A Primer on Real Analysis

Alan Sultan

Queens College

Contents

1 Preliminaries   **1**
  1.1 Preliminary Facts For Sets and Functions 1
  1.2 Preliminary Facts about the Real Numbers 15
  1.3 Logic Preliminaries 19

2 Bounds and Sequences of Real Numbers   **23**
  2.1 Preliminaries 23
  2.2 Basic Theorems 26
  2.3 Sequences 32
  2.4 Subsequences 50
  2.5 Limit Superior and Limit Inferior 53
  2.6 Cauchy Sequences 61

3 Metric Spaces   **65**
  3.1 Basic Results and Examples 65
  3.2 Sequences in Metric Spaces 71
  3.3 Open and Closed Sets 77

4 Limits, Compactness and Uniform Continuity   **91**
  4.1 Limits 91
  4.2 Limits of Functions Defined On Metric Spaces 95
  4.3 Continuity 100
  4.4 Compactness 112
  4.5 Some Useful Theorems Related to Compactness 124
  4.6 Uniform Continuity 127

5 Sequences of Functions   **137**

6 Countability (An Informal Approach)   **151**
  6.1 Sets of Measure 0 155

7 The Riemann Integral — 159
- 7.1 Preliminaries — 159
- 7.2 Intuitive Review of the Riemann Integral — 167
- 7.3 Lebesgue's Theorem — 170
- 7.4 Basic Results on Riemann Integrals — 174
- 7.5 Other Approaches to the Riemann Integral — 182
- 7.6 The Equivalence of... — 185
- 7.7 Riemann-Stieltjes Integrals — 194
- 7.8 Evaluation of Riemann-Stieltjes Integrals — 197

8 Introduction to The Lebesgue Integral — 203
- 8.1 Overview — 203
- 8.2 The Modern Approach to the Lebesgue Integral — 216

Appendix — 225

Bibliography — 231

Solutions to Most Exercises — 233

Preface

This book is a primer on analysis. Its sole purpose it to give those who want a quick and somewhat intuitive introduction to analysis, a resource. Once the student goes through this book, he or she should be ready to tackle the more popular books on the subject. As a primer, one will see certain things missing, and certain assumptions made that the student will be able to see developed in other texts at the time he or she is ready. So, one will not see Dedekind cuts, nor proofs of certain properties of real numbers, for example, the Completeness Theorem. Rather, these will be assumed. This allows us to get into the main concepts very quickly. Also, I have not included a chapter on derivatives (since what one needs can easily be found in any calculus book). This book assumes a basic knowledge of calculus for example, the calculation of derivatives and standard integrals.

While mathematicians often prefer to use symbols rather than words to make things more concise, this book goes pretty much in the opposite direction, becoming quite informal at times and using more words than symbols (but makes an effort to balance the two). I feel that words make the transition to the symbols easier. I also often use pictures to illustrate the concepts, since I feel this works best for a beginner. The pictures facilitate the understanding and are in fact the intuitive link that mathematicians often use in constructing proofs in analysis. Sometimes a picture is drawn that makes it clear that something must be true, and that often suggests a way to prove the theorem in a more abstract setting. These pictures are meant to be a prop only to help remember the theorems and possibly gain insight into how the proofs of the theorems might have been constructed.

Because this is a book for beginners, there is a lot of detail in many of the proofs and we will at times appear to be quite repetitive. This is done purposely to remind students of definitions and theorems and is primarily a pedagogical technique that some of my own teachers used which I found to be helpful. Sometimes when I prove a theorem with two parts, and the proofs of the two parts are similar, I leave one half of the proof to the student as I feel this is good for his or her training.

Most analysis books begin with preliminary material on sets and functions, and this text is no different. While one may not use some of these theorems right away,

I will use them later in the book. The results and their proofs are an essential part of the students' training, and many of the results are used heavily in more advanced subject areas like topology, complex analysis, and functional analysis.

I have provided solutions to virtually all the problems in this book. This way the beginning student or self learner gets a chance to check his or her work. This also will aid any teacher who uses this book as it gives the teacher the freedom to cover more ground while feeling secure that the students have answers. Also, this gives the teacher time to discuss student solutions which are different from those given to determine if an error has been made. Although there are answers at the end, any serious student knows that to learn a subject one must try to work the problems. So looking at the solutions without trying the problem will usually work against oneself.

The numbering in the book is consecutive within each chapter to make things easy to locate. Thus, if in Chapter 2 we start with a definition, then follow it by an example, and follow that by a theorem, these will appear as Definition 2.1, Example 2.2, and Theorem 2.3, respectively.

I assume that the reader has a knowledge of direct proof, proof by contradiction, and proof by induction. The reader is also expected to know the basic laws of logic, and how to negate statements with quantifiers, some of which I will review.

As in many books, typos and errors sometimes creep in. Please let me know soon if you find any, as I will be updating this manuscript from time to time and I will make the corrections that are needed. E-mail me at asultan956@aol.com, and let me know in the subject area of the message what the e-mail is about. And thank you for taking the time to do this.

Finally, I would like to thank Jeremy Miller, David Miller, and Nikolina Dabovic who read through a preliminary version of this manuscript and made good suggestions. Special thanks go to Mona Ali who carefully read through the first full draft of this manuscript and made many very fine suggestions. She is amazing!

Chapter 1

Preliminaries

1.1 Preliminary Facts For Sets and Functions

In this first chapter, we lay the groundwork for much of what follows. The reader who knows most of this material should just peruse this chapter to see what definitions are being used since definitions vary from book to book. The approach taken here is somewhat informal, as this is all that is needed for our purposes.

Definition 1.1 *A **set** is simply a collection of objects.*

Sets are often indicated by braces, { }. So, when we write $\{1, 2, 3\}$, we can read this as "The set consisting of the objects $1, 2$, and 3." Another way to describe sets is by some kind of verbal description of the objects. So, we might write {real numbers $x|\ 1 \leq x \leq 3$}, which is read as "the set of all real numbers x such that x is between 1 and 3 inclusive of 1 and 3." We see that the vertical bar in this description of the set is read as "such that." When the set is infinite, listing the elements is really impossible unless there is a pattern to the elements listed. So, if we wanted to describe the set of natural numbers, we could write $\{1, 2, 3, ...\}$ to indicate the pattern, or we might wish to use some letter to represent the set of natural numbers (which we will soon do since we will be using this set fairly often).

Definition 1.2 *The objects making up a set are called the **elements** of the set.*

Thus, the elements of the set $\{1, 2, 3\}$ are 1, 2, and 3. We use the notation $\in$ to indicate that an object is in the set. Thus $1 \in \{1, 2, 3\}$, but $4 \notin \{1, 2, 3\}$, where the symbol $\notin$ means "is not an element of".

Sets are often denoted by capital letters. There are several sets we will need throughout this primer which are highlighted below.

Z is the set of integers. That is, $Z = \{0, \pm 1, \pm 2, \pm 3, ...\}$.
Z^+ is the set of positive integers, also known as the natural numbers. That is $Z^+ = \{1, 2, 3...\}$.
Q is the set of rational numbers. That is, $Q = \{\frac{p}{q}|\ p, q \varepsilon Z,$ and $q \neq 0\}$.
R is the set of all real numbers.
R^+ is the set of all positive real numbers.
$[a, b]$ stands for the closed interval $[a, b]$ with endpoints a and b. That is $[a, b] = \{x \in R|\ a \leq x \leq b\}$.
(a, b) stands for the interval $[a, b]$ without endpoints a and b. That is $(a, b) = \{x \in R|\ a < x < b\}$.
$(-\infty, a) = \{x \in R|\ x < a\}$.
$(-\infty, a] = \{x \in R|\ x \leq a\}$.
$(a, \infty) = \{x \in R|\ x > a\}$.
$[a, \infty) = \{x \in R|\ x \geq a\}$.

Definition 1.3 *If A and B are sets, we say that A is a **subset** of B, and write $A \subset B$, if every element of A is an element of B.*

Example 1.4 *If $A = \{1, 2\}$ and $B = \{1, 2, 3\}$, then since every element of A is an element of B, A is a subset of B.*

We should mention that this notation for subset is not standard, and in some books the symbol $\subseteq$ is used for subset, while $\subset$ is used to mean that A is a subset of B, but is not the same as B. In this primer, however, the notation $A \subset B$ allows A and B to be the same.

Observe that for any set A, $A \subset A$.

It is convenient to discuss a set having no elements. Such a set is denoted by $\emptyset$. It is also convenient to agree that $\emptyset$ is considered to be a subset of every set.

Definition 1.5 *If A and B are sets, then*

*(1) the **union** of A and B, denoted by $A \cup B$, is the set of all elements that belong to A or B or both, and*

*(2) the **intersection** of A and B denoted by $A \cap B$, is the set of elements that belong to both A and B.*

*(3) The **difference** of A and B, denoted by $A - B$, is the set of those elements of A which are not in B.*

1.1. PRELIMINARY FACTS FOR SETS AND FUNCTIONS

Example 1.6 *If $A = \{1, 2, 3, 4\}$ and $B = \{4, 5, 6, 7\}$, then $A \cup B = \{1, 2, 3, 4, 5, 6, 7\}$. Notice that we don't write the element 4 twice when writing the union, since we are interested only in the elements that are in both sets and not how many times they occur. $A \cap B = \{4\}$, and $A - B = \{1, 2, 3\}$.*

The definition of union and intersection extend to any collection of sets. So if $\{A_\alpha\}$ denotes a collection of sets, then $\cup A_\alpha$ represents the collection of those things that are in one or more of the sets, while $\cap A_\alpha$ represents the collection of all things that are common to all the sets.

Example 1.7 *For each real number, x, let $A_x = (x, x+1)$. Thus $A_{\frac{1}{2}} = (\frac{1}{2}, \frac{3}{2})$, $A_{\sqrt{\pi}} = (\sqrt{\pi}, \sqrt{\pi}+1)$, etc. Here the index x comes from the set of real numbers. In this case we can write $\cup A_x$ for the union of the sets, where the index is understood since it was stated, or, if we wish to specify where x comes from, we can write $\underset{x \in R}{\cup} A_x$. We prefer the first form, though we will use both forms in this book. In any event, in this case, $\cup A_x = \cup(x, x+1) = R$, the set of all real numbers, while $\cap A_x = \cap A_x = \emptyset$. If the index x came from the natural numbers instead of R, then $\cup A_x = (1, 2) \cup (2, 3) \cup (3, 4) \cup ...$; if the index, x, came from the set $\{-3, 0, 4\}$, then $\cup A_x = (-3, -2) \cup (0, 1) \cup (4, 5)$. Usually when talking about a collection of sets, $\{A_\alpha\}$, the index set (the set from which α comes) is stated, but if it isn't, then we just assume that we have a collection of sets whose indices come from some set which we are not specifying.*

Definition 1.8 *When all sets under discussion are subsets of a fixed set U, we call U the **universal set**.*

Definition 1.9 *If U is the universal set, then by the **complement** of A, denoted by A^c, we mean $U - A$. That is, A^c means everything (in U) which is not in A.*

Example 1.10 *If $U = \{1, 2, 3, 4, 5\}$ and $A = \{1, 2\}$, then $A^c = \{3, 4, 5\}$.*

When we just write A^c, we are assuming that all sets under discussion are coming from some universal set U, and the context will usually make it clear what U is.

Venn diagrams often can be used to represent relationships among sets. For example, in Figure 1.1 the universal set is represented by a rectangle. Subsets of U can be represented by circles.

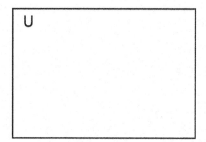

Figure 1.1. A depiction of a universal set

In Figure 1.2 we see how we can represent $A \subset B$.

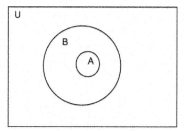

Figure 1.2. $A \subset B$.

In Figure 1.3 we see how to represent $A \cap B$. It is the shaded portion of the diagram.

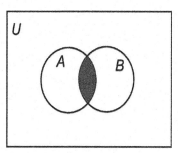

Figure 1.3. $A \cap B$

Similarly, in Figure 1.4 we see how to represent $A \cup B$.

1.1. PRELIMINARY FACTS FOR SETS AND FUNCTIONS 5

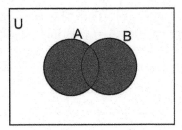

Figure 1.4. $A \cup B$.

In Figure 1.5 we see $A - B$.

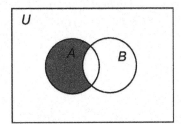

Figure 1.5. $A - B$.

We observe at once from the definitions, that for any sets A and B it must be true that $A \cap \emptyset = \emptyset$, $A \cup \emptyset = A$, $A \subset A \cup B$, $A \cap B \subset A$ and that $A \cap B \subset B$. The last 3 relationships are suggested by looking at Venn Diagrams but also follow immediately from the definitions of subset, intersection, and union. Venn diagrams also suggest relationships like $(A - B) \cup (B - A) = (A \cup B) - (A \cap B)$, which to the beginner is often not obvious. (See Exercise 2).

Definition 1.11 *A collection of sets $\{E_i\}$ is called* **pairwise disjoint** *if $E_i \cap E_j = \emptyset$, when $i \neq j$.*

Example 1.12 *If $E_1 = (0, 1)$, $E_2 = (1, 2)$ and in general, $E_i = (i - 1, i)$, where $i = 1, 2, 3, ...$, the sets E_i are pairwise disjoint.*

Definition 1.13 *Two sets are* **equal** *if they have the same elements.*

This last definition seems so obvious that it makes us think there is no need to state it. For example, if $A = \{x|\ x = 2n, n \in Z^+\}$ and B is the set of positive even integers, it may be clear that A and B are the same. But in analysis, we sometimes are given two equal sets that are described in very different ways, and it is not immediately clear they are the same. For example, we might let $A = \{x \in Z|\ |x^2 - 8x + 12|$ is a prime number} and $B = \{1, 3, 5, 7\}$. That these are really the same set requires some work to establish. The fact that $A = \{$distinct positive integers $x, y,$ and z such that $x^3 + y^3 = z^3\}$ and $B = \varnothing$ are the same set is not so easy to prove. (This is a special case of Fermat's Last Theorem.) And if we let T be the set of triangles in the Euclidean plane with the property that $\dfrac{a^3 + b^3 + c^3}{a + b + c} = c^2$, where a, b and c are the sides of the triangle, and $S = \{$triangles in the Euclidean plane where angle C is 60 degrees$\}$, then $T = S$, but it takes some work to show.

In general, when we wish to prove that two sets A and B are the same, we need to prove two things. We need to show that both $A \subset B$ and that $B \subset A$. This simply says that everything in A is in B and everything in B is in A. If both of these are true, then $A = B$. Let us begin by showing something that many of you have probably seen, but is not so obvious.

Theorem 1.14 *(DeMorgan's Laws)*
 (1) $(\cap A_\alpha)^c = \cup A_\alpha^c$ *(In words: The complement of the intersection of a collection of sets, is the union of the complements of the individual sets.)*
 (2) $(\cup A_\alpha)^c = \cap A_\alpha^c$ *(In words: The complement of the union of a collection of sets, is the intersection of the complements of the individual sets.)*

Proof. (of 1) We will first show that the set on the left side of the equal sign is a subset of the set on the right side of the equal sign, and then show the reverse.

To show that $(\cap A_\alpha)^c \subset \cup A_\alpha^c$, we pick an arbitrary $x \in (\cap A_\alpha)^c$, and show that it is in $\cup A_\alpha^c$. Using definitions is critical in proving this (and throughout this book). Remembering that the complement of a set means everything not in the set, we have the following:

If $x \in (\cap A_\alpha)^c$, then $x \notin \cap A_\alpha$.
But if x is not in the intersection, $\cap A_\alpha$, then x is not in at least one of the sets making up this intersection. So x is not in some specific set, call it A_{α_0}.
Since x is not in A_{α_0}, then by definition of the complement, $x \in A_{\alpha_0}^c$.
But x being in the set $A_{\alpha_0}^c$ implies that $x \in \cup A_\alpha^c$,
since the union consists of all things that are in one or more of the sets.

We have shown that if we start with an *arbitrary* $x \in (\cap A_\alpha)^c$, it will be in $\cup A_\alpha^c$. This shows that every element of $(\cap A_\alpha)^c$ is in $\cup A_\alpha^c$, or said another way, that

$$(\cap A_\alpha)^c \subset \cup A_\alpha^c. \tag{1.1}$$

1.1. PRELIMINARY FACTS FOR SETS AND FUNCTIONS

To show the reverse inclusion, we essentially reverse the steps. Here are the details:

Suppose that $x \in \cup A_\alpha^c$.
Then $x \in A_{\alpha_0}^c$ for some specific set $A_{\alpha_0}^c$ in the union, by definition of union.
So $x \notin A_{\alpha_0}$ by definition of $A_{\alpha_0}^c$.
Since $x \notin A_{\alpha_0}$, $x \notin \cap A_\alpha$.
But if $x \notin \cap A_\alpha$, then $x \in (\cap A_\alpha)^c$.

We have just shown that any $x \in \cup A_\alpha^c$ must be contained in $(\cap A_\alpha)^c$. So we have

$$\cup A_\alpha^c \subset (\cap A_\alpha)^c. \tag{1.2}$$

Together, (1.1) and,(1.2) show that

$$(\cap A_\alpha)^c = \cup A_\alpha^c.$$

The proof of (2) is similar and is left for you. ∎

Example 1.15 *We illustrate part (2) of this theorem. Suppose that our universal set is $U = \{1, 2, 3, ...10\}$, that $A_1 = \{1, 2\}$, $A_2 = \{2, 3, 4\}$, and that $A_3 = \{5, 6, 7\}$. Then $\cup A_\alpha = A_1 \cup A_2 \cup A_3 = \{1, 2, 3, ..., 7\}$, and $(\cup A_\alpha)^c = \{8, 9, 10\}$. On the other hand $A_1^c = \{3, 4, 5, ..., 10\}$, $A_2^c = \{1, 5, 6, 7, 8, 9, 10\}$, and $A_3^c = \{1, 2, 3, 4, 8, 9, 10\}$. Now, $(\cap A_\alpha^c) = \{8, 9, 10\}$. So this example illustrates that $(\cup A_\alpha)^c = (\cap A_\alpha^c)$.*

The following theorem is also referred to as DeMorgan's Law.

Theorem 1.16 *If $\{B_\alpha\}$ (where α runs through some index set) is a collection of sets, then for any set A, $A \cap (\cup B_\alpha) = \cup(A \cap B_\alpha)$. In particular, if B_1 and B_2 are sets, then $A \cap (B_1 \cup B_2) = (A \cap B_1) \cup (A \cap B_2)$.*

Proof. Suppose that $x \in A \cap (\cup B_\alpha)$. Using the definition of intersection, $x \in A$ and $x \in \cup B_\alpha$. So $x \in A$, and (by the definition $\cup B_\alpha$) x is in at least one set B_{α_0} in the union. So $x \in A \cap B_{\alpha_0}$, which implies that $x \in \cup(A \cap B_\alpha)$. To recap, we started with an arbitrary $x \in A \cap (\cup B_\alpha)$, and showed it was in $\cup(A \cap B_\alpha)$. So

$$A \cap (\cup B_\alpha) \subset \cup(A \cap B_\alpha). \tag{1.3}$$

All the steps are completely reversible. So we can start from the last step, $x \in \cup(A \cap B_\alpha)$, and argue the same way moving back to the first step: If $x \in \cup(A \cap B_\alpha)$, then $x \in A \cap B_{\alpha_0}$ for some α_0. Thus, $x \in A$ and $x \in B_{\alpha_0}$. So, $x \in A$ and $x \in \cup B_\alpha$, or said differently, $x \in A \cap (\cup B_\alpha)$. Since we started with an arbitrary x in $\cup(A \cap B_\alpha)$ and showed it was in $A \cap (\cup B_\alpha)$, we have

$$\cup(A \cap B_\alpha) \subset A \cap (\cup B_\alpha). \tag{1.4}$$

From (1.3) and (1.4) we have that $A \cap (\cup B_\alpha) = \cup(A \cap B_\alpha)$. ∎

Definition 1.17 *If A and B are sets, then $A \times B = \{(a,b) |\ a \in A$ and $b \in B\}$. If A_1, $A_2, A_3, ..., A_n$ are sets, then $A_1 \times A_2 \times A_3 \times ... A_n = \{n\text{-tuples } (a_1, a_2, a_3, ... a_n) |\ a_1 \in A_1, a_2 \in A_2, a_3 \in A_3, ..., a_n \in A_n\}$.*

Example 1.18 *Let $A = \{1,2\}$ and $B = \{3,4,5\}$. Then $A \times B = \{(1,3), (1,4), (1,5), (2,3), (2,4), (2,5)\}$. Notice that $A \times B \neq B \times A$ since $(3,1)$ is in $B \times A$, but not in $A \times B$.*

So if R is the real line, then $R \times R = \{(a,b) |\ a, b$ are real numbers$\}$. This set of ordered pairs is of course the ordinary Cartesian plane that we study in high school, and is abbreviated R^2, while $R \times R \times R$, abbreviated R^3, is $\{(a,b,c) |\ a, b$ and c are real numbers$\}$. This is what we usually call 3-dimensional space while $R^n = R \times R \times R... \times R$ (where R occurs n times) is called n-dimensional space.

Definition 1.19 *(Informal) A **function from a set** A **to a set** B is a rule which associates with each element $a \in A$, one and only one element, b in B. The element, b in B, associated with a, is called the **image** of b under this function. A is called the **domain** of the function.*

Notation 1.20 *If f is a function from A to B we write $f : A \to B$. This can be read as "f is a function from A to B," or "f maps A to B." The image of an element $a \in A$ under this function is denoted by $f(a)$.*

Example 1.21 *Let $f : R \to R$ be defined by $f(x) = x^2$. (So f maps the real numbers to the real numbers, and the image of any real number x is x^2.) This is the kind of function you studied in high school. We have that $f(3) = 3^2 = 9$. $f(x+h) = (x+h)^2$, and so on.*

Example 1.22 *A less familiar example to a beginner might be: Let A be the set of 2×2 matrices, and let B be R. The mapping from A to B which associates with each 2×2 matrix $\begin{pmatrix} a & b \\ c & d \end{pmatrix}$ its determinant, $ad - bc$, is a function since each matrix in A has one and only one determinant. So, for example, $f\left(\begin{pmatrix} 1 & -2 \\ 3 & 4 \end{pmatrix}\right) = (1)(4) - (-2)(3) = 10$.*

Example 1.23 *As a final illustration, if we let A be the set of all people in the United States who have been assigned social security numbers, and let B be the set of assigned social security numbers, then we can form the function $f : A \to B$ which associates with each person in set A, his or her social security number. Assuming that each person is on the up and up and has only one social security number assigned to him or her, this associates with each person in A one and only one social security number in B.*

1.1. PRELIMINARY FACTS FOR SETS AND FUNCTIONS

Definition 1.24 *If $f: A \to B$ and $S \subset A$, we define $f(S)$ to be the set $\{f(x)|x \in S\}$. We call this set **the image of S under** f, or when the function is clear, just the **image** of S. The **range** of the function is defined to be $f(A)$, the image of the domain.*

Example 1.25 *In Example 1.21, the range of f is the set of nonnegative numbers. The image of the set $\{-1, 1, 2, 3\}$, under f, is the set $\{1, 4, 9\}$.*

Example 1.26 *In Example 1.22, the range of f is the set of all real numbers, since every real number r, is $f(A)$ for some A. More specifically, for any real number r, $r = f(A)$ when $A = \begin{pmatrix} r & -2 \\ 0 & 1 \end{pmatrix}$. Also notice that $r = f(B)$ where $B = \begin{pmatrix} r & 4 \\ 0 & 1 \end{pmatrix}$. So in a function, many elements can have the same image. But given any fixed element in the domain, there can only be one image of it.*

Example 1.27 *In Example 1.23, the range of f is the set of social security numbers that have been assigned.*

Figure 1.6 shows how we can picture a function from a set A to a set B.

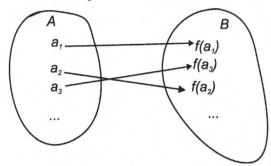

Figure 1.6. Depicting a function from A to B.

The arrows go from points in the domain to their images in the range. In Figure 1.7 you see a depiction of the image of a subset S, of A under a function.

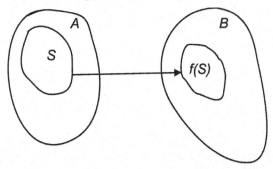

Figure 1.7. The image of a set S under a function

Definition 1.28 *If f and $g : A \to R$, we can form new functions from A to R. We can form the functions $(f+g)$, $(f-g)$, fg, and f/g whose values for any $x \in A$, are respectively, $f(x) + g(x)$, $f(x) - g(x)$, $f(x)g(x)$, and $f(x)/g(x)$, where in the last expression, $g(x) \neq 0$.*

Example 1.29 *If $f(x) = x^2$ and $g(x) = \sqrt{x}$, then $(f+g)(x) = f(x)+g(x) = x^2+\sqrt{x}$. Since $g(x)$ limits x to nonnegative values, we consider both f and g to be defined only for nonnegative values when talking about $f+g$. Similarly, $(f-g)(x) = f(x)-g(x) = x^2 - \sqrt{x}$, $(fg)(x) = x^2\sqrt{x}$ and $(f/g)(x) = \dfrac{x^2}{\sqrt{x}} = \sqrt{x}$. In f/g we assume that both f and g are defined only for $x > 0$, in order to compute the quotient. Said another way, the domain of f/g is $\{x|\ x > 0\}$.*

Definition 1.30 *If $f : A \to B$, and T is a subset of B, then $f^{-1}(T) = \{x \in A|\ f(x) \in T\}$ and is called the **pre-image** of T under f or simply the pre-image of T, (when it is clear what function we are talking about).*

The pre-image of a set T in B is the collection of everything in A that maps into T. In Figure 1.8 we see a depiction of the pre-image of T under a function.

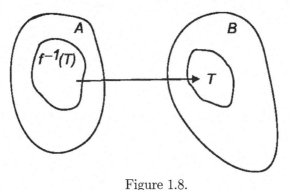

Figure 1.8.

Example 1.31 *In Example 1.21, $f^{-1}(\{0,1,5\}) = \{0, \pm 1, \pm\sqrt{5}\}$.*

Example 1.32 *In Example 1.22, $f^{-1}(2)$ is the set of all 2×2 matrices with determinant 2. $f^{-1}(0)$ is the set of all 2×2 matrices with determinant 0. You may remember from linear algebra, that these matrices are precisely the non-invertible 2×2 matrices.*

1.1. PRELIMINARY FACTS FOR SETS AND FUNCTIONS

Example 1.33 *In Example 1.23, the pre-image T of social security numbers, is the set of people who have been assigned the social security numbers in T.*

One should observe right away that if S and T are subsets of B, where $f : A \to B$, then if $S \subset T$, $f^{-1}(S) \subset f^{-1}(T)$. Figure 1.9 illustrates this. You will try to explain why this is true in the exercises.

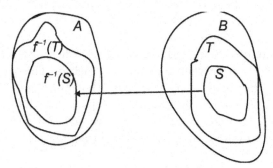

Figure. 1.9. If $S \subset T$, $f^{-1}(S) \subset f^{-1}(T)$.

The following results will be needed later in this book.

Theorem 1.34 *If $f : A \to B$, and $\{E_\alpha\}$ is a collection of subsets of B, then*
  (a) $f^{-1}(\cup E_\alpha) = \cup f^{-1}(E_\alpha)$, and
  (b) $f^{-1}(\cap E_\alpha) = \cap f^{-1}(E_\alpha)$.

Proof. (a) Suppose that $x \in f^{-1}(\cup E_\alpha)$. Since the pre-image of $\cup E_\alpha$ consists of all those things from A that map into $\cup E_\alpha$, $f(x) \in \cup E_\alpha$. So, $f(x) \in E_{\alpha_0}$ some α_0. This implies that $x \in f^{-1}(E_{\alpha_0})$ since the image of x is in E_{α_0}, and this in turn implies that $x \in \cup f^{-1}(E_\alpha)$. We started with an arbitrary $x \in f^{-1}(\cup E_\alpha)$, and showed that $x \in \cup f^{-1}(E_\alpha)$. So
$$f^{-1}(\cup E_\alpha) \subset \cup f^{-1}(E_\alpha). \tag{1.5}$$

To get the reverse inclusion, we reverse the steps: Suppose that $x \in \cup f^{-1}(E_\alpha)$. Then $x \in f^{-1}(E_{\alpha_0})$ for some α_0. So $f(x) \in E_{\alpha_0}$, which implies that $f(x) \in \cup E_\alpha$. And since the image of x is in $\cup E_\alpha$, $x \in f^{-1}(\cup E_\alpha)$. We chose an arbitrary $x \in \cup f^{-1}(E_\alpha)$, and showed that $x \in f^{-1}(\cup E_\alpha)$. So, it follows that $\cup f^{-1}(E_\alpha) \subset f^{-1}(\cup E_\alpha)$. This, coupled with (1.5) gives us that $f^{-1}(\cup E_\alpha) = \cup f^{-1}(E_\alpha)$. Although we have written out everything in detail, all we had to say for the second half of the proof we gave was "reverse the steps." We will do that occasionally from here on.

Part (b) is done similarly and is left to the reader. ∎

Theorem 1.35 *If $f : A \to B$, and $\{E_\alpha\}$ is a collection of subsets of A, then*
(1) $f(\cup E_\alpha) = \cup f(E_\alpha)$, and
(2) $f(\cap E_\alpha) \subset \cap f(E_\alpha)$.

Proof. (1): Suppose that $x \in f(\cup E_\alpha)$. Then $x = f(z)$ for some $z \in \cup E_\alpha$ and since $z \in \cup E_\alpha$, $z \in E_{\alpha_0}$ for some α_0. Since x is the image of z, which is in E_{α_0}, we have that $x \in f(E_{\alpha_0})$. So $x \in \cup f(E_\alpha)$. We started with an arbitrary $x \in f(\cup E_\alpha)$ and showed it was contained in $\cup f(E_\alpha)$. So, $f(\cup E_\alpha) \subset \cup f(E_\alpha)$. To show the set inclusion the other way, we ask you to reverse the steps.

(2) Suppose that $x \in f(\cap E_\alpha)$. Then $x = f(z)$, where $z \in \cap E_\alpha$. Since $z \in \cap E_\alpha$, z is in each E_α. So $x = f(z)$ where $z \in E_\alpha$ for each α. This says that $x \in \cap f(E_\alpha)$. Since every $x \in f(\cap E_\alpha)$ is in $\cap f(E_\alpha)$, we have that $f(\cap E_\alpha) \subset \cap f(E_\alpha)$.

Now, many people will say "Reverse the steps to get the set inclusion the other way." But the steps are not reversible, and we ask you to determine why in the exercises. The next example will show this inclusion does not go the other way. ∎

Example 1.36 *Let $f : R \to R$ be the constant function $f(x) = 5$. Let $E_1 = \{1\}$ and let $E_2 = \{2\}$. Then $f(E_1 \cap E_2) = \varnothing$, but $f(E_1) \cap f(E_2) = \{5\}$. So $f(E_1 \cap E_2) \neq f(E_1) \cap f(E_2)$.*

Theorem 1.37 *If $f : A \to B, E \subset A$, and $F, S \subset B$, then we have*
(1) $E \subset f^{-1}(f(E))$
(2) $f(f^{-1}(F)) \subset F$, and
(3) $f^{-1}(S^c) = (f^{-1}(S))^c$.

Proof. (1): If $x \in E$ then of course $f(x) \in f(E)$, and it follows since $f(x)$ is in $f(E)$, that $x \in f^{-1}f(E)$. We have shown that any $x \in E$ is in $f^{-1}f(E)$, so $E \subset f^{-1}f(E)$.

A very simple way of seeing this is to realize that $f^{-1}(f(E))$ consists of *all* things that map into $f(E)$. And since E maps into $f(E)$, this set contains E.

(2): Suppose that $x \in f(f^{-1}(F))$, then $x = f(z)$ for some $z \in f^{-1}(F)$. By definition of the pre-image of F, this tells us that $f(z) \in F$. But $f(z) = x$, so $x \in F$. Since we took an arbitrary $x \in f(f^{-1}F)$ and showed it was in F we conclude that $f(f^{-1}F) \subset F$.

Again, a simple way of seeing this is to realize that $f^{-1}(F)$ consists of all things whose images are in F. So the image of this set, which is $f(f^{-1}(F))$, must be in F.

(3): Suppose $p \in f^{-1}(S^c)$. This implies that $f(p) \in S^c$ and that $f(p) \notin S$, which in turn implies that $p \notin f^{-1}(S)$ and that $p \in (f^{-1}(S))^c$. We started with an arbitrary $p \in f^{-1}(S^c)$ and showed that is was in $(f^{-1}(S))^c$, so $f^{-1}(S^c) \subset (f^{-1}(S))^c$. All of our steps are reversible, which means that $(f^{-1}(S))^c \subset f^{-1}(S^c)$. These two set inclusions tell us that $f^{-1}(S^c) = (f^{-1}(S))^c$. ∎

1.1. PRELIMINARY FACTS FOR SETS AND FUNCTIONS

Example 1.38 *We illustrate parts (1) and (2) of the above theorem. Let $f : R \to R$ where $f(x) = x^2$. Let $E = \{1\}$. Then $f(E) = 1$, and $f^{-1}f(E) = \{-1, 1\}$. So $E \subset f^{-1}f(E) = \{-1, 1\}$. And if we let $F = \{-4, 1\}$, then $f^{-1}(F) = \{1, -1\}$, $f(f^{-1}(F)) = \{1\}$, and we have that $f(f^{-1}(F)) \subset F$.*

Definition 1.39 *If $f : A \to B$, and $g : B \to C$, then the **composition** of g and f, denoted by $g \circ f$, is a function from $A \to C$, and the image of any $x \in A$ under $g \circ f$ is $g(f(x))$.*

Figure 1.10 illustrates the notion of composition.

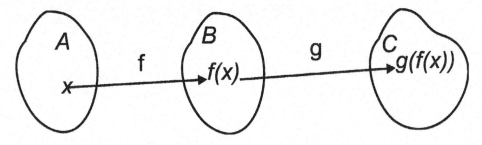

Example 1.40 *Suppose that $f, g : R \to R$, where $f(x) = x + 1$ and $g(x) = \sin x$. Then $g(f(x)) = \sin(x + 1)$.*

EXERCISES

When you are asked to prove a set inclusion or identity, assume all sets under consideration are subsets of some universal set U.

1. Draw a Venn diagram that shows 3 mutually disjoint sets.

2. Draw a Venn diagram that shows that $(A - B) \cup (B - A) = (A \cup B) - (A \cap B)$.

3. Find $B \times A$ and $A \times A$ if $A = \{1, 2\}$ and $B = \{2, 4, 5\}$.

4. Suppose that for each x in $[0, 1]$, $O_x = (x - .5, x + .5)$. What are $\cup O_x$ and $\cap O_x$ equal to? Would the intersection be different if $O_x = [x - .5, x + .5]$?

5. $S_n = (2 - 1/n, 2 + 1/n)$, where $n = 1, 2, 3, \ldots$. What is the intersection of the S_n equal to?

6. Prove that $A - B = A \cap B^c$.

7. Prove that if $A \subset B$, then $B^c \subset A^c$.

8. Express in set notation, those elements that are in exactly two of the sets A, B, and C.

9. A function $f : A \to B$ is said to be $1-1$ (read "one to one") if whenever $a_1 \neq a_2$, it follows that $f(a_1) \neq f(a_2)$.

    (a) Write the contrapositive of this statement.

    (b) Show that any linear function $f(x) = mx + b$ (from R to R), where $m \neq 0$, is $1-1$.

    (c) Show that if $f : A \to B$ is $1-1$ and $g : B \to C$ is $1-1$, then $g \circ f$ is $1-1$.

    (d) Suppose that $f : A \to B$, and that E_α are subsets of A. We proved on page 12 that $f(\cap E_\alpha) \subset \cap f(E_\alpha)$. If we tried reversing the steps, where would the reversal break down?

    (e) True or False: If $f : A \to B$ is $1-1$, then $f(\cap E_\alpha) = \cap f(E_a)$, where the E_α are subsets of A. Prove your result.

10. A function $f : A \to B$ is said to be onto B if for any $b \in B$, there is an $a \in A$ such that $f(a) = b$.

    (a) Show that any function is onto its range.

    (b) Show that any linear function $f(x) = mx + b$ (from R to R), where $m \neq 0$, is onto R.

    (c) Find a function $f : (0,1) \to R^+$ which is onto R^+.

    (d) True or False: If $f : A \to B$ is onto B and $g : B \to C$ is onto C, then $g \circ f$ is onto C. Explain.

    (e) True or False: If $f : A \to B$ is $1-1$, then $f^{-1}(f(E)) = E$. Prove your result.

    (f) True or False: If $f : A \to B$ is onto B, then $f^{-1}(f(E)) = E$ for any subset E of A. Prove your result.

11. If T is a subset of A, and $f : A \to B$, such that f is onto B, give a condition that will guarantee that $(f(T))^c = f(T^c)$.

1.2 Preliminary Facts about the Real Numbers

There are certain facts that we will accept throughout this primer which are probably familiar to most people. The proofs of some of them are outlined in the exercises. One need only recall that a rational number is any number that can be written in the form $\frac{p}{q}$, where p and q are integers and $q \neq 0$. That is, a rational number is an element of Q.

Here are some of the things with which we assume the reader is familiar.

Proposition 1.41 *The sum, difference, product, and quotient, of rational numbers is rational. Of course, in the quotient statement, we cannot divide by zero.*

Proposition 1.42 *If $a < b$, then*
  (1) $a - x < b - x$ for any number x, and
  (2) $a + x < b + x$ for any number x.
  (3) If $c > 0$, then $ca < cb$.
  (4) If $c < 0$, then $ca > cb$.
  (5) If $a, b > 0$ and $a < b$, then $1/a > 1/b$.
  (6) If $a \leq b$ and $b \leq a$, then $a = b$.

Example 1.43 *So from $-3x > 5$, we get all of the following: $x < \frac{-5}{3}$, $-6x > 10$, and $-3x - 7 > -2$. Also, by part (5) of this proposition, if $a > 2$ then $\frac{1}{a} < \frac{1}{2}$ and from part (6) of this proposition, if $a \geq 2$ and $a \leq 2$, then $a = 2$.*

Proposition 1.44 $\sqrt{2}$ *is irrational.*

Proposition 1.45 *Given any positive number ε, one can always find a positive number N such that for $n > N$, $\frac{1}{n} < \varepsilon$.*

Proposition 1.46 *If $n > 0$, and you add $\frac{1}{n}$ to itself enough times, your sum can be made as large as you want. Said another way, if b is any positive number, then there exists a positive integer m such that $m \cdot \frac{1}{n} \geq b$, or more simply, such that $\frac{m}{n} \geq b$.*

Using these propositions we can prove several things that we will need in our development of analysis.

Theorem 1.47 *Between any two real numbers a and b, one can find a rational number.*

Proof. The first case we prove is for $b > 0$. Suppose that $a < b$, and that $b > 0$. Since $a < b$, $b - a > 0$. And so, according to Proposition 1.45, there is an n such that $\frac{1}{n} < b - a = \varepsilon$, from which it follows (by Proposition 1.42 part (4)) that

$$-\frac{1}{n} > a - b. \tag{1.6}$$

By Proposition 1.46, there is a positive integer m such that $m \cdot \frac{1}{n} \geq b$, or said differently, such that

$$b \leq \frac{m}{n}. \tag{1.7}$$

Let m be the *smallest* positive integer which makes $b \leq \frac{m}{n}$. Then

$$\frac{m-1}{n} < b. \tag{1.8}$$

Now,

$$\begin{aligned} a &= b + (a - b) \\ &< \frac{m}{n} + \left(-\frac{1}{n}\right) \text{ (by (1.7) and (1.6))} \\ &= \frac{m-1}{n}, \end{aligned}$$

which of course yields

$$a < \frac{m-1}{n}. \tag{1.9}$$

From (1.9) and (1.8) we get that

$$a < \frac{m-1}{n} < b.$$

Thus, we have found a rational number, $\frac{m-1}{n}$, between a and b when $b > 0$. Now, to show the general case where $a < b$, we add a positive integer p to both sides of this inequality so that

$$a + p < b + p,$$

and $b + p > 0$. Now, from the first part of the proof, there is a rational number r such that

$$a + p < r < b + p.$$

It follows from this that

$$a < r - p < b.$$

1.2. PRELIMINARY FACTS ABOUT THE REAL NUMBERS

Since $r-p$ is rational by Proposition (1.41), we have found a rational number between a and b, and we are done. ∎

The following is probably also familiar to the readers.

Theorem 1.48 *If $r = \dfrac{p}{q}$ is rational, and i is irrational, then $r + i$ is irrational.*

Proof. Suppose this is not the case. Then $r + i$ is rational, say $\dfrac{p'}{q'}$. Since $r = \dfrac{p}{q}$, this yields $\dfrac{p}{q} + i = \dfrac{p'}{q'}$, or $i = \dfrac{p'}{q'} - \dfrac{p}{q}$, which tells us (by Proposition 1.41) that i is rational. But we started with i being irrational! We have our contradiction. Since our contradiction arose from assuming that the sum of a rational number and an irrational number was rational, it must be true that the sum of a rational number and an irrational number is irrational. ∎

Theorem 1.49 *Between any two real numbers a and b, there is an irrational number.*

Proof. If a and b are real numbers such that $a < b$, then $a - \sqrt{2} < b - \sqrt{2}$. So by Theorem 1.47, there is a rational number such that

$$a - \sqrt{2} < r < b - \sqrt{2}.$$

By adding $\sqrt{2}$ to these inequalities, it follows that

$$a < r + \sqrt{2} < b.$$

Since $r + \sqrt{2}$ is irrational by the previous theorem, we have produced an irrational number between a and b. ∎

The following results will also be used routinely.

Proposition 1.50 *If a and b are real numbers, then*
(1) $|ab| = |a|\,|b|$
(2) $\left|\dfrac{a}{b}\right| = \dfrac{|a|}{|b|}$ *provided $b \neq 0$, and*
(3) $-|a| \leq a \leq |a|$.
(4) *If $c > 0$, then the inequality $|x| < c$ is equivalent to $-c < x < c$.*
(5) *The inequality $-c \leq x \leq c$, is equivalent to $|x| \leq c$, even when $c = 0$.*

Theorem 1.51 (Triangle Inequality) *If a and b are real numbers, then $|a + b| \leq |a| + |b|$.*

Proof. We know by part (3) of the previous proposition, that $-|a| \leq a \leq |a|$, and that $-|b| \leq b \leq |b|$. Adding these inequalities we get that $-(|a|+|b|) \leq a+b \leq |a|+|b|$, and this, by part (5) of the previous proposition (with $x = a + b$, and $c = |a| + |b|$), tells us that $|a + b| \leq |a| + |b|$. ∎

By induction, we can prove the following corollary.

Corollary 1.52 *If $a_1, a_2, a_3, ...a_n$, are real numbers, then $|a_1 + a_2 + ... + a_n| \leq |a_1| + |a_2| + ... + |a_n|$.*

Remark 1.53 *The name "triangle inequality" comes from the fact that if a and b are vectors in the plane which form sides of a triangle, then $c = a + b$ is the third side of the triangle. And if we think of $|a|$ and $|b|$ as the lengths of the sides of the triangle (which is common to do in calculus), the triangle inequality tells us that the length of the third side of the triangle is less than or equal to the sum of the lengths of the other two sides (a fact we know from geometry).*

Proposition 1.54 *(Reverse Triangle Inequality) If a and b are real numbers, then $||a| - |b|| \leq |a - b|$.*

We will guide you through the proof of this in the exercises.

EXERCISES

1. Prove that the sum, product, difference, and quotient, of rational numbers is rational provided that in the case of quotient we are not dividing by zero.

2. Show that between any two real numbers there are infinitely many rational numbers, and infinitely many irrational numbers.

3. Prove or disprove (by giving a counterexample): The sum of two irrational numbers is irrational.

4. Show there is only one null set.

5. Prove that if S and T are subsets of B, where $f : A \to B$, then if $S \subset T$, $f^{-1}(S) \subset f^{-1}(T)$.

6. Prove that if a and b are real numbers then $|a - b| \geq |a| - |b|$.

7. Prove the reverse triangle inequality. [Hint: Use the previous exercise and then reverse the roles of a and b. Finally, use Proposition 1.50 part (5).

8. The definition of a is less than b is that there is a positive number c such that $a + c = b$. Prove each of the following using this definition. You may assume that the product of two positive numbers is positive, and that the product of a positive number times a negative number is a negative number, as well as the usual rules for solving equations.

    (a) If $a < b$ and $b < c$ then $a < c$.
    (b) If $a < b$, and c is positive, then $ca < cb$.
    (c) If $a < b$ and c is negative, then $ca > cb$. (Hint: If $c < 0$, then $-c$ is positive.)
    (d) Show that if $0 < a < b$, then $a^2 < b^2$. Show that if $a < b$, it does not follow that $a^2 < b^2$.

9. We define $|a|$ to be a when $a \geq 0$, and $-a$, when $a < 0$.

    (a) Show $|a| \geq 0$.
    (b) Show that $|ab| = |a||b|$ by taking the cases when a and b have the same sign, and when they have opposite signs.
    (c) The geometric interpretation of $|a|$ is the distance a is from the origin. Using this interpretation, explain why if $c > 0$, the inequality $|x| < c$ and $-c < x < c$ are equivalent.

1.3 Logic Preliminaries

The statement "If p is true then q is true," is called a conditional statement and is written as $p \implies q$. The statement that p is true if and only if q is true is written as $p \iff q$. When we want to prove the statement $p \iff q$, there are two things we need to prove. One is the forward direction, that $p \implies q$; that is, we assume that p is true and try to show that q is true. In proofs, this forward direction will be abbreviated ($\implies$). The other thing we need to show is the reverse direction, that $q \implies p$, where we assume that q is true and then attempt to prove that p is true. This direction of proof is abbreviated ($\impliedby$). To illustrate, suppose we want to prove the theorem: p is a limit point of S if and only if there is a sequence of distinct points from S, none equal to p, converging to p. (Never mind what the words mean.) In the forward direction, ($\implies$), we have to prove that IF p is a limit point of S, THEN there is a sequence of distinct points from S, none equal to p, converging to p. In the backward direction, ($\impliedby$), we have to prove that IF there is a sequence of distinct points from S, none equal to p, converging to p, THEN p is a limit point of S.

We will often prove things by contradiction in this book, and many times that requires negating statements. Recall that the negation of a statement like, "Today is Tuesday," is the statement that says, "It is not true that Today is Tuesday." Said another way, the negation of "Today is Tuesday," is "Today is not Tuesday." The negation of "$x \geq 3$," is "It is not true that $x \geq 3$," or more simply, "$x < 3$." Given a statement p, the negation of p is denoted by $\sim p$.

Many times the statements we wish to negate have quantifiers like, "For each x," or "There exists a y." Another way of saying, "There exists a y" is, "For some y." Suppose that we wish to negate the statement, "For each $x, x \geq 3$." The negation of this is, "It is not true that for each x, $x \geq 3$." We can also say this as, "There exists an x for which $x < 3$," or, "For some x, $x < 3$." In general, the negation of the statement "For each x, p is true" is "For some x, p is false, or equivalently, "For some $x, \sim p$ is true." Similarly, the negation of "For some x, p is true," is "For all $x, \sim p$. is true, or equivalently, "For all x, p is false."

Example 1.55 *The negation of, "For some x, $x + y = 5$" is "For all x, $x + y \neq 5$." The negation of "For all x, $x > 0$," is "For some x, $x \leq 0$."*

If a statement has several quantifiers, we negate each one using the rules above.

Example 1.56 *The negation of, "For each x, there exists a y such that $x + y = 7$", would be, "There is some x, such that for all y, $x + y \neq 7$." The negation of "For some x, and for all y, $xy > 5$," would be "For each x, there is some y, such that $xy \leq 5$."*

Now let us take a few more complicated statements.

People will often negate an expression like, "For each $x > 0$, p is true," incorrectly. They will say, "There exists an $x \leq 0$, such that p is false." This is not correct. The negation would be "It is not true that for every $x > 0$, p is true". Said another way, "There is an $x > 0$, such that $\sim p$ is the case" or more simply, "There is some $x > 0$ for which p is false." So our point is that the negation of, "For each x *of a certain type*, p is true" is "For some x *of this type*, p is false."

Example 1.57 *The negation of, "There exists an $x < 0$ such that $x < -3$" is, "For each $x < 0$, $x \geq -3$." The negation of the statement, "For each $\varepsilon > 0$ there is a $\delta > 0$, such that $\varepsilon + \delta < 4$" is, "For some $\varepsilon > 0$, and for each $\delta > 0, \varepsilon + \delta \geq 4$."*

The contrapositive of the statement $p \implies q$ is $\sim q \implies \sim p$. Thus, the contrapositive of, "If today is Tuesday, then tomorrow is Wednesday" is, "If tomorrow is not Wednesday, then today is not Tuesday." Every conditional statement has the same truth value as its contrapositive, so if we wish to prove a theorem that begins with "If p then q", we need only prove the contrapositive, "If not q then not p," and we will have proved our theorem.

1.3. LOGIC PRELIMINARIES

EXERCISES

1. Write the negation of "For all x with $|x| < 2$, $|f(x)| < 4$."

2. Write the negation of "For all $x \in E$, there exists an r, such that the set $B_r \subset E$."

3. Write the negation of "For all $\varepsilon > 0$, there is a $\delta > 0$, such that if $x \in E$ and $|x| < \delta$, $|f(x)| < \varepsilon$."

4. Write the negation of "There exists an $M > 0$ such that for all n, $|a_n| < M$."

Chapter 2

Bounds and Sequences of Real Numbers

2.1 Preliminaries

In this chapter we will only be dealing with sets of real numbers.

Definition 2.1 *Suppose A is a set of real numbers. We say that a real number u is an **upper bound** for A, if $u \geq a$ for all $a \in A$. When a set A has an upper bound u, we say that A is **bounded above**, or more specifically, that A is bounded above by u.*

Example 2.2 *Suppose that A is the interval $(1,3]$. This set is bounded above. The number 3 is an upper bound for the set, since 3 is greater than or equal to every element in the set. But A is also bounded above by $4, 5.1, 10+\sqrt{2}$, and so on. In fact, if a set of real numbers has an upper bound, then it has infinitely many upper bounds. Any number larger than an upper bound for A is also an upper bound for A.*

Definition 2.3 *We say that l is a **lower bound** for a set, A, of real numbers, if $l \leq a$ for every $a \in A$. If a set A has a lower bound, l, then we say that A is **bounded below**, or more specifically, that A is bounded below by l.*

Example 2.4 *The set $(1,3]$ is bounded below by 1, since 1 is less than or equal to all elements of this set. It is also bounded below by $0, -1, -2, -\pi$, and so on. Notice that upper and lower bounds of a set may or may not be contained in the given set.*

Definition 2.5 *A set is said to be **bounded**, if it is bounded above and below.*

Example 2.6 *The set $A = (1,3]$ is bounded, since it is bounded above by 3 and below by 1. The set $B = (-\infty, 3)$ is bounded above by 3, but is not bounded below, and the set $C = [3, \infty)$ is bounded below by 3, but is not bounded above. The set of integers, Z, is neither bounded above nor below.*

The following is an alternate way of defining a bounded set, which is quite useful:

Definition 2.7 *(Alternate) We say that a set A is **bounded**, if for some positive number M, $|a| \leq M$ for all $a \in A$. This is equivalent to saying that the set A can be enclosed by the interval $[-M, M]$, or that A is a subset of $[-M, M]$ for some positive M. If $A \subset [-M.M]$, then obviously it is bounded below by $-M$ and above by M. Conversely, if a set A is bounded above by u, and below by l, then, then $|a| \leq M$ where M is the larger of $|u|$ and $|l|$. So the two definitions of bounded are equivalent (in that either one implies the other), and we use whichever is most convenient at a given time.*

Definition 2.8 *We say that a number, s, is the **least upper bound** or **supremum** of a set A, if (1) s is an upper bound for A, and (2) s is less than or equal to any other upper bound for A.*

We abbreviate the supremum of a set A by sup A. Alternately, we can denote the supremum of A by LUB(A).

Example 2.9 *3 is the least upper bound, or supremum, of the finite set $A = \{1, 2, 3\}$. It is also the least upper bound for the closed interval $[1, 3]$, for the open interval $(1, 3)$, as well as for the set of rational numbers between 1 and 3.*

Definition 2.10 *We say that a number, i, is a **greatest lower bound** or **infimum** of a set A, if (1) i is a lower bound for A, and (2) i is greater than or equal to any other lower bound for A.*

The greatest lower bound for a set A is abbreviated inf A or $GLB(A)$.

Example 2.11 *The greatest lower bound for any finite set of real numbers is always the smallest number in the set. The greatest lower bound or infimum of the set A of positive numbers is 0. The infimum of the set of rational numbers whose squares are greater than 2 is, as intuition suggests, $\sqrt{2}$.*

Example 2.12 *If $A = \{1 + 1/n |\ n \in Z^+\} = \{2, 3/2, 4/3, 5/4, ...\}$, then inf $A = 1$ while sup $A = 2$. Notice that sup A is contained in A, but inf A is not. So the supremum or infimum of a set may or may not be in the set.*

2.1. PRELIMINARIES

Definition 2.13 *If A is not bounded above we write $\sup A = \infty$, and if A is not bounded below we write $\inf A = -\infty$. A set is called **unbounded** if either $\sup A = \infty$ or $\inf A = -\infty$.*

Example 2.14 *If A is the set of integers, then $\sup A = \infty$ and $\inf A = -\infty$. Thus, the set of integers is unbounded. If B is the set of real numbers greater than 2, then $\inf B = 2$, but $\sup B = \infty$, so B is also unbounded.*

Remark 2.15 *One of the things in analysis that one sees routinely is a statement like this: Since the number N is greater than or equal to all $s \in S$, N is greater than or equal to $\sup S$. This statement follows from the definition of least upper bound. For, if $N \geq s$ for all s in S, then N is an upper bound for S. And since the least upper bound or supremum of S is less than or equal to all upper bounds for S, $\sup S \leq N$, or equivalently, $N \geq \sup S$. In a similar manner, if $N \leq s$ for all $s \in S$, then $N \leq \inf S$. We will use these results routinely, so it is important to internalize them for a better reading of this primer.*

Throughout this primer we will accept the following theorem, (known as the Completeness Theorem). A proof of this is beyond the scope of this book.

Theorem 2.16 *(**Completeness Theorem**) Any set of real numbers which is bounded above, has a least upper bound. Any set of real numbers which is bounded below, has a greatest lower bound.*

One can easily deduce some non trivial results from this theorem.

Example 2.17 *It can be shown that the set $\{(1+1/n)^n | \, n \in Z^+\}$, is bounded above by 3. (We will guide you through a proof of this in an exercise in the next section.) So it has a least upper bound, which turns out to be the number e, which is approximately 2.71828. This fact is hardly obvious to the novice.*

Example 2.18 *It is clear that the set of numbers of the form $\dfrac{2^n}{n!}$ where n is a positive integer, is bounded below by 0. Intuition tells us (since the factorial grows so big so fast), that these fractions approach 0 as n gets large. So it seems that 0 is the greatest lower bound of this set. And in fact, it is true. Proving it, however, requires some cleverness. We guide you with one way to do it in the exercises in the next section.*

EXERCISES

1. Show that a set can only have one supremum and one infimum.

2. By writing out the elements of the set, determine the greatest lower bound and least upper bound of the following sets. No proofs are necessary. Determine if the sets are bounded or not.

   (a) $A = \{(-1)^n | \; n = 1, 2, 3, ...\}$

   (b) $B = \{\left(\frac{1}{2}\right)^n | \; n = 1, 2, 3, ...\}$

   (c) $C = \left\{n - \dfrac{2}{3+n^2} | \; n = 1, 2, 3, ...\right\}$

   (d) $D = \{(-1)^n n | \; n = 1, 2, 3, ...\}$

   (e) $E = \left\{\dfrac{2n+3}{3n+2} | \; n = 1, 2, 3, ...\right\}$

   (f) $F = \{$rational numbers between 0 and 1$\}$

   (g) $G = \{\sqrt{n+1} - \sqrt{n} | \; n = 1, 2, 3, ...\}$

3. Show that if B is bounded, and $A \subset B$, then A is bounded.

4. Show that if $a \leq b$ for each $a \in A$ and $b \in B$, then $\sup A \leq \inf B$.

5. Is 3 the least upper bound of the set of irrational numbers strictly between 1 and 3? Prove your answer.

6. Show that the supremum of the set of rationals less than $\sqrt{2}$, is $\sqrt{2}$.

7. Show that if we assume that every set of numbers which is bounded above has a least upper bound, then it follows that every set of real numbers which is bounded below has a greatest lower bound.

2.2 Basic Theorems

In this section we give some of the fundamental facts about the supremum and infimum of a set.

Theorem 2.19 *(a) If $A \subset B$, and B is bounded above, then $\sup A \leq \sup B$.*
(b) If $A \subset B$, and B is bounded below, then $\inf A \geq \inf B$.

(You can refer to Figure 2.1, which will help you to remember this theorem. There we have taken B to be an interval and A to be a subinterval. Of course, this theorem is true for sets that are not intervals.)

2.2. BASIC THEOREMS

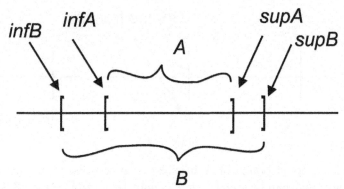

Figure 2.1. When $A \subset B$, sup A is less than or equal to sup B and inf A is greater than or equal to inf B.

Proof. (a): We know from the Completeness Theorem (Theorem 2.16), that since B is bounded above, B has a least upper bound, which we call x. Now, by definition of upper bound, $x \geq b$ for all $b \in B$. But since everything in A is in B by virtue of A being a subset of B, $x \geq a$ for all $a \in A$. But if $x \geq a$ for all $a \in A$, then

$$x \geq \sup A \tag{2.1}$$

by our Remark 2.15 on page 25. Since by definition $x = \sup B$, we can substitute into (2.1) to get that $\sup B \geq \sup A$. ∎

Now, as an important part of your training, you should try to prove part (b). You might stumble initially, which is not unusual, but if you do, keep trying. When you need to check your answer, you can always find the solution on the Internet.

The next theorem is a very important one in analysis, and is used continuously in this primer.

Theorem 2.20 *(1) Suppose that A is bounded above, and $y = \sup A$. Then if $b < y$, there will always exist an element $a \in A$, such that*

$$b < a \leq y. \tag{2.2}$$

(2) Suppose that A is bounded below, and that $z = \inf A$. Then if $b > z$, there will always exist an element $a \in A$, such that

$$z \leq a < b. \tag{2.3}$$

Proof. (1) Figure 2.2 illustrates part (1) of this theorem.

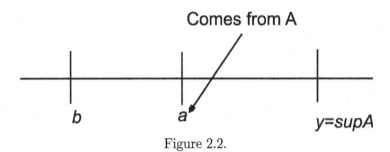

Figure 2.2.

Since y is the least upper bound of A, it is less than or equal to any upper bound for A. Since y is not less than or equal to b, b cannot be an upper bound for A. Since b is not an upper bound for A, b is not greater than or equal to everything in A. Thus, there must be an $a \in A$ greater than b. This proves half of what we want. As for the other half, we know $a \leq y$ since y is an upper bound for A, and $a \in A$. This completes the proof.

We ask you to write out the proof of (2). It essentially amounts to changing the words "upper bound" to "lower bound" and reversing the inequalities. ∎

Corollary 2.21 *If A is a bounded set, then both (2.2) and (2.3) are true.*

Proof. This follows since a bounded set is bounded above and below. ∎

Example 2.22 *Here are two examples to illustrate part (1) of the theorem. Suppose $A = (0, 1)$. Then $\sup A = 1$. The theorem tells us that if we consider any number less than 1, there is something in A greater than it. This is pretty clear.*

As another example, suppose that we adjoin the number 2 to the set A. That is, let $B = (0, 1) \cup \{2\}$. Then $\sup B = 2$. Suppose we consider something smaller than 2, say 1. The theorem asserts there is something in the set greater than 1 but ≤ 2. Of course, 2 is such a number.

The following lemma is also used a lot in analysis. The result is very intuitive.

Lemma 2.23 *(1) If for any $\varepsilon > 0$, $y \geq N - \varepsilon$, then $y \geq N$. (Said another way, if y is $\geq$ everything less than N, than y itself must be greater than or equal to N.)*
(2) Similarly, if $y \leq N + \varepsilon$ for all $\varepsilon > 0$, then $y \leq N$.

Proof. (of (1)): We do a proof by contradiction. Suppose that $y < N$. Then by Theorem 2.20, there is a number r such that $y < r < N$. Since $y < r$, this contradicts that y is greater than or equal to everything less than N. This contradiction, which arose from assuming that $y < N$, shows that $y < N$ is false. So, y must be $\geq N$. We leave the proof of the second part to you. ∎

2.2. BASIC THEOREMS

Definition 2.24 *If A and B are sets of real numbers, then $A + B = \{a + b|\ a \in A$ and $b \in B\}$.*

Example 2.25 *If $A = \{1, 2, 3\}$ and B is the set $\{4, 5, 6\}$, then $A + B = \{1+4, 1+5, 1+6, 2+4, 2+5, 2+6, 3+4, 3+5, 3+6\} = \{5, 6, 7, 8, 9\}$. (Recall that we don't write elements of a set more than once.)*

The next result is a bit surprising to beginners.

Theorem 2.26

(1) If A and B are sets of real numbers which are bounded above, then
$$\sup(A + B) = \sup A + \sup B.$$

(2) If A and B are sets of real numbers which are bounded below, then
$$\inf(A + B) = \inf A + \inf B.$$

Proof. (1) We know that
$$\begin{aligned} a &\leq \sup A \text{ for all } a \in A, \text{ and that} \\ b &\leq \sup B \text{ for all } b \in B. \end{aligned}$$

If we add these two inequalities, we get
$$a + b \leq \sup A + \sup B, \text{ for all } a \in A \text{ and } b \in B.$$

This says that the fixed number, $\sup A + \sup B$, is greater than or equal to everything in $A + B$. So, by Remark 2.15 we have,
$$\sup(A + B) \leq \sup A + \sup B. \tag{2.4}$$

(Here we are taking N in that remark to be $\sup A + \sup B$, and S to be $A + B$.)

Now, to get the inequality going the other way we observe, that by Theorem 2.20, there is an $a \in A$ and a $b \in B$ such that
$$\begin{aligned} (\sup A) - \varepsilon/2 &< a, \text{ and} \\ (\sup B) - \varepsilon/2 &< b. \end{aligned}$$

Adding these two inequalities and noting that the supremum of a set is greater than or equal to every element of the set, we get
$$(\sup A) + (\sup B) - \varepsilon \leq a + b \leq \sup(A + B). \tag{2.5}$$

Now we appeal to Lemma 2.23 (with $N = \sup A + \sup B$, and $y = \sup(A + B)$), to conclude that
$$(\sup A) + (\sup B) \leq \sup(A + B). \tag{2.6}$$
Now, from (2.4) and (2.6) we have that
$$\sup A + \sup B = \sup(A + B).$$

We encourage you to write out the proof to part two of the theorem. The proof essentially switches all the inequalities, and uses the word "inf" instead of "sup." ∎

Since bounded sets are bounded above and below, we have:

Corollary 2.27 *If A and B are bounded sets of real numbers, then $\sup(A + B) = \sup A + \sup B$ and $\inf(A + B) = \inf A + \inf B$.*

Later in this primer we will need the following definition:

Definition 2.28 *If A is a set of real numbers and c is a constant, then $cA = \{ca|\ a \in A\}$.*

Theorem 2.29 *If A is a bounded set of real numbers, then*

(1) $\sup cA = c \sup A$ when $c > 0$, and
(2) $\sup cA = c \inf A$ when $c < 0$. In particular, when $c = -1$, we have that $\sup(-A) = -\inf A$.

Proof. (1) Where do we begin? Deciding that is always part of the learning process. Often, it is trial and error. We may try something, and it doesn't work. So we try something else, and we might make some progress. Where to begin is not always clear. Here we begin with:
$$a \leq \sup A \text{ for each } a \in A, \tag{2.7}$$
since the supremum of a set is greater than or equal to everything in the set. Multiplying inequality (2.7) by c, won't flip the inequality, since we are assuming c is positive. So we have:
$$ca \leq c \sup A \text{ for each } a \in A. \tag{2.8}$$
Now, the elements on the left side of (2.8) are the elements in the set cA, so by Remark 2.15,
$$\sup cA \leq c \sup A. \tag{2.9}$$
We need to get the inequality in (2.9) going the other way in order to finish the proof. We know that
$$ca \leq \sup cA \text{ for each } a \in A. \tag{2.10}$$

2.2. BASIC THEOREMS

If we divide (2.10) by c, which is positive, we get

$$a \leq \frac{\sup cA}{c} \text{ for each } a \in A. \tag{2.11}$$

Now again by Remark 2.15, taking N to be $\frac{\sup cA}{c}$ and S to be A, we get that

$$\sup A \leq \frac{\sup cA}{c}. \tag{2.12}$$

If we multiply (2.12) by c, we get

$$c \sup A \leq \sup cA \tag{2.13}$$

From (2.9) and (2.13) we get that

$$\sup cA = c \sup A,$$

which is what we were trying to prove.

Again, we ask you to prove the second part of the theorem. It is very instructive to do so. Note that if $c = 0$, then trivially $c \sup A = \sup cA$. ∎

When people first see this theorem, they often think part (2) of the theorem is incorrect. We illustrate both parts of the theorem now.

Example 2.30 *Suppose $A = [-3, 5]$ and $c = 2$. Then $\sup A = 5$ and $c \sup A = 10$. Since $cA = [-6, 10]$, $\sup cA = 10$ which equals $c \sup A$. Now, start over with the same A but let $c = -5$. Then $\inf A = -3$ and $c \inf A = 15$. But $cA = [-25, 15]$ and $\sup cA = 15$, which equals $c \inf A$.*

EXERCISES

1. A maximum element of a set, A, is defined to be an element of A which is greater than or equal to every element of A. Show that if $\sup A \in A$, then $\sup A$ is the maximum element of A.

2. Show that if a is a real number and B is a bounded set, then $\sup\{a + b | b \in B\} = a + \sup B$.

3. Prove the second part of Theorem 2.29.

4. True or False: If $f(x) = 3x + 2$ is defined on a bounded subset E of the real line, and $\sup E = 2$, then $\sup f(E) = 8$.

2.3 Sequences

Intuitively, a sequence of real numbers is a listing of numbers, with or without a pattern. So, for example, each of the listings, $1, 3, 5, 7, ...$ and $2, \pi, e, \sqrt{2}, ...$, are sequences. Only, in the first case the three dots, "...", seem to indicate that we are listing the odd positive integers and continuing. In the second case it is not clear what the three dots mean, except that there is more to the listing.

The first number that appears in the listing of a sequence is called the first term of the sequence, the second number in the listing is called the second term, etc. So a general sequence can be written as $a_1, a_2, a_3, ...$, and of course a_n represents the nth term of the sequence. The formal definition of a sequence of real numbers is as follows:

Definition 2.31 *A sequence of real numbers is a function $f : Z^+ \to R$. We call $f(1)$ the first term, and $f(2)$ the second term, and so on. So $f(1)$ is what we referred to as a_1 earlier, and $f(2) = a_2$, and so on.*

While this may be the formal definition, it is much easier to think of a sequence of numbers as a listing, rather than as a function.

There are several notations we use to represent the sequence $a_1, a_2, a_3, ...$ One such notation is $\{a_n\}$. Obviously, this can cause confusion since we use the same notation for sets. A better notation might be $\{a_n\}_{n=1}^{\infty}$, which now clearly distinguishes this sequence from a set. We will use the former notation when it is clear we are talking about a sequence. For example, if we say "the sequence $\{a_n\}$" or later on when we talk about the "convergence of $\{a_n\}$", it will be clear that $\{a_n\}$ refers to a sequence. Sometimes, we will just write the listing of the sequence, and sometimes we will write $\{a_n\} = a_1, a_2, a_3, ...$ This is not very standard notation, but it certainly clarifies that we are talking about a sequence and not a set. One thing to take note of right away is that when talking about sequences in this book, n can only be a positive integer.

Sequences are a fundamental concept in analysis and are used in many areas related to analysis. They are as fundamental as the notion of a variable is in algebra. Probably the most important concept related to sequences is the notion of convergence. Intuitively, we say that a sequence of real numbers, $\{a_n\}$, converges to a real number L, if as you go farther and farther out in the sequence, the terms of the sequence get closer and closer to L. In fact, the terms can be made as close as we want to L by going far enough out in the sequence. If we picture the terms of the sequence (as well as L), as points on the number line (see Figure 2.3), we want the *distance* between the terms of the sequence and L to get smaller and smaller as we go farther and farther out in the sequence.

In Figure 2.3, one sees three different ways a sequence can converge to L. We have indicated in that figure that the terms after a_{100} are starting to get closer to L. That

2.3. SEQUENCES

is, a_{102} is closer to L than a_{101} is, and a_{103} is closer to L than a_{102} is, and so on. Although this picture shows three different ways a sequence can converge, there are other ways this can happen. In the example we have just given, it need not be that a_{102} is closer to L than a_{101} is and that a_{103} is closer to L than a_{102} is. They just need "to be sufficiently close" to L after a certain point. Obviously, this informal approach will not do. We need to be more precise, and soon, we will be.

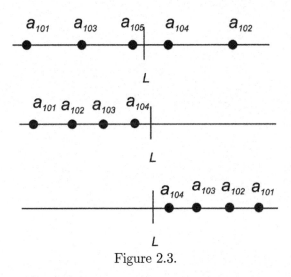

Figure 2.3.

Suppose we want to show that the sequence $\left\{\dfrac{n}{n+1}\right\}$ converges to 1. We have to show that, as we go farther out in the sequence, the terms get close to 1; that is, that the distance between the numerical values of the terms of the sequence and the number 1, can be made as small as we want. Now, how do we measure the distance between two numbers, a and b, on the number line? We measure it by the absolute value of their difference. So either $|a - b|$ or $|b - a|$ will represent this distance. Thus, to show that $\left\{\dfrac{n}{n+1}\right\}$ gets close to 1 as n gets larger, we need to show that $\left|\dfrac{n}{n+1} - 1\right|$ can be made as small as we want when we take a large enough n. The notion of smallness, however, is relative. To one person, saying the distance between $\dfrac{n}{n+1}$ and 1 is small might mean $\left|\dfrac{n}{n+1} - 1\right| < \dfrac{1}{10}$. To another, it might mean that $\left|\dfrac{n}{n+1} - 1\right| < \dfrac{1}{100}$, and to another, it might mean $\left|\dfrac{n}{n+1} - 1\right| < \dfrac{1}{100{,}000{,}000}$. In order to satisfy everyone's notion of closeness, we don't use a numerical value after the $<$

34 CHAPTER 2. BOUNDS AND SEQUENCES OF REAL NUMBERS

sign; rather, we use a letter, ε, which can represent anyone's measure of closeness. This discussion leads to the following definition of the limit of a sequence.

Definition 2.32 *We say that the sequence $\{a_n\}$ **converges** to L, or L **is the limit of the sequence** $\{a_n\}$, if for any $\varepsilon > 0$ (for any measure of closeness), there is an $N > 0$, such that $n > N$ implies $|a_n - L| < \varepsilon$. When a sequence doesn't converge to a finite number, we say that the sequence **diverges**.*

Notation 2.33 *When the sequence $\{a_n\}$ converges to L, we either write $a_n \to L$, $\lim_{n \to \infty} a_n = L$, or more simply, $\lim a_n = L$.*

Example 2.34 *Show that the sequence $\left\{\dfrac{1}{n}\right\}$ converges to the number 0.*

Solution: We want to show that we can make the terms of the sequence as close to 0 as we want, by going far enough out in the sequence. That is, we want to make the distance between $\dfrac{1}{n}$ and 0, which is $\left|\dfrac{1}{n} - 0\right| = \dfrac{1}{n}$, to be less than ε after a certain point. But if we want $\dfrac{1}{n}$ to be less than ε, we need only let n be any integer $> \dfrac{1}{\varepsilon}$. Here, we take our N to be $\dfrac{1}{\varepsilon}$. Notice that $\left|\dfrac{1}{n}\right| = \dfrac{1}{n}$, since n, being the term number in the sequence, is always positive.

Remark 2.35 *The N in the definition above, usually depends on ε, as we have just seen. Some books require that N be an integer in the definition. This is unnecessary, though n will always be a positive integer since it represents which term in a sequence we are talking about (first, second, etc.). So n can only be an integer. Also, some books state the latter part of the definition as $n \geq N$ implies $|a_n - L| < \varepsilon$, instead of $n > N$. All of these definitions are equivalent and all yield the same theory.*

Example 2.36 *Show that the sequence $\left\{\dfrac{n}{n+1}\right\}$ converges to 1.*

Solution: We begin by picking any $\varepsilon > 0$. According to the definition of the limit of a sequence, we have to find an N such that $n > N$ implies that $\left|\dfrac{n}{n+1} - 1\right| < \varepsilon$, or equivalently (by combining the terms in the absolute value), that $\left|\dfrac{-1}{n+1}\right| = \left|\dfrac{1}{n+1}\right| < \varepsilon$. Also, since n is a positive integer in all the sequences we are talking about, $\dfrac{1}{n+1}$ is positive, so we don't need the absolute values. Thus, showing that $\left|\dfrac{n}{n+1} - 1\right| < \varepsilon$

2.3. SEQUENCES

really amounts to showing that $\frac{1}{n+1} < \varepsilon$ *"after a certain point in the sequence"; which is more formally described as* $\frac{1}{n+1} < \varepsilon$ *for* $n > N$ *for some N. We need only multiply both sides of* $\frac{1}{n+1} < \varepsilon$ *by* $n+1$, *divide by* ε *and subtract 1, to get that for* $n > N = \frac{1}{\varepsilon} - 1$, *it will be true that* $\left|\frac{n}{n+1} - 1\right| = \frac{1}{n+1} < \varepsilon$. *Since we have found an N for every possible* ε, *we are done.*

Example 2.37 *Show that* $\left\{\frac{2n^2}{3n^2 + 3n + 1}\right\}$ *converges to* $\frac{2}{3}$.

Solution: *We examine*

$$\left|\frac{2n^2}{3n^2 + 3n + 1} - \frac{2}{3}\right|$$
$$= \left|\frac{(-6n - 2)}{3(3n^2 + 3n + 1)}\right|$$
$$= \frac{6n + 2}{3(3n^2 + 3n + 1)}$$
$$< \frac{6n + 2}{3n^2}$$
$$< \frac{12n}{3n^2}$$
$$= \frac{4}{n}.$$

Here we have made use of the fact that since $-6n - 2$ *is negative, its absolute value is its negative, which is* $6n + 2$, *and also that since* $3(3n^2 + 3n + 1)$ *is positive, we don't need the absolute value notation around it. We have also made use of the fact that* $\frac{6n+2}{3(3n^2+3n+1)} < \frac{6n+2}{3n^2}$, *since the denominator of the fraction on the left side of the inequality, is bigger than the denominator of the fraction on the right side of the inequality, but the numerators are the same. Finally, since* $n \geq 1$ *in any sequence,* $6n + 2 < 6n + 6n = 12n$. *Getting back to the main thrust of the problem, we wish to show that for any* $\varepsilon > 0$, *we can make* $\left|\frac{2n^2}{3n^2+3n+1} - \frac{2}{3}\right| < \varepsilon$ *for* $n > N$, *for some N. But* $\left|\frac{2n^2}{3n^2+3n+1} - \frac{2}{3}\right| < \frac{4}{n}$, *as we have seen, and we can make* $\frac{4}{n} < \varepsilon$ *by making* $n > \frac{4}{\varepsilon}$. *Thus, we need only take* $N = \frac{4}{\varepsilon}$ *to be sure that* $\left|\frac{2n^2}{3n^2+3n+1} - \frac{2}{3}\right| < \varepsilon$, *for* $n > N$.

Remark 2.38 *The key idea behind the notion of a sequence converging is that for **any** $\varepsilon > 0$, we can find a point in the sequence after which $|a_n - L| < \varepsilon$. Since ε represents **any** positive real number, we can, if it is more convenient, replace ε by $\varepsilon/2$, $\varepsilon/3$, or anything else we might wish. Thus, if $\{a_n\}$ converges to L, we can find an N_2 such that $|a_n - L| < \varepsilon/2$ for $n > N_2$. We can find an N_3 such that for $n > N_3$, $|a_n - L| < \varepsilon/3$, and we can also find an N_4 such that $n > N_4$ implies that $|a_n - L| < \varepsilon |M|/2\pi$, where M is any nonzero number.*

Remark 2.39 *When we talk about a sequence $\{a_n\}$ converging to L and want to discuss ideas intuitively, we will use expressions like "we can make $|a_n - L|$ small after some point." This is just a verbal description of "for any $\varepsilon > 0$ there is an $N > 0$ such that $n > N$ implies $|a_n - L| < \varepsilon$," and often helps in understanding the intuitive explanations.*

Remark 2.40 *There is a geometric interpretation of $\{a_n\}$ converging to L. When a sequence converges to L, then for any $\varepsilon > 0$ the terms of $\{a_n\}$ are eventually contained in the interval $(L - \varepsilon, L + \varepsilon)$; that is, for any $\varepsilon > 0$ there is an N such that $n > N$ implies that $a_n \in (L - \varepsilon, L + \varepsilon)$. See Figure 2.4, where we indicate that for the given ε, the terms after a_{99} are in the interval $(L - \varepsilon, L + \varepsilon)$.*

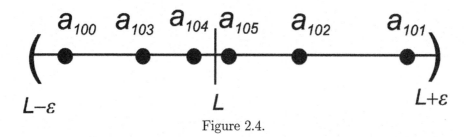

Figure 2.4.

This geometric interpretation follows since $|a_n - L| < \varepsilon$ is equivalent to $-\varepsilon < a_n - L < \varepsilon$, which in turn is equivalent to $L - \varepsilon < a_n < L + \varepsilon$. The more complicated a convergent sequence is, the more difficult it becomes to use the definition of a limit to show that the limit exists. So we need to prove some theorems that will make our study of convergence somewhat easier. The formal definition of a limit is all we need to establish the basic results. What we must establish right away (to give us the control we need), is that if a sequence converges, it can only converge to one limit. To do that, we need a lemma, whose conclusion seems pretty clear. But in analysis, what seems intuitively clear is not always rigorously clear, and more to the point, sometimes things that seem clear, are not even true! So we will take the time to prove the obvious. This is also good training in how to construct proofs.

2.3. SEQUENCES

Lemma 2.41 *If $|a| \leq \varepsilon$ for all $\varepsilon > 0$, then $a = 0$.*

Proof. We do a proof by contradiction. Suppose that a is not zero. Then $|a|$ is cannot equal zero, and so the number $\varepsilon = \dfrac{|a|}{2}$ is positive. Now, since we are told that $|a|$ is less than *any* $\varepsilon > 0$, we have that $|a| < \dfrac{|a|}{2}$, and this is a contradiction, of course, since a positive number cannot be less than half of itself. Since our contradiction arose from assuming $a \neq 0$, a must be 0. ∎

Theorem 2.42 *The limit of a convergent sequence is unique.*

Proof. Suppose that L and M are both limits of the sequence $\{a_n\}$. We will prove that $L = M$, thus establishing the uniqueness of the limit. Suppose then, that ε is any positive number. Since we are assuming that $\{a_n\}$ converges to both L and M, we have, from the definition of a limit, that there is an N_1 such that

$$|a_n - L| < \frac{\varepsilon}{2} \text{ for } n > N_1, \qquad (2.14)$$

and there is an N_2 such that

$$|a_n - M| < \frac{\varepsilon}{2} \text{ for } n > N_2. \qquad (2.15)$$

Now, let N be any number bigger than both N_1 and N_2. Then we have

$$\begin{aligned}
|L - M| &= |L - a_n + a_n - M| \\
&\leq |L - a_n| + |a_n - M| \quad \text{(triangle inequality)} \\
&= |a_n - L| + |a_n - M| \\
&\leq \frac{\varepsilon}{2} + \frac{\varepsilon}{2} \quad \text{(by (2.14) and (2.15))} \\
&= \varepsilon.
\end{aligned}$$

This string of inequalities shows that we can make $|L - M| \leq \varepsilon$ for any $\varepsilon > 0$ by going far enough out in the sequence. Hence by the previous lemma (with $a = L - M$), $L - M = 0$, so that $L = M$. ∎

Theorem 2.43 *If $\{a_n\}$ is a sequence converging to L, then $\{|a_n|\}$ converges to $|L|$.*

Proof. Figure 2.5 shows part of a sequence converging to L. The terms a_{101}, a_{102}, etc., are getting closer and closer to L. It appears that the absolute values of the terms are getting closer and closer to $|L|$.

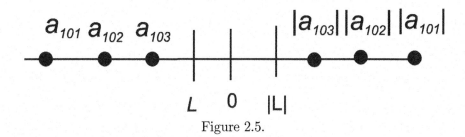

Figure 2.5.

The following is the formal way to establish this theorem. Our goal is to show that for any $\varepsilon > 0$, we can make $||a_n| - |L|| < \varepsilon$ when $n > N$, for some N. But we know that since $a_n \to L$, we can make

$$|a_n - L| < \varepsilon \text{ when } n > N, \text{ for some } N. \tag{2.16}$$

This same N will work for the sequence $\{|a_n|\}$. Here are the steps using the reverse triangle inequality (Proposition 1.54):

$$||a_n| - |L|| \leq |a_n - L| < \varepsilon, \text{ which is true for } n > N \text{ by (2.16)}.$$

Since we have shown that we can make $||a_n| - |L|| < \varepsilon$ when $n > N$ for some N, we have shown that $|a_n|$ converges to $|L|$. ∎

Theorem 2.44 *If $\{a_n\}$ is a sequence converging to L, then the set $\{a_n\}$ is bounded.*

Proof. Intuitively this is clear if we look at Figure 2.6,

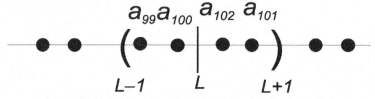

Figure 2.6. A convergent sequence is bounded

since if the sequence converges, the terms of the sequence will eventually lie in $(L-1, L+1)$, and the finite number of terms that don't lie in that interval can be enclosed by *some* interval. So all the terms of the sequence can be enclosed by one big interval, making the range of the sequence, or the set $\{a_n\}$, bounded. Here is the formal proof: Since $\{a_n\}$ converges to L, we know we can surely find a positive integer N, such that $n > N$ implies that

$$|a_n - L| < 1. \tag{2.17}$$

2.3. SEQUENCES

Now it follows that for $n > N$,

$$\begin{aligned} |a_n| &= |a_n - L + L| \\ &\leq |a_n - L| + |L| \text{ (by the triangle inequality (Theorem 1.51))} \\ &< 1 + |L| \quad \text{(by 2.17)} \\ &\leq 1 + |L|. \end{aligned}$$

We have shown that $|a_n| \leq M_1$ where $M_1 = 1 + |L|$, for $n > N$. But we want all the terms to have an absolute value less than or equal to some number M, not just those a_n with $n > N$. So we let M be the largest of $|a_1|, |a_2|, ... |a_N|$ and M_1. Now we have that $|a_n| \leq M$ for all n, and this is the definition of the set $\{a_n\}$ being bounded. (See page 24.) ∎

The next theorem is a very fundamental one in the study of sequences. Before giving the next theorem, we clarify the meaning of some of its terms. When we talk about $\{a_n + b_n\}$, we mean the sequence $\{c_n\}$ where $c_1 = a_1 + b_1, c_2 = a_2 + b_2, c_3 = a_3 + b_3$, etc. (That is, $\{c_n\}$ is the sequence obtained by adding the corresponding terms of the sequences $\{a_n\}$ and $\{b_n\}$.) So if $\{a_n\} = -1, 1, -1, 1, ...$, and $\{b_n\} = 1, -1, 1, -1, ...$, the sequence $\{a_n + b_n\} = 0, 0, 0, ...$. The sequences $\{a_n - b_n\}, \{a_n b_n\}, \left\{\dfrac{1}{b_n}\right\}$, and $\left\{\dfrac{a_n}{b_n}\right\}$, have similar meanings. (In the latter two sequences, $b_n \neq 0$.)

Theorem 2.45 *If $\{a_n\}$ and $\{b_n\}$ are sequences of real numbers that converge to L and M respectively, then*
  (1) $\{a_n + b_n\}$ converges to $L + M$,
  (2) $\{a_n - b_n\}$ converges to $L - M$,
  (3) $\{a_n b_n\}$ converges to LM,
  (4) $\left\{\dfrac{1}{b_n}\right\}$ converges to $\dfrac{1}{M}$, if $M \neq 0$, and
  (5) $\left\{\dfrac{a_n}{b_n}\right\}$ converges to $\dfrac{L}{M}$ if $M \neq 0$.

Proof. (1): According to Remark 2.38, since $\{a_n\}$ and $\{b_n\}$ converge to L and M respectively, we can make the following statement: For any given $\varepsilon > 0$, there exists an N_1 such that

$$|a_n - L| < \frac{\varepsilon}{2} \text{ for } n > N_1, \tag{2.18}$$

and an N_2 such that

$$|b_n - M| < \frac{\varepsilon}{2} \text{ for } n > N_2. \tag{2.19}$$

If we let N be any number larger than N_1 and N_2, then both (2.18) and (2.19) hold when $n > N$. Now, suppose $\varepsilon > 0$ is given. To show that $\{a_n + b_n\}$ converges to $L + M$, we need to show that we can make

$$|a_n + b_n - (L + M)| < \varepsilon$$

after a certain point. We take n to be $> N$, where N is as above. Then we have

$$\begin{aligned}|a_n + b_n - (L + M)| &= |a_n + b_n - L - M| \\ &= |a_n - L + b_n - M| \\ &\leq |a_n - L| + |b_n - M| \quad \text{(by the triangle inequality)} \\ &< \frac{\varepsilon}{2} + \frac{\varepsilon}{2} = \varepsilon. \quad \text{(using (2.18) and (2.19))}.\end{aligned}$$

Since we have shown that for any $\varepsilon > 0$ we can make $|a_n + b_n - (L + M)| < \varepsilon$ after a certain point, we have shown that $\{a_n + b_n\}$ converges to $L + M$. The proof of (2) is similar, and we leave it to you.

Proof of (3): This is a bit trickier than you might think. Our goal is to show that $|a_n b_n - LM|$ gets small after a certain point (that is, can be made less than ε for all $n > N$, some N.). But,

$$\begin{aligned}|a_n b_n - LM| &= |a_n b_n - L b_n + L b_n - LM| \\ &= |b_n(a_n - L) + L(b_n - M)| \\ &\leq |b_n(a_n - L)| + |L(b_n - M)| \quad \text{(triangle inequality)} \\ &= |b_n||a_n - L| + |L||b_n - M| \quad \text{(Proposition 1.50 part (1))}.\end{aligned}$$

Now, because $|b_n - M|$ gets small as n gets large, the term $|L||b_n - M|$ should also get small. But can we say the same thing for $|b_n||a_n - L|$? If $|b_n|$ is not bounded, then it is not so clear. But if we knew that $|b_n|$ is bounded, then since $|a_n - L|$ can be made small, the answer would be yes. Fortunately, we know that $|b_n|$ is bounded by Theorem 2.44, so intuitively, we should be able to make $|b_n||a_n - L|$ small, and therefore make $|a_n b_n - LM|$ small. Now, let's formalize all this: Let $\varepsilon > 0$ be given. Since $\{b_n\}$ converges, we know it is bounded. So $|b_n| \leq p$ for some $p > 0$ and for all n. Since $\{a_n\}$ converges to L, we know that we can make $|a_n - L| < \dfrac{\varepsilon}{p + |L|}$ and that the same is true for $|b_n - M|$ when $n > N$, for some N. Now we can put on the finishing touches. When $n > N$,

$$\begin{aligned}|a_n b_n - LM| &\leq |b_n||a_n - L| + |L||b_n - M| \quad \text{(from our last display)} \\ &< p \frac{\varepsilon}{p + |L|} + |L| \frac{\varepsilon}{p + |L|} \quad \text{(as we just pointed out)} \\ &= \frac{\varepsilon(p + |L|)}{p + |L|} = \varepsilon.\end{aligned}$$

2.3. SEQUENCES

Having shown that for any $\varepsilon > 0$ we can find an N such that when $n > N$ we have $|a_n b_n - LM| < \varepsilon$, we have proven that $\{a_n b_n\}$ converges to LM.

Proof of (4): Once again, we need to be a bit creative with the proof. First, lets look at this intuitively. We want to show that we can make $\left|\dfrac{1}{b_n} - \dfrac{1}{M}\right|$ small after a certain point. But this absolute value simplifies to $\left|\dfrac{M - b_n}{b_n M}\right| = \dfrac{|M - b_n|}{|b_n||M|}$. Now, for n large, the numerator gets small while the denominator approaches the fixed number $|M|^2$, since $|b_n|$ approaches $|M|$ by Theorem 2.43. Thus, we should expect to be able to make $\dfrac{|M - b_n|}{|b_n||M|}$ small. Formalizing all of this is trickier than it looks. We observe one last thing: Since $\{|b_n|\}$ converges to $|M|$, we can make $|b_n|$ close to $|M|$ after a certain point. But if $|b_n|$ is close enough to $|M|$ after a certain point, then the terms have to be bigger than $|M|/2$ after that point, since $|M| > 0$. We will need this. Now, to the formal proof.

Since $\{b_n\}$ converges to M, $\{|b_n|\}$ converges to $|M|$ by Theorem 2.43. Thus, we can find an N_1 such that $||b_n| - |M|| < \dfrac{|M|}{2}$, when $n > N_1$. From this inequality we get $-\dfrac{|M|}{2} < |b_n| - |M| < \dfrac{|M|}{2}$, when $n > N_1$, and adding $|M|$ to all the inequalities we get that $\dfrac{|M|}{2} < |b_n| < 3\dfrac{|M|}{2}$. From this we can extract the fact that $|b_n| > \dfrac{|M|}{2}$ when $n > N_1$. Taking reciprocals we get that for $n > N_1$,

$$\frac{1}{|b_n|} < \frac{2}{|M|}.$$

Multiplying both sides of this inequality by $\dfrac{1}{|M|}$, we get

$$\frac{1}{|b_n||M|} < \frac{2}{|M|^2}. \tag{2.20}$$

Also, since $\{b_n\}$ converges to M, there is an N_2 such that $n > N_2$ implies that

$$|b_n - M| < \frac{|M|^2 \varepsilon}{2}. \tag{2.21}$$

Now, let N be any number greater than N_1 and N_2, so that both (2.20) and (2.21) hold. Then for $n > N$, we have

$$\begin{aligned}
\left|\frac{1}{b_n} - \frac{1}{M}\right| &= \left|\frac{M - b_n}{b_n M}\right| \\
&= \frac{|M - b_n|}{|b_n||M|} \\
&= \frac{1}{|b_n||M|} \cdot |M - b_n| \\
&< \frac{2}{|M|^2} \cdot \frac{|M|^2 \varepsilon}{2} \quad \text{(by (2.20) and (2.21))} \\
&= \varepsilon.
\end{aligned}$$

We have just shown that $\left|\dfrac{1}{b_n} - \dfrac{1}{M}\right| < \varepsilon$ for $n > N$, and so we have proven that $\left\{\dfrac{1}{b_n}\right\}$ converges to $\dfrac{1}{M}$ if $M \neq 0$.

Proof of (5) The proof of this goes quickly once we use parts (3) and (4) of the theorem and note that $\dfrac{a_n}{b_n} = a_n \cdot \dfrac{1}{b_n}$. From here, $\lim a_n = L$ and $\lim \dfrac{1}{b_n} = \dfrac{1}{M}$, $\lim \dfrac{a_n}{b_n} = \lim \left(a_n \cdot \dfrac{1}{b_n}\right) = \lim a_n \cdot \lim \dfrac{1}{b_n} = L \cdot \dfrac{1}{M} = \dfrac{L}{M}$. ∎

Now we come to some other basic theorems about limits. Drawing a picture for the next theorem convinces us that it is true. But in analysis, we don't use pictures to prove things, only to guide us.

Theorem 2.46 *If a sequence $\{a_n\}$ converges to L and all the terms of $\{a_n\}$ are greater than or equal to zero, then $L \geq 0$.*

Proof. Suppose not. Then L is less than zero. So $-L$ is greater than 0. Now, by the definition of convergence, we can make $|a_n - L| < \varepsilon$ for any $\varepsilon > 0$. So we can take ε to be $-L$. That yields $|a_n - L| < -L$ (for $n > N$ and for some N), which is equivalent to $L < a_n - L < -L$. Now, if we add L to the inequalities, we get that $2L < a_n < 0$. The inequality on the right, $a_n < 0$, gives us a contradiction, since we are given that all the terms of the sequence are greater than or equal to zero. Since our contradiction arose from assuming that L was less than zero, L must be greater than or equal to zero. ∎

Remark 2.47 *It is a very common error to assume in a proof that if all the terms of a convergent sequence are positive, that the limit is positive. This is false, as the sequence $\{a_n\} = \{1/n\}$ shows. Here, all the terms are positive, but the limit is zero.*

2.3. SEQUENCES

Theorem 2.48 *If $\{a_n\}$ and $\{b_n\}$ are convergent sequences, and if $a_n \leq b_n$ for each n, then $\lim a_n \leq \lim b_n$.*

Proof. Consider the sequence $c_n = b_n - a_n$. Then since $a_n \leq b_n$, $c_n \geq 0$ for all n. So by the previous theorem, $\lim c_n \geq 0$. But $\lim c_n = \lim b_n - \lim a_n$ by Theorem 2.45 part (2). So $\lim b_n - \lim a_n \geq 0$, and it follows that $\lim b_n \geq \lim a_n$ or equivalently, $\lim a_n \leq \lim b_n$. ∎

Corollary 2.49 *If $\{a_n\}$ is a convergent sequence such that $a_n \leq M$ after a certain point, then $\lim a_n \leq M$. Similarly, if $\{b_n\}$ is a convergent sequence, and if $b_n \geq M$ after a certain point, then $\lim b_n \geq M$.*

Proof. For the first part we take $\{b_n\}$ in the theorem to be the constant sequence each of whose terms is equal to M. Clearly, $\lim b_n = M$. (See Exercise 1.) For the second part, we take $\{a_n\}$ to be the sequence all of whose terms are M. ∎

Corollary 2.50 *If $\{a_n\}$, $\{b_n\}$, and $\{c_n\}$ are convergent sequences and if $a_n \leq b_n \leq c_n$ for all n after some point, then $\lim a_n \leq \lim b_n \leq \lim c_n$.*

Proof. It follows immediately from the previous theorem that $\lim a_n \leq \lim b_n \leq \lim c_n$. ∎

Corollary 2.51 *(Squeeze Theorem) If $\{a_n\}$, $\{b_n\}$, and $\{c_n\}$ are sequences, with $a_n \leq b_n \leq c_n$ for all n after some point, and if $\lim a_n = \lim c_n = L$, then $\lim b_n$ exists and is also equal to L.*

Proof. Choose any $\varepsilon > 0$. Since $\{a_n\}$ and $\{c_n\}$ converge to L, there is an N_1 such that $n > N_1$ implies that
$$L - \varepsilon < a_n < L + \varepsilon, \tag{2.22}$$
and there is an N_2 such that
$$L - \varepsilon < c_n < L + \varepsilon. \tag{2.23}$$
Let N be larger than both N_1 and N_2. Then we have from (2.22), (2.23) and the given, that for $n > N$,
$$L - \varepsilon < a_n \leq b_n \leq c_n < L + \varepsilon.$$
From this we can extract that for $n > N$,
$$L - \varepsilon < b_n < L + \varepsilon.$$
This last inequality is equivalent to $|b_n - L| < \varepsilon$ for $n > N$, and this is what it means for $\{b_n\}$ to converge to L. ∎

Remark 2.52 *The intuitive way of seeing this is to say that if $\{a_n\}$ and $\{c_n\}$ both converge to L, then for any $\varepsilon > 0$, the terms of these sequences are eventually in the interval $(L-\varepsilon, L+\varepsilon)$, and since b_n is between them, b_n will also eventually be in that interval, which means that b_n converges to L.*

Remark 2.53 *Sometimes when a beginner is given the information that $\{a_n\}$ and $\{c_n\}$ are convergent sequences and that $\{b_n\}$ is a sequence such that $a_n \leq b_n \leq c_n$, they immediately conclude that $\lim a_n \leq \lim b_n \leq \lim c_n$. This step would be an error since we are not given information about b_n converging; so it makes no sense to talk about $\lim b_n$.*

The following is a theorem that is used a lot in analysis.

Theorem 2.54 *Let A be a set of real numbers. If $L = \sup A$, then there is a sequence of points from A converging to L.*

Proof. By Theorem 2.20, we know that there exists a point a_1 in A such that
$$L - 1 < a_1 \leq L,$$
and there is a point a_2 in A such that
$$L - 1/2 < a_2 \leq L,$$
and so on. In this way we generate a sequence of points a_n from the set A satisfying
$$L - \frac{1}{n} < a_n \leq L,$$
which we can write as
$$L - \frac{1}{n} < a_n \leq L < L + \frac{1}{n}.$$
From this we extract that $L - 1/n < a_n < L + 1/n$, from which we get that $-1/n < a_n - L < 1/n$, and this is equivalent to
$$|a_n - L| < \frac{1}{n}. \tag{2.24}$$
Now, for any given $\varepsilon > 0$, we choose N big enough to make $1/N < \varepsilon$. So if $n > N$, $\frac{1}{n} < \frac{1}{N} < \varepsilon$. Together, this observation and (2.24) yield
$$|a_n - L| < \varepsilon \text{ for } n > N,$$
and this is the definition of $\{a_n\}$ converging to L. ∎

2.3. SEQUENCES

Definition 2.55 *(1) A sequence, $\{a_n\}$, is said to be **increasing** if $a_1 \leq a_2 \leq a_3 \leq ....$ A sequence, $\{a_n\}$, is said to be **strictly increasing** if $a_1 < a_2 < a_3 < ....$*

*(2) A sequence, $\{a_n\}$, is said to be **decreasing** if $a_1 \geq a_2 \geq a_3 \geq ....$ A sequence, $\{a_n\}$, is said to be **strictly decreasing** if $a_1 > a_2 > a_3 > ....$*

*(3) A sequence is said to be **monotonic** if it is either increasing or decreasing.*

Example 2.56 *The sequence $1, 1, 2, 2, 3, 3, ...,$ is increasing but not strictly increasing. The sequence $1, \frac{1}{2}, \frac{1}{3}, ...,$ is strictly decreasing. The constant sequence, $1, 1, 1, ...,$ is both increasing and decreasing. Thus, all of these sequences are monotone. The sequence $1, 2, \frac{1}{2}, 3, 4, 5, 6, 7, ...,$ is not increasing because the third term, $\frac{1}{2}$, is smaller than the second term, nor is it decreasing, since the second term is greater than the first. However, if we are to assume that the pattern of terms being successive integers continues, then after the third term, this sequence is increasing*

The following theorem is a critical result in analysis and becomes intuitively clear once you draw some pictures of sequences on the number line.

Theorem 2.57 *A bounded monotonic sequence converges. More specifically, if the sequence is increasing, it converges to its supremum. If it is decreasing, it converges to its infimum.*

Proof. We will prove the case for when the sequence is increasing, leaving the other half to you.

If the sequence is increasing, then $a_1 \leq a_2 \leq a_3 \leq ...$, and since it is bounded, all the terms are less than or equal to the least upper bound of the sequence, L, which we know exists by Theorem 2.16. By Theorem 2.20, if $\varepsilon > 0$ is given, then there is a term of the sequence, a_N, such that

$$L - \varepsilon < a_N \leq L.$$

Since the sequence is increasing, $a_n \geq a_N$, for $n \geq N$. So we have that

$$L - \varepsilon < a_N \leq a_n \leq L, \text{ for } n > N.$$

Of course, from this we have

$$L - \varepsilon < a_N \leq a_n \leq L < L + \varepsilon \quad \text{for } n > N,$$

and from this we can extract,

$$L - \varepsilon < a_n < L + \varepsilon \quad \text{for } n > N,$$

which can be written as

$$|a_n - L| < \varepsilon, \quad \text{for } n > N \text{ (by Proposition 1.50 part 5).}$$

This of course, is the definition of $\{a_n\}$ converging to L. ∎

Example 2.58 *It can be shown (and in the exercises we will ask you to show it) that the sequence $\{(1 + 1/n)^n\}$ is a bounded increasing sequence. So it converges, and the limit happens to be e, which is not obvious, unless you know this fact, or if you have seen this expression as the definition of e. (There are many possible ways to define e.)*

Example 2.59 *As early as 1500 BC the following algorithm for computing the square root of a positive number, a, was known. Begin with any positive number s_1 and generate a sequence of numbers, s_n, as follows: $s_2 = \frac{1}{2}\left(s_1 + \frac{a}{s_1}\right)$, $s_3 = \frac{1}{2}\left(s_2 + \frac{a}{s_2}\right)$, and so on with $s_n = \frac{1}{2}\left(s_{n-1} + \frac{a}{s_{n-1}}\right)$. Since we are starting with a and $s_1 > 0$, s_n is positive for all $n \in Z^+$. It turns out that this sequence converges (and rather rapidly), to $\sqrt{a}$. (You should check this on a calculator for some specific values of a.)*

We will show the convergence of the sequence $\{s_n\}$ to $\sqrt{a}$. The first step in this process is to show that $s_n \geq \sqrt{a}$ for each positive integer n. We do that by showing that for any positive s, $\frac{1}{2}\left(s + \frac{a}{s}\right) \geq \sqrt{a}$. For if this were not true, then $\frac{1}{2}\left(s + \frac{a}{s}\right) < \sqrt{a}$. By multiplying both sides by $2s$ and bringing everything over to one side we get $s^2 - 2s\sqrt{a} + a < 0$, which simplifies to $(s - \sqrt{a})^2 < 0$. This is impossible, of course, since the square of any real number must be ≥ 0. Since our contradiction arose from assuming that $s_n < \sqrt{a}$, we have that

$$s_n \geq \sqrt{a}. \tag{2.25}$$

So $\{s_n\}$ is bounded below by $\sqrt{a}$. Squaring both sides of (2.25) we get $s_n^2 \geq a$, and dividing this by s_n, we get that $s_n \geq \frac{a}{s_n}$. Adding s_n to both sides of this inequality, we get $2s_n \geq s_n + \frac{a}{s_n}$. Finally, dividing this last inequality by 2, we get $s_n \geq \frac{1}{2}\left(s_n + \frac{a}{s_n}\right)$. Notice that the right side of this inequality is s_{n+1}. So we have that

$$s_n \geq s_{n+1}, \tag{2.26}$$

and we have therefore shown that $\{s_n\}$ is decreasing. To summarize, (2.26) and (2.25) show that $\{s_n\}$ is decreasing and bounded below by $\sqrt{a}$, so by Theorem 2.57, the sequence $\{s_n\}$ converges.

2.3. SEQUENCES

We now show that the sequence $\{s_n\}$ converges to $\sqrt{a}$. To see this, take the limit of both sides of the equation $s_n = \frac{1}{2}(s_{n-1} + \frac{a}{s_{n-1}})$ as $n \to \infty$, and realize that $\lim s_n = \lim s_{n-1}$ (since the sequences $\{s_n\}$ and $\{s_{n-1}\}$ are the same when $n \geq 2$, except that $\{s_{n-1}\}$ has one extra term). Calling the limit of s_n, L (which we know is greater than or equal to zero by Theorem 2.46), we get from $s_n = \frac{1}{2}(s_{n-1} + \frac{a}{s_{n-1}})$ that $L = \frac{1}{2}(L + \frac{a}{L})$. Multiplying both sides by $2L$, then subtracting L^2, we get $L^2 = a$, and solving for the positive value of L, we get $L = \sqrt{a}$. So the limit of the sequence is $\sqrt{a}$.

EXERCISES

1. Prove that a constant sequence converges.

2. True or False: If $\{a_n\}$ and $\{b_n\}$ are sequences such that $a_n < b_n$ for all n, then $\lim a_n < \lim b_n$. If this is false, give an example to show it is false, and then correct the statement. If it is true, prove it.

3. Give an example of a bounded sequence which is not convergent.

4. Give an example (if possible) of a convergent sequence which is not bounded.

5. If $\{a_n\}$ and $\{b_n\}$ are sequences, then $\sup\{a_n + b_n\} \leq \sup\{a_n\} + \sup\{b_n\}$. Prove this and then give an example to show that they are not equal in general. Does this contradict Theorem 2.26?

6. Find the limits of each of the following sequences using your knowledge of calculus, or by guessing, and then prove that these are the limits, using the definition of convergence of a sequence.

    (a) $\left\{\left(\frac{1}{2}\right)^n\right\}$

    (b) $\left\{\dfrac{-2}{3+n^2}\right\}$

    (c) $\left\{\dfrac{2n+3}{3n+2}\right\}$

    (d) $\left\{\dfrac{3n^2+4n+1}{4n^3+2n+3}\right\}$

    (e) $\{\sqrt{n+1} - \sqrt{n}\}$

    (f) $\{\sqrt{n^2+n} - n\}$

7. Using theorems in the chapter, show that $\lim \dfrac{3n}{n+1} = 3$. Then prove that the limit of quotients of polynomials of the same degree is the ratio of the coefficients of the highest powered terms in the numerator and denominator. [Hint: For the first case, divide numerator and denominator by n.]

8. Give an example of a sequence, $\{a_n\}$, which doesn't converge, but where $\{|a_n|\}$ does converge.

9. Prove that if $\{a_n\}$ converges to 3, then $\{\sqrt{a_n}\}$ converges to $\sqrt{3}$. State and prove an analogous result for a nonnegative sequence which converges to L.

10. Show that the sequence $\{a_n\}$ has a limit of 0 if and only if $\{|a_n|\}$ has a limit of 0.

11. Prove that for any constant, c, the sequence $\left\{\dfrac{c}{n!}\right\}$ converges to 0. [Hint: Show that $n! \geq 2^{n-1}$.]

12. Use the squeeze theorem to show that the sequence $\left\{\dfrac{\sin 3n}{n}\right\}$ converges to zero.

13. Show that if $0 < a < 1$, then the sequence $\{a^n\}$ is decreasing and bounded below by 0, and hence has a limit, L. Then show, using the fact that $a^n = a(a^{n-1})$ and taking the limit of both sides, that L is zero.

14. Using the definition of limit, show that $\lim a^n = 0$ when $0 < a < 1$.

15. Show that the set of numbers of the form $\dfrac{2^n}{n!}$, where n is a positive integer, has an infimum of zero. (Hint: Show by induction that after a certain point $n! > 3^n$.)

16. Suppose that $\{x_n\}$ is a sequence of real numbers and that $\{a_n\}$ is a sequence of nonnegative real numbers with $\lim a_n = 0$. Suppose further that $|x_n - x| \leq Ca_n$ for some positive constant C. Show that x_n converges to x.

17. Prove that for any real number, r, there is an increasing sequence of rational numbers converging to r.

18. We say that the sequence $\{a_n\}$ diverges to infinity, and write $\lim a_n = \infty$, if for every positive M, there is an N such that $n > N$ implies $a_n > M$. Show that if $\{a_n\}$ and $\{b_n\}$ diverge to infinity, so do $\{a_n + b_n\}$ and $\{a_n b_n\}$. Also, show that if $a_n \leq b_n$ and a_n diverges to infinity, then so does b_n.

2.3. SEQUENCES

19. Show that if $\lim a_n = L$ and $\lim b_n = \infty$, then $\lim(a_n/b_n) = 0$.

20. If $c > 1$, then $c = 1 + b$ where $b > 0$. Use this and the binomial theorem to show that $\lim c^n = \infty$.

21. Give a definition of $\lim\{a_n\} = -\infty$ and state analogous results to those in Exercise 18.

22. Show that the sequence $\{(1 + \frac{1}{n})^n | n \text{ is a positive integer}\}$, is between 2 and 3 and so is bounded. Also, show it is monotonic. [Hint: By the binomial theorem,
$$a_n = \left(1 + \frac{1}{n}\right)^n = 1 + n \cdot \frac{1}{n} + \frac{n(n-1)}{2!} \cdot \frac{1}{n^2} + \frac{n(n-1)(n-2)}{3!} \cdot \frac{1}{n^3} + \ldots +$$
$\frac{n(n-1)(n-2)\ldots 1}{n!} \cdot \frac{1}{n^n} \leq 1 + 1 + \frac{1}{2!} + \frac{1}{3!} + \ldots + \frac{1}{n!}$. Now use the fact that $n! \geq 2^{n-1}$, so that $\frac{1}{n!} \leq \frac{1}{2^{n-1}}$ and use the formula for the sum of a geometric series (which you can get from the internet if you don't know it) to show that $\frac{1}{2} + \frac{1}{2^2} + \ldots + \frac{1}{2^{n-1}} < 1$. To show the sequence is increasing, show that each term of a_n when expanded by the binomial theorem, is less than or equal to the corresponding term of a_{n+1} (when expanded) and a_{n+1} has one more term.]

23. Using the fact that for any positive x, $x + \frac{1}{x} \geq 2$, show that the sequence defined by $a_1 = 2$, and $a_{n+1} = 2 - \frac{1}{a_n}$ is decreasing, by showing that $a_n - a_{n+1} > 0$ for each positive integer n. Then show that $\{a_n\}$ is bounded, and find $\lim a_n$. Do different values of $a_1 > 1$ give us different limits for the sequence?

24. Suppose that $a_1 < 2$ and we define successive terms of a sequence by the formula, $a_{n+1} = \sqrt{2 + a_n}$. Show that $\{a_n\}$ is bounded above by 2 and is monotonic. To what number does this sequence converge?

25. Show that the sequence $x_n = \frac{1}{1^2} + \frac{1}{2^2} + \frac{1}{3^2} + \ldots + \frac{1}{n^2}$ is bounded and monotonic, and hence converges. [Hint: For $k \geq 2$, $\frac{1}{k^2} < \frac{1}{k(k-1)} = \frac{1}{k-1} - \frac{1}{k}$.] An interesting fact is that this sequence converges to $\frac{\pi^2}{6}$. Many very fine mathematicians tried to find the limit, but it took over 50 years for the answer to be found.

2.4 Subsequences

Intuitively, a subsequence of a sequence is a sequence obtained by eliminating a finite or infinite number of terms from the original sequence. What we are left with is the subsequence.

Example 2.60 *The sequence of positive integers, $\{n\}$, has infinitely many subsequences. For example, we have the subsequence of even integers, $\{2n\}$, obtained by eliminating all the odd integers; we have the subsequence of odd integers, $\{2n+1\}$, obtained by eliminating all the even integers; we also have the subsequence $\{3n-2\} = 1, 4, 7, 10, ...$; and we have the subsequence $\{n\}$, where $n \geq 3$. Making up examples of subsequences is quite easy. Notice, however, that the sequence $3, 2, 5, 7, 8, 9...$ is not a subsequence of $\{n\}$ since we have rearranged some terms of $\{n\}$. In a subsequence, we never rearrange terms of the original sequence. We only eliminate terms.*

More formally, we have:

Definition 2.61 *A **subsequence** of a sequence $\{a_n\}$ is a sequence of the form $\{a_{n_k}\}$ where $n_1 < n_2 < n_3$, etc. Said another way, it is a sequence of the form $\{a_{f(n)}\}$ where $f(n)$ is a strictly increasing function from the natural numbers to the natural numbers.*

Remark 2.62 *Since the subsequence $\{a_{n_k}\}$ of $\{a_n\}$ is obtained by eliminating certain terms, $n_1 \geq 1$, and since the subscripts are strictly increasing, $n_2 \geq 2$, $n_3 \geq 3$, etc. So a_{n_k} is either in the same place, as a_k in the sequence $\{a_n\}$, or farther out. Said another way, $n_k \geq k$, for all k.*

Remark 2.63 *Every subsequence of a sequence is itself a sequence since the original sequence is a listing, and knocking out terms just leaves another listing.*

Theorem 2.64 *If a sequence, $\{a_n\}$, of real numbers converges to L, then any subsequence of $\{a_n\}$ also converges to L.*

Proof. Since $\{a_n\}$ converges to L, we know that for any positive ε, there is a positive integer N such that
$$|a_n - L| < \varepsilon, \text{ for } n \geq N. \tag{2.27}$$

By Remark 2.62 it follows that $n_k > N$ when $k > N$, since the indices in a subsequence are increasing. So by (2.27)
$$|a_{n_k} - L| < \varepsilon \text{ for } k > N,$$

and this is what it means to say that the subsequence $\{a_{n_k}\}$ converges to L. ∎

We are coming up to a major result, known as the Bolzano-Weierstrass theorem. To prove that theorem, we will need the following:

2.4. SUBSEQUENCES

Definition 2.65 *A **peak** of a sequence $\{a_n\}$ is a term, a_k, of the sequence, such that $a_k \geq a_n$ when $n \geq k$. That is, a peak of a sequence is a term whose numerical value is greater than or equal to the numerical values of all the terms following it in the sequence.*

Example 2.66 *For the sequence $\left\{\dfrac{1}{n}\right\} = 1, \dfrac{1}{2}, \dfrac{1}{3}, ...$, every term is a peak. For the sequence $\{n-1\} = 0, 1, 2, 3, ...$, no term is a peak. For the sequence $0, 1, 0, \dfrac{1}{2}, ...$, where we insert terms of the sequence $\left\{\dfrac{1}{n}\right\}$ between successive zeros, all the even numbered terms are peaks, and for the sequence $10, 9, 8, 1, -2, 1, -2, 1, -2, ...$, the first 3 terms are peaks as well as all the terms equal to 1.*

Theorem 2.67 *Every sequence of real numbers, $\{a_n\}$, has a monotone subsequence.*

Proof. There are two cases to consider.

Case 1: The sequence has infinitely many peaks.

In order to find the monotonic subsequence we start at the first term in the sequence, and work through the terms of the sequence in order. Suppose that we find that a_{n_1} is the first peak we come upon and a_{n_2} is the second peak etc. Then by the definition of a peak $a_{n_1} \geq a_{n_2} \geq a_{n_3} \geq ...$, and this subsequence, $\{a_{n_k}\}$, is our monotone subsequence, which happens to be decreasing.

Case 2: There are only a finite number of peaks.

Suppose that the peaks are $a_{n_1}, a_{n_2}, ..., a_{n_k}$ and let $s_1 = n_k + 1$, the index right after the last peak. Since a_{s_1} is not a peak, there is some $s_2 \geq s_1$ such that $a_{s_2} > a_{s_1}$. Similarly, since a_{s_2} is not a peak, there is an $s_3 \geq s_2$ such that $a_{s_3} > a_{s_2}$, and so on. The subsequence, $a_{s_1}, a_{s_2}, a_{s_3}, ...$, is our monotone subsequence since $a_{s_1} < a_{s_2} < a_{s_3} < ...$, and this subsequence is increasing. ∎

We now arrive at a very important theorem.

Theorem 2.68 *(Bolzano Weierstrass) Every bounded sequence of real numbers, $\{a_n\}$, has a convergent subsequence.*

Proof. By the previous theorem, $\{a_n\}$ has a monotone subsequence. This subsequence also must be bounded since the original sequence is. But by Theorem 2.57, this monotone subsequence must converge. ∎

Example 2.69 *Consider the bounded sequence where a_n is defined by,*

$$a_n = \begin{cases} 1 + \dfrac{1}{n} & \text{if } n \text{ is odd} \\ 2 - \dfrac{4}{n} & \text{if } n \text{ is even} \end{cases}$$

whose first few terms are $2, 0, \frac{4}{3}, 1, \frac{6}{5}, \ldots$. Any subsequence of this sequence, whose terms are eventually all of the form $1 + \frac{1}{n}$, will converge to 1. Similarly, any subsequence all of whose terms are eventually of the form $2 - \frac{4}{n}$, will converge to 2. So, certainly this sequence has at least one convergent subsequence.

The following will be useful later on.

Theorem 2.70 Let $\{a_n\}$ be a bounded sequence with the property that every convergent subsequence of $\{a_n\}$ converges to the same number, L. Then $\{a_n\}$ must also converge to L.

Proof. We will do a proof by contradiction. Suppose that the sequence $\{a_n\}$ doesn't converge to L. Using the geometric interpretation of a convergent sequence (see Remark 2.40), it is not true that the sequence $\{a_n\}$ is eventually in the interval $(L - \varepsilon, L + \varepsilon)$, for every $\varepsilon > 0$. Thus, there must be some fixed $\varepsilon_0 > 0$, such that $(L - \varepsilon_0, L + \varepsilon_0)$ doesn't include all the terms of the sequence after a certain point. Said another way, no matter which N you choose, there will be a term of the sequence whose index is greater than N and which is not in the interval $(L - \varepsilon_0, L + \varepsilon_0)$. In particular, if you choose N to be 1, there will be a term, a_{n_1}, of the sequence, such that $a_{n_1} \notin (L - \varepsilon_0, L + \varepsilon_0)$. If you take $N > n_1$, there will be a term, a_{n_2}, of the sequence, whose index n_2 is greater than N (and hence greater that n_1), such that $a_{n_2} \notin (L - \varepsilon_0, L + \varepsilon_0)$. Similarly, if we now take $N > n_2$, there will be a term of the sequence, a_{n_3}, whose index n_3 is greater than N (and hence greater than n_2), such that $a_{n_3} \notin (L - \varepsilon_0, L + \varepsilon_0)$. Continuing in this manner with we get a subsequence $a_{n_1}, a_{n_2}, a_{n_3}, \ldots$ from the original sequence (since the indices are increasing), such that

none of the terms of this subsequence $\{a_{n_k}\}$ are contained in $(L-\varepsilon_0, L+\varepsilon_0)$. (2.28)

But since the terms of this subsequence, $\{a_{n_k}\}$, are bounded (because the original sequence is), this subsequence (by virtue of being a sequence) will have a convergent subsequence (by the Bolzano Weierstrass theorem (Theorem 2.68)). Call this subsequence, S. By the conditions of the theorem, S has to converge to L. That means that **all** the terms of S have be within $(L - \varepsilon_0, L + \varepsilon_0)$ after a certain point. But all the terms of S come from $\{a_{n_k}\}$. This contradicts (2.28).

Since our assumption that $\{a_n\}$ didn't converge to L led to a contradiction, $\{a_n\}$ must converge to L. ∎

EXERCISES

1. Give an example of a sequence which is not bounded, but which has a convergent subsequence.

2. Give an example of a sequence with exactly three subsequential limits.

3. Give a quick proof using subsequences to show that the sequence $\{(-1)^n\}$ diverges.

4. Show that if $\{a_n\}$ converges to a nonzero number c, the sequence $\{(-1)^n a_n\}$ diverges, but if $\{a_n\}$ converges to 0, the sequence also converges to 0.

5. If $\{a_n\}$ is unbounded, show there is a subsequence $\{a_{n_k}\}$ such that $\left\{\dfrac{1}{a_{n_k}}\right\}$ converges to 0.

6. True or False: A subsequence of a subsequence of a sequence, $\{a_n\}$, is a subsequence of $\{a_n\}$.

7. If $\{a_n\}$ and $\{b_n\}$ are bounded sequences, then by the Bolzano Weierstrass theorem, $\{a_n\}$ has a convergent subsequence $\{a_{n_k}\}$. Give an example to show that every $\{b_{n_k}\}$ need not converge. But show that such a $\{b_{n_k}\}$ has a convergent subsequence $\{b_{n_{k_l}}\}$ and it does follow that the subsequence $\{a_{n_{k_l}}\}$ converges.

8. If $c > 1$, the sequence $\{c^{\frac{1}{n}}\}$ is a decreasing sequence, bounded below by 1. So it has a limit. Show that the limit is 1 by considering the subsequence $\{c^{\frac{1}{2n}}\}$, which must have the same limit.

2.5 Limit Superior and Limit Inferior

According to the Bolzano Weierstrass theorem, every bounded sequence has a convergent subsequence. But a bounded sequence $\{a_n\}$, may have many convergent subsequences. We will denote by $\mathcal{L}$, the collection of limits of all subsequences of $\{a_n\}$.

Definition 2.71 *(1) If $\{a_n\}$ is a bounded sequence, and if $\mathcal{L}$ represents the set of all limits of subsequences of $\{a_n\}$, then $\sup \mathcal{L}$ is called the **limit superior** of $\{a_n\}$, while $\inf \mathcal{L}$ is called the **limit inferior** of the sequence $\{a_n\}$. The limit superior of a sequence $\{a_n\}$ is denoted by $\limsup\{a_n\}$, or by $\overline{\lim}\{a_n\}$, while the limit inferior of the sequence is denoted by $\liminf\{a_n\}$, or $\underline{\lim}\{a_n\}$.*

(2) If a sequence $\{a_n\}$ is not bounded above, we say that $\overline{\lim}\{a_n\} = \infty$, while if the sequence is not bounded below, we say that $\underline{\lim}\{a_n\} = -\infty$.

Example 2.72 *(1) Consider the sequence* $0, 1, 0, 1, \ldots$. *Since all the terms of the sequence are between 0 and 1, all subsequential limits are between 0 and 1, by Corollary 2.49. So* $\inf \mathcal{L}$ *and* $\sup \mathcal{L}$ *are between 0 and 1. The subsequence of odd numbered terms has a limit of 0, while the subsequence of even numbered terms has a limit of 1. So* $\overline{\lim}\{a_n\} = \sup \mathcal{L} = 1$ *while* $\underline{\lim}\{a_n\} = \inf \mathcal{L} = 0$. *This result is easy to intuit since the given sequence has only two subsequential limits namely, 0 and 1.*

(2) Consider the sequence $\left\{\dfrac{1}{n}\right\} = 1, \dfrac{1}{2}, \dfrac{1}{3}, \ldots$. *This sequence converges to 0. So by Theorem 2.64, every subsequence converges to 0. So* $\mathcal{L} = \{0\}$, *and* $\overline{\lim}\{a_n\} = \sup \mathcal{L} = 0$, *and* $\underline{\lim}\{a_n\} = \inf \mathcal{L} = 0$.

(3) The sequence $-1, 1, -2, 2, -3, 3, \ldots$, *has a limit superior of* ∞ *since it is not bounded above, and a limit inferior of* $-\infty$ *since it is not bounded below.*

(4) The sequence $0, 1, 0, 2, 0, 3, \ldots$, *has a limit superior of* ∞ *since it is not bounded above, and a limit inferior of 0. To see this latter fact, note that 0 is the limit of the subsequence of odd numbered terms, and so is a subsequential limit. But no subsequential limit can be less than zero since all the terms of this sequence are nonnegative.*

(5) We will see in Chapter 6 Section 1 that the set of rational numbers in $[0, 1]$ *can be written as a sequence,* $\{s_n\}$, *which we call* S. *We claim that for this sequence,* $\mathcal{L}$ *consists of all numbers between 0 and 1. To see that* $\mathcal{L} = [0, 1]$, *take any real number* $r \in [0, 1]$. *We construct a sequence of rationals converging to* r *by making use of the fact that between any two real numbers one can find infinitely many rational numbers, and hence infinitely many elements of* S. *Let* s_{n_1} *be a rational number in* S *between* $r - 1$ *and* r. *Now take a rational number in* S, s_{n_2}, *such that* n_2 *is greater than* n_1, *which is between* $r - 1/2$, *and* r. *Next, pick the rational number* s_{n_3} *with* $n_3 > n_2$ *between* $r - 1/3$ *and* r, *etc., making sure that* $r - 1/k < s_{n_k} < r$ *and that the subscripts on the* s_{n_k} *are increasing. The sequence* $\{s_{n_k}\}$ *is a subsequence of our original sequence, which converges to* r. *Since* r *can be any real number in* $[0, 1]$, *and we have shown that* r *is the limit of a subsequence of this sequence, the collection of all subsequential limits is* $[0, 1]$. *So* $\overline{\lim}\{a_n\} = \sup \mathcal{L} = 1$ *while* $\underline{\lim}\{a_n\} = \inf \mathcal{L} = 0$.

Here is our first important theorem in this section:

Theorem 2.73 *A bounded sequence* $\{a_n\}$ *converges to* L *if and only if* $\overline{\lim}\{a_n\} = \underline{\lim}\{a_n\} = L$.

Proof. ($\Longrightarrow$) Suppose that the sequence $\{a_n\}$ converges to L. Then by Theorem 2.64 all subsequences converge to L. Thus, $\mathcal{L}$, the set of all subsequential limits, consists of only one number, L. Thus, $\sup \mathcal{L} = \overline{\lim}\{a_n\} = L$ and $\inf \mathcal{L} = \underline{\lim}\{a_n\} = L$.

($\Longleftarrow$) Now we are assuming that $\overline{\lim}\{a_n\} = \underline{\lim}\{a_n\} = L$, and want to show that that the sequence $\{a_n\}$ converges. Since all subsequential limits are between

2.5. LIMIT SUPERIOR AND LIMIT INFERIOR

$\overline{\lim}\{a_n\} = \sup \mathcal{L}$ and $\underline{\lim}\{a_n\} = \inf \mathcal{L}$, and $\overline{\lim}\{a_n\} = \underline{\lim}\{a_n\} = L$, all subsequential limits are L. But by Theorem 2.70 if all convergent subsequences of $\{a_n\}$ converge to L, so does $\{a_n\}$. ∎

The next theorem is somewhat sophisticated.

Theorem 2.74 *If $\{a_n\}$ is a bounded sequence of real numbers, then $\overline{\lim}\{a_n\}$ and $\underline{\lim}\{a_n\}$ are themselves subsequential limits of $\{a_n\}$.*

Proof. Figure 2.6 gives us an idea of why this might be true, and how to go about proving it.

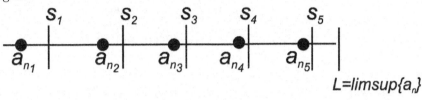

Figure 2.6.

Let us explain how to read the figure. We know, from Theorem 2.54, that since $L = \overline{\lim}\{a_n\} = \sup \mathcal{L}$, there is a sequence of points from $\mathcal{L}$ that converges to L. Label the terms of the sequence as s_1, s_2, s_3, etc. Since $s_1, s_2, s_3, ...$, are elements of $\mathcal{L}$, they are themselves limits of subsequences of $\{a_n\}$. Now, since s_1 is the limit of a subsequence of $\{a_n\}$, there is an element a_{n_1} of the sequence $\{a_n\}$ within a distance of 1 from s_1. Similarly, since s_2 is the limit of a subsequence of $\{a_n\}$, there is an element a_{n_2} of $\{a_n\}$ within $\frac{1}{2}$ unit from s_2, and we can take $n_2 > n_1$ since there are infinitely many elements of the subsequence within $\frac{1}{2}$ unit of s_2. Similarly, we generate a_{n_3}, a_{n_4}, etc., where the distance between a_{n_k} and s_k is less than $\frac{1}{k}$. Since the s_n are converging to L and the a_{n_k} are close behind, they too should converge to L. Now we just have to write out these observations formally. We do that now.

Recapping, since L is the supremum of $\mathcal{L}$, there is a sequence $\{s_n\}$ of elements of $\mathcal{L}$, converging to L. Since s_1 is the limit of a subsequence of $\{a_n\}$, we can find an a_{n_1} in $\{a_n\}$ such that $|a_{n_1} - s_1| < 1$. Since each s_n is a subsequential limit of $\{a_n\}$, we can find $n_2 > n_1$ such that $|a_{n_2} - s_2| < \frac{1}{2}$, and an $n_3 > n_2$ such that $|a_{n_3} - s_3| < \frac{1}{3}$, where in general,

$$|a_{n_k} - s_k| < \frac{1}{k}. \tag{2.29}$$

We now show that $\{a_{n_k}\}$ converges to L. So, let $\varepsilon > 0$ be given. Since $1/k$ can be made as small as we want by making k large enough, we can choose an N_1 such that

$$k > N_1 \text{ makes } \frac{1}{k} < \frac{\varepsilon}{2}. \tag{2.30}$$

Since $\{s_n\}$ converges to L, there is an N_2 such that

$$k > N_2 \text{ makes } |s_k - L| < \frac{\varepsilon}{2}. \tag{2.31}$$

Pick N to be larger than both N_1 and N_2. We now have, for $k > N$,

$$\begin{aligned}
|a_{n_k} - L| &= |a_{n_k} - s_k + s_k - L| \\
&\leq |a_{n_k} - s_k| + |s_k - L| \quad \text{(triangle inequality)} \\
&< \frac{1}{k} + |s_k - L| \quad \text{(by (2.29))} \\
&< \frac{\varepsilon}{2} + \frac{\varepsilon}{2} \quad \text{(by (2.30) and (2.31))} \\
&= \varepsilon.
\end{aligned}$$

We have shown with this set of inequalities that for $k > N$, $|a_{n_k} - L| < \varepsilon$, and this is the definition of $\{a_{n_k}\}$ converging to L. Thus L, the limit superior of $\{a_n\}$, is the limit of the subsequence $\{a_{n_k}\}$ of $\{a_n\}$, which means that $L \in \mathcal{L}$. The proof for the limit inferior is done similarly and is therefore left to the reader. ∎

Remark 2.75 *The previous theorem tells us that* $\sup \mathcal{L} \in \mathcal{L}$, *and since* $\sup \mathcal{L}$ *is greater than or equal to everything in* $\mathcal{L}$, $\sup \mathcal{L}$ *is the maximum element of* $\mathcal{L}$. *Said another way,* **the limit superior of a sequence is the largest subsequential limit of the sequence.** *In a similar manner,* **the limit inferior of a sequence is the smallest subsequential limit of the sequence.** *Thinking of* $\overline{\lim}\{a_n\}$ *and* $\underline{\lim}\{a_n\}$ *this way, as the largest and smallest subsequential limits, makes proving certain things very easy.*

The following are some properties of the limit inferior and limit superior of a sequence.

Theorem 2.76 *If $\{a_n\}$ is a bounded sequence of real numbers, then for any $\varepsilon > 0$,*

(1) there are only a finite number of terms greater than $\overline{\lim}\{a_n\} + \varepsilon$, and
(2) there are only a finite number of terms less than $\underline{\lim}\{a_n\} - \varepsilon$.

Proof. We will prove (1) leaving the proof of (2) to you. Our proof is a proof by contradiction. Suppose that (1) is not true. Then there are infinitely many terms of $\{a_n\} > \overline{\lim}\{a_n\} + \varepsilon$. If we list these in order of occurrence, this infinite set of terms forms a subsequence, $\{a_{n_k}\}$, of $\{a_n\}$, all of whose terms are greater than $\overline{\lim}\{a_n\} + \varepsilon$. Since the original sequence is bounded, $\{a_{n_k}\}$ is also bounded, so by the Bolzano Weierstrass theorem (Theorem 2.68) $\{a_{n_k}\}$ has a convergent subsequence, which we call $\{b_n\}$. Since all the terms of $\{b_n\}$ come from $\{a_{n_k}\}$, they are all $> \overline{\lim}\{a_n\} + \varepsilon$.

2.5. LIMIT SUPERIOR AND LIMIT INFERIOR

Hence the limit of $\{b_n\}$ is $\geq \overline{\lim}\{a_n\} + \varepsilon$ (by Corollary 2.49 with $M = \overline{\lim}\{a_n\} + \varepsilon$). And since $\overline{\lim}\{a_n\} + \varepsilon$ is greater than $\overline{\lim}\{a_n\}$, we have established the existence of a subsequence of $\{a_n\}$, (namely $\{b_n\}$), whose limit is greater than the largest subsequential limit, $\overline{\lim}\{a_n\}$. This is our contradiction. Our contradiction arose from assuming that (1) was not true. Thus, (1) must be true. ∎

Remark 2.77 *Part (1) of Theorem 2.76 can be stated as, "All but a finite number of the terms of the sequence are less than or equal to $\overline{\lim}\{a_n\} + \varepsilon$."*

Theorem 2.78 *If $\{a_n\}$ is a bounded sequence of real numbers, then for any given $\varepsilon > 0$,*

(1) there are infinitely many terms of $\{a_n\}$ greater than $\overline{\lim}\{a_n\} - \varepsilon$, and

(2) there are infinitely many terms of $\{a_n\}$ less than $\underline{\lim}\{a_n\} + \varepsilon$.

Proof. (1) Again, we do a proof by contradiction. Suppose that (1) is not true. Then there are only a finite number of terms of $\{a_n\} > \overline{\lim}\{a_n\} - \varepsilon$. Therefore, all terms of $\{a_n\}$ after a certain point are less than or equal to the fixed number $\overline{\lim}\{a_n\} - \varepsilon$, which means that *all* subsequential limits of $\{a_n\}$ are less than or equal to $\overline{\lim}\{a_n\} - \varepsilon$ (refer to Theorem 2.48, where we take b_n to be the constant sequence $\overline{\lim}\{a_n\} - \varepsilon$). Since $\overline{\lim}\{a_n\} - \varepsilon < \overline{\lim}\{a_n\}$, our previous sentence tells us that *all* subsequential limits are $< \overline{\lim}\{a_n\}$. But by Theorem 2.74, $\overline{\lim}\{a_n\}$ is itself a subsequential limit of $\{a_n\}$, so this would imply that $\overline{\lim}\{a_n\} < \overline{\lim}\{a_n\}$, a contradiction. Since we arrived at a contradiction by assuming (1) was false, it must be that (1) is true. We leave the similar proof of (2) to you. ∎

Figure 2.7 summarizes the results of the last two theorems. In this picture $\overline{L}$ and $\underline{L}$ represent $\limsup\{a_n\}$ and $\liminf\{a_n\}$, respectively.

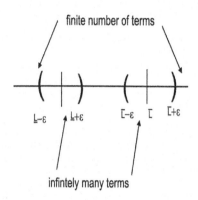

Figure 2.7.

We are headed toward a different way of describing limit superiors and limit inferiors which has some uses in higher analysis. We begin with a bounded sequence, $\{a_n\}$, and define the following quantities:

$$s_n = \sup\{a_n, a_{n+1}, a_{n+2}, ...\} = \sup_{k \geq n}\{a_k\}$$
$$i_n = \inf\{a_n, a_{n+1}, a_{n+2}, ...\} = \inf_{k \geq n}\{a_k\}.$$

So, $s_1 = \sup\{a_1, a_2, a_3, ...\}$, $s_2 = \sup\{a_2, a_3, a_4, ...\}$, $s_3 = \sup\{a_3, a_4, a_5, ...\}$, ..., while $i_1 = \inf\{a_1, a_2, a_3, ...\}$, $i_2 = \inf\{a_2, a_3, a_4, ...\}$, $i_3 = \inf\{a_3, a_4, a_5, ...\}$, ... Since $s_1, s_2, s_3, ...$, are suprema over smaller and smaller sets, $\{s_n\}$ decreases, while the infima sequence, $\{i_n\}$, increases (Theorem 2.19). Since $\{s_n\}$ decreases, and is bounded, it converges to its infimum, s, while $\{i_n\}$, being bounded and increasing, converges to its supremum, i (Theorem 2.57). The following theorem gives us a new way to describe the largest subsequential limit of the sequence $\{a_n\}$ (the limit superior of $\{a_n\}$), and the smallest subsequential limit of $\{a_n\}$ (the limit inferior of $\{a_n\}$), and also suggests why we call the limits superior and inferior by their respective names.

Theorem 2.79 *Using the notation from the previous paragraph, if $\{a_n\}$ is a bounded sequence, then*

(1) $\limsup\{a_n\}$ *equals* $\lim s_n = \lim \sup_{k \geq n}\{a_k\}$. *Similarly,*

(2) $\liminf\{a_n\}$ *equals* $\lim i_n = \lim \inf_{k \geq n}\{a_k\}$.

2.5. LIMIT SUPERIOR AND LIMIT INFERIOR

Proof. Pick any $\varepsilon > 0$. Then by Theorem 2.76, there are only finitely many terms of the sequence $\{a_n\}$ which are greater than $\overline{\lim}\{a_n\} + \varepsilon$. This implies that there is an integer N such that for all $n \geq N$,

$$a_n \leq \overline{\lim}\{a_n\} + \varepsilon.$$

It follows that
$$\sup_{n \geq N}\{a_n\} \leq \overline{\lim}\{a_n\} + \varepsilon,$$

(by Remark 2.15) or, said another way,

$$s_N \leq \overline{\lim}\{a_n\} + \varepsilon.$$

But as we pointed out, the sequence $\{s_n\}$ is decreasing, so for all $n > N$

$$s_n \leq s_N \leq \overline{\lim}\{a_n\} + \varepsilon,$$

which implies (by Corollary 2.49 with $M = \overline{\lim}\{a_n\} + \varepsilon$) that,

$$\lim s_n \leq \overline{\lim}\{a_n\} + \varepsilon$$

or, that
$$s \leq \overline{\lim}\{a_n\} + \varepsilon, \tag{2.32}$$

where $s = \lim s_n$. Since this is true for all $\varepsilon > 0$, we have (by Lemma 2.23 with $y = s$ and $N = \overline{\lim}\{a_n\}$) that

$$s \leq \overline{\lim}\{a_n\}. \tag{2.33}$$

We now proceed to show that $s \geq \overline{\lim}\{a_n\}$, so that with (2.33) we will have $s = \overline{\lim}\{a_n\}$. Now, by Theorem 2.78, there are infinitely many terms of $\{a_n\} > \overline{\lim}\{a_n\} - \varepsilon$. So pick any

$$a_{n_1} > \overline{\lim}\{a_n\} - \varepsilon. \tag{2.34}$$

Since $s_{n_1} = \sup\{a_{n_1}, a_{n_1+1}, a_{n_1+2}, ...\}$, $s_{n_1} \geq a_{n_1}$. Using this together with (2.34) we have that

$$s_{n_1} > \overline{\lim}\{a_n\} - \varepsilon.$$

Again, since there are infinitely many terms of $\{a_n\}$ greater than $\overline{\lim}\{a_n\} - \varepsilon$, there is an $n_2 > n_1$ such that

$$a_{n_2} > \overline{\lim}\{a_n\} - \varepsilon.$$

And since $s_{n_2} \geq a_{n_2}$, we have that

$$s_{n_2} > \overline{\lim}\{a_n\} - \varepsilon.$$

Continuing in this manner, we generate a subsequence $\{s_{n_k}\}$ of $\{s_n\}$ such that

$$s_{n_k} > \overline{\lim}\{a_n\} - \varepsilon,$$

and hence it follows that

$$\lim s_{n_k} \geq \overline{\lim}\{a_n\} - \varepsilon. \qquad (2.35)$$

But we already pointed out that the sequence $\{s_n\}$ was decreasing and bounded, and therefore converges to s, its limit. Thus, this subsequence, $\{s_{n_k}\}$, also converges to s by Theorem 2.64. So, we have from (2.35) that

$$s \geq \overline{\lim}\{a_n\} - \varepsilon. \qquad (2.36)$$

From this it follows, as it did previously (since ε is arbitrary), that

$$s \geq \overline{\lim}\{a_n\}. \qquad (2.37)$$

It follows from (2.33) and (2.37) that $s = \overline{\lim}\{a_n\}$, or said another way, that $s = \lim s_n = \lim \sup_{k \geq n}\{a_k\}$, which is what we wanted to prove. We leave the similar proof of (2) to you. ∎

Remark 2.80 *Since the sequence $\{s_n\}$ decreases, $\lim s_n = \inf\{s_n\}$ by Theorem 2.57. So, using the definition of s_n, we see that another way of describing the limit superior of a sequence is $\overline{\lim}\{a_n\} = \lim s_n = \inf\{s_n\} = \inf \sup_{k \geq n}\{a_k\}$. Since the sequence $\{i_n\}$ increases, $\lim i_n = \sup\{i_n\}$. So, another way of describing the limit inferior of a sequence is $\underline{\lim}\{a_n\} = \lim i_n = \sup\{i_n\} = \sup \inf_{k \geq n}\{a_k\}$.*

Example 2.81 *Consider the sequence $\{a_n\} = 0, 1, 0, 1, \ldots$. The set $\{a_k, a_{k+1}, a_{k+2}, \ldots\} = \{0, 1\}$, for all $k = 1, 2, 3, \ldots$. So $s_1 = \sup\{a_1, a_2, a_3, \ldots\} = \sup\{0, 1\} = 1$. Similarly, $s_2 = s_3 = \ldots = s_n = 1$. Thus, $\overline{\lim}\{a_n\} = \lim s_n = 1$. Similarly, $i_1 = i_2 = \ldots = i_n = 0$ for all n, so $\underline{\lim}\{a_n\} = \lim i_n = 0$. If $\{b_n\} = 1, 1/2, 1/3, \ldots$ then $s_1 = \sup\{1, 1/2, 1/3\ldots\} = 1$, $s_2 = \sup\{1/2, 1/3, 1/4\ldots\} = 1/2$, $s_3 = \sup\{1/3, 1/4, 1/5\ldots\} = 1/3$, and so on, so that $s_n = 1/n$. And $\overline{\lim}\{b_n\} = \lim s_n = 0$. In a similar manner, $i_n = 0$ for $n = 1, 2, 3, \ldots$, so $\underline{\lim}\{b_n\} = \lim i_n = 0$.*

EXERCISES

1. Suppose that $\{a_n\}$ is a bounded sequence of real numbers. Show that if $\{a_{n_k}\}$ is a subsequence of $\{a_n\}$, then $\overline{\lim}\{a_{n_k}\} \leq \overline{\lim}\{a_n\}$.

2. Show that if $\{a_n\}$ is a bounded sequence of real numbers, then $\overline{\lim}\{a_n\} = -\underline{\lim}\{-a_n\}$.

3. Let $A_n = \sup_{k \geq n}\{a_k\}$ and let $B_n = \sup_{k \geq n}\{b_k\}$. Using the idea that $\overline{\lim}\{a_n\} = \lim A_n$ and that $\overline{\lim}\{b_n\} = \lim B_n$, prove that if $\{a_n\}$ and $\{b_n\}$ are bounded sequences of real numbers, and $a_n \leq b_n$, then $\overline{\lim}\{a_n\} \leq \overline{\lim}\{b_n\}$. What is the corresponding result for limit inferiors?

4. Using the ideas from the previous exercise, show that $\overline{\lim}\{a_n + b_n\} \leq \overline{\lim}\{a_n\} + \overline{\lim}\{b_n\}$. Are they equal? If not, give a counterexample.

5. Show that if $c > 0$, $\overline{\lim}\, ca_n = c \cdot \overline{\lim} a_n$.

6. Show that it is not true that $\overline{\lim}\{a_n b_n\} \leq \left(\overline{\lim}\{a_n\}\right)\left(\overline{\lim}\{b_n\}\right)$.

7. Show that when $\{a_n\}$ and $\{b_n\}$ are nonnegative sequences, that $\overline{\lim}\{a_n b_n\} \leq \left(\overline{\lim}\{a_n\}\right)\left(\overline{\lim}\{b_n\}\right)$.

8. Suppose we write the rational numbers in $[0, 1]$ as a sequence, $\{a_n\}$, as we pointed out is possible in the text. Instead of finding all subsequential limits as we did in the text, show that $\limsup\{a_n\} = 1$ another way.

2.6 Cauchy Sequences

To prove that a sequence converges, we have to know the limit. Is it possible to tell if a sequence converges without knowing its limit? This section will answer this question.

Definition 2.82 *A sequence of real numbers, $\{a_n\}$, is called a **Cauchy sequence** if for any $\varepsilon > 0$ there exists an N (which usually depends on ε), such that $m, n > N$ implies that*
$$|a_n - a_m| < \varepsilon.$$

Intuitively, this says that the terms of the sequence get close to each other as you go farther out into the sequence. That is, they bunch up.

Cauchy sequences turn out to be critical objects in analysis. If a sequence converges to L, the terms will get close to L after a certain point, and hence be close to each other. So, it appears that a convergent sequence is Cauchy. Also, after drawing pictures on the real line, one would suppose that if the terms of a sequence eventually bunch up, they must bunch up near some limit value. That is, every Cauchy sequence converges. But is our intuition correct here? If it is, then the notion of convergent sequences of real numbers and Cauchy sequences of real numbers, are one in the same. And that means, we can use whichever notion suits us best to prove things. This is done regularly in analysis. Let's examine these ideas. First we look at some examples.

Example 2.83 *Consider the sequence $\{1/n\}$. We claim it is Cauchy. To see this, pick any $\varepsilon > 0$. We know there is an N such that $1/N < \varepsilon/2$. Suppose that $m, n > N$. Then $1/n$ and $1/m$ are both less than $1/N < \varepsilon/2$. Now we consider $|a_n - a_m| = |1/n - 1/m|$. This, by the triangle inequality, is less than or equal to $|1/n| + |1/m| = 1/n + 1/m < \varepsilon/2 + \varepsilon/2 = \varepsilon$. We have shown that for $m, n > N$, that $|a_n - a_m| < \varepsilon$, which means that this sequence is Cauchy.*

Example 2.84 *The sequence $1, 0, 1, 0, \ldots$, is not Cauchy because there is no way to make $|a_n - a_m| < 1/2$, for all n after a certain point. To see this notice that the difference of successive terms always has absolute value 1.*

Example 2.85 *Let $x_n = 1 - \dfrac{1}{3!} + \dfrac{1}{5!} - \ldots + \dfrac{(-1)^{n+1}}{(2n-1)!}$. It can be shown that this sequence is Cauchy. We will guide you through a proof of this in the exercises.*

We now move on to some basic results for Cauchy sequences. The first is to be expected.

Theorem 2.86 *If a sequence $\{a_n\}$ is convergent, then it is Cauchy.*

Proof. Suppose that $\{a_n\}$ converges to L. Then, we know that for any $\varepsilon > 0$ we can find an N such that $|a_n - L| < \varepsilon/2$, for $n > N$. Now, let both m and n be $> N$. Then we have

$$\begin{aligned} |a_n - a_m| &= |a_n - L + L - a_m| \\ &\leq |a_n - L| + |L - a_m| \quad \text{(triangle inequality)} \\ &< \varepsilon/2 + \varepsilon/2 \\ &= \varepsilon. \end{aligned}$$

Since this set of relationships implies that $|a_n - a_m| < \varepsilon$ for $m, n > N$, we have shown that the definition of sequence being Cauchy holds. So this sequence is Cauchy. ∎

Theorem 2.87 *If the sequence $\{a_n\}$ is Cauchy, then it is bounded.*

Proof. This is really no surprise since the terms bunching up after a certain point imply that these terms can be contained in an interval, making this collection bounded. The remaining terms not in this interval are finite in number and therefore, can certainly be enclosed by an interval, and so are also bounded. It follows that the whole sequence is bounded. Now, here is the formal proof: Since the sequence is Cauchy,

2.6. CAUCHY SEQUENCES

we can make $|a_n - a_m| < 1$ for m, n greater than some integer N. In particular, $|a_n - a_{N+1}| < 1$. Now, by the triangle inequality we have that, for $n > N$

$$\begin{aligned} |a_n| &= |a_n - a_{N+1} + a_{N+1}| \\ &\leq |a_n - a_{N+1}| + |a_{N+1}| \\ &\leq 1 + |a_{N+1}|. \end{aligned}$$

Now, let $M = \max\{|a_1|, |a_2|, |a_3|, ..., |a_N|, 1 + |a_{N+1}|\}$. Then $|a_n| \leq M$ for all n, which is one of the equivalent definitions of bounded. ∎

Theorem 2.88 *If a Cauchy sequence $\{a_n\}$ has a subsequence which converges to L, then $\{a_n\}$ converges to L.*

Proof. Suppose $\{a_{n_k}\}$ is $\{a_n\}$ which converges to L. To show that $\{a_n\}$ converges to L, we need to show that for any $\varepsilon > 0$ we can find an N, such that $n > N$ implies that $|a_n - L| < \varepsilon$. By the definition of Cauchy, there exists an N_1 such that

$$m, n > N_1 \text{ implies that } |a_n - a_m| < \varepsilon/2. \quad (2.38)$$

Since the sequence $\{a_{n_k}\}$ converges to L, there is an N_2 such that

$$n_k > N_2 \text{ implies that } |a_{n_k} - L| < \varepsilon/2. \quad (2.39)$$

Now, let N be bigger than both N_1 and N_2, and let n and n_k both be greater than N, while letting n_k be fixed. Since N is bigger than both N_1 and N_2, both (2.38) and (2.39) hold. So we have that for $n > N$,

$$\begin{aligned} |a_n - L| &= |a_n - a_{n_k} + a_{n_k} - L| \\ &\leq |a_n - a_{n_k}| + |a_{n_k} - L| \\ &< \varepsilon/2 + \varepsilon/2 \quad \text{(by (2.38) and (2.39))} \\ &= \varepsilon. \end{aligned}$$

We have shown that for $n > N$, $|a_n - L| < \varepsilon$, so the sequence converges to L. ∎

An intuitive proof of the above theorem could go like this: Since the sequence is Cauchy, after a certain point all the terms of the original sequence are close to each other. In particular, they are close to the latter terms of the known convergent subsequence, which are close to L. So the terms of the original sequence must also get close to L.

We now come to a theorem which is considered a major result in analysis. It says (together with Theorem 2.86), that for sequences of real numbers, the notions of convergence and Cauchy are one in the same.

CHAPTER 2. BOUNDS AND SEQUENCES OF REAL NUMBERS

Theorem 2.89 *Every Cauchy sequence of real numbers converges.*

Proof. Let us call our Cauchy sequence $\{a_n\}$. Then by Theorem 2.87, our sequence is bounded. But by the Bolzano Weierstrass theorem (Theorem 2.68), this sequence has a convergent subsequence which converges to some number L. Now, by Theorem 2.88, the original sequence converges to L. ∎

The way this theorem is used is as follows: We are given a sequence whose convergence we are unsure of. But we are able to show it is Cauchy. This then tells us the sequence converges.

Example 2.90 *We pointed out in Example 2.85 that the sequence $\{x_n\}$ where $x_n = 1 - \frac{1}{3!} + \frac{1}{5!} - .... + \frac{(-1)^{n+1}}{(2n-1)!}$, is Cauchy. (See also Exercise 5.) Therefore it converges, but to what? The student who remembers calculus, will know this sequence converges to sine of 1 radian, which to the uninitiated, is hardly obvious.*

EXERCISES

1. Give an example of a sequence of real numbers which is not Cauchy.

2. Show that if a sequence $\{x_n\}$ is Cauchy, then so is every subsequence of $\{x_n\}$.

3. Show, using the definition of Cauchy, that if $\{a_n\}$ and $\{b_n\}$ are Cauchy, then so is $\{a_n + b_n\}$. Then, without using the definition, but using theorems from the chapter, show that $\{a_n^2 b_n^3\}$ is Cauchy. If $\lim b_n \neq 0$, is $\left\{\frac{a_n}{b_n}\right\}$ Cauchy?

4. Show the sequence $\{a_n\}$ where $a_n = 1 + \frac{1}{2} + \frac{1}{3} + ... + \frac{1}{n}$, is not Cauchy by showing that $a_{2n} - a_n$ is always greater than or equal to $\frac{1}{2}$.

5. Show that the sequence $\{x_n\}$ where $x_n = 1 - \frac{1}{3!} + \frac{1}{5!} - ... + \frac{(-1)^{n+1}}{(2n-1)!}$, is Cauchy, using the definition of Cauchy. [Hint: Show that $n! \geq 2^{n-1}$ when $n \geq 1$ (so that $\frac{1}{n!} \leq \frac{1}{2^{n-1}}$) and then compare $|x_m - x_n|$ to a geometric series.]

Chapter 3

Metric Spaces

3.1 Basic Results and Examples

In this section we discuss the notion of a metric space. Intuitively, a metric space is a set, together with a notion of distance between points. This distance does not have to bear any relationship to the ordinary notion of distance as we will soon see. The distance is simply a function with certain properties that mimic the properties that we ordinarily observe with distance in Euclidean space.

Definition 3.1 *Suppose M is any set. A **metric**, d, on M, is a function from $M \times M$ to R which satisfies the following properties:*
  (1) $d(x,y) \geq 0$ for all $x,y \in M$ and $d(x,y) = 0$ if and only if $x = y$.
  (2) $d(x,y) = d(y,x)$ for all $x,y \in M$.
  (3) (triangle inequality) $d(x,y) \leq d(x,z) + d(y,z)$ for all $x,y,z \in M$.

The first property of a metric tells us that the distance between any two points, x and y, can never be negative, and is only zero when the two points are the same. The second property tells us that the order in which you compute the distance doesn't matter, while the third property is essentially saying the distance between two points, x and y, is really the shortest distance between the two points. Intuitively, if you want to travel from x to y, any path through an intermediate point, z, will be at least as long as the direct path from x to y. (Just to be clear, the word "path" here has no formal meaning. The set M is an abstract set and the notion of a path is not defined in an abstract set.)

Definition 3.2 *A **metric space** is a set M, together with a metric d defined on M.*

Notation 3.3 *We denote the metric space by the ordered pair (M, d).*

Remark 3.4 *A set can have many metrics defined on it. If d_1, d_2 and d_3 are three different metrics defined on a set M, then the metric spaces $(M, d_1), (M, d_2)$, and (M, d_3) are all considered to be different.*

Example 3.5 *Let our set, M, be the set of real numbers, R. Define the following metric on R: $d(x, y) = |x - y|$. This is known as the **usual metric** on R. We now show that it is a metric.*

It is clear that for any real numbers, $d(x, y) = |x - y| \geq 0$ and $d(x, y) = |x - y| = 0$ if and only if $x = y$. So property (1) in the definition of a metric holds. It is also clear that $d(x, y) = d(y, x)$ since $|x - y| = |y - x|$. Finally, the third condition of a metric space, $d(x, y) \leq d(x, z) + d(y, z)$, in the context of this example says that

$$|x - y| \leq |x - z| + |z - y|,$$

and we know this is true by the ordinary triangle inequality for real numbers, (Theorem 1.51, with $a = x - z$ and $b = z - y$).

We have verified all three conditions for a metric space, and so the set of real numbers with the usual metric, is a metric space.

Example 3.6 *Let M be any set. The function $d : M \times M \to R$ defined by*

$$d(x, y) = \begin{cases} 0 & \text{if } x = y \\ 1 & \text{if } x \neq y \end{cases}$$

*where x and y are any two points in M, is a metric on M. This is known as the **trivial** or **discrete** metric.* It is clear that conditions (1) and (2) of the definition of a metric hold. Let us examine condition (3). We observe that if $d(x, y) = 0$, then certainly condition 3 holds since $d(x, z)$ and $d(y, z)$ are both nonnegative. Also, if either $d(x, z)$ or $d(y, z) = 1$ then (3) holds since $d(x, y)$ is always ≤ 1. Therefore the only case we have to consider is when $d(x, y) = 1$, and both $d(x, z)$ and $d(y, z) = 0$. But if $d(x, z) = 0$, then $x = z$ by condition (1). Similarly, if $d(y, z) = 0$, then $y = z$. So if both $d(x, z)$ and $d(y, z) = 0$, then $x = y = z$ which means that $d(x, y) = 0$, not 1, as we originally hypothesized. Said another way, this last case can never happen. So in all legitimate cases, the triangle inequality holds, and we have that (M, d) is a metric space.

Example 3.7 *Anyone who is reading this book is probably familiar with the notion of a continuous function. We will discuss this concept in some detail later on in this book. But for now, let us present an example of a metric space using them. This is an important metric space in higher analysis.*

Let M be the set of continuous real valued functions with domain $[0, 1]$. We denote this set by $C[0, 1]$. So $M = C[0, 1]$ consists of functions, and our metric will have to

3.1. BASIC RESULTS AND EXAMPLES

be defined on pairs of functions. Suppose that $f(x)$ and $g(x)$ are in $C[0,1]$. Define $d(f,g)$ to be the maximum of $|f(x) - g(x)|$, on $[0,1]$. We abbreviate this maximum as $\max|f(x) - g(x)|$. (That this maximum exists follows from Corollary 4.51 in Chapter 4.) It is clear that conditions (1) and (2) of a metric space hold. To verify condition (3), suppose that $h(x)$ is in $C[0,1]$. We observe from the ordinary triangle inequality for real numbers, that if x is any fixed real number in $[0,1]$,

$$\begin{aligned} |f(x) - g(x)| &\leq |f(x) - h(x)| + |h(x) - g(x)| \\ &\leq \max|f(x) - h(x)| + \max|h(x) - g(x)| \\ &= d(f,h) + d(h,g). \end{aligned}$$

So we have shown that $|f(x) - g(x)|$ is less than or equal to the fixed number $d(f,h) + d(h,g)$ for each $x \in [0,1]$, from which it follows that $\max|f(x) - g(x)| \leq d(f,h) + d(h,g)$. Said another way, $d(f,g) \leq d(f,h) + d(h,g)$, and this, of course, is property (3) of the definition of a metric. So (M,d) is a metric space.

We have other examples to give, but first, there is a rather important result we will need known as the **Cauchy-Schwartz Inequality** (which some books call the **Cauchy-Schwartz-Bunyakovski**, or **CBS** inequality).

Theorem 3.8 *If $p = (p_1, p_2, p_3, ..., p_n)$, and $q = (q_1, q_2, q_3, ..., q_n)$ are points in R^n, and not all the components of p are zero, then*

$$\left| \sum_{i=1}^{n} p_i q_i \right| \leq \left(\sqrt{\sum_{i=1}^{n} p_i^2} \right) \cdot \left(\sqrt{\sum_{i=1}^{n} q_i^2} \right).$$

Proof. For any real number x, it is clearly true that $(p_i x + q_i)^2 \geq 0$. So it follows that

$$\sum_{i=1}^{n} (p_i x + q_i)^2 \geq 0.$$

If we expand this it becomes,

$$\sum_{i=1}^{n} \left(p_i^2 x^2 + 2 p_i q_i x + q_i^2 \right) \geq 0,$$

and distributing the summation, this can be written as

$$\left(\sum_{i=1}^{n} p_i^2 \right) x^2 + \left(2 \sum_{i=1}^{n} p_i q_i \right) x + \sum_{i=1}^{n} q_i^2 \geq 0.$$

Now, call $A = \sum_{i=1}^{n} p_i^2$, $B = 2\sum_{i=1}^{n} p_i q_i$, and $C = \sum_{i=1}^{n} q_i^2$. The above display now reads,

$$Ax^2 + Bx + C \geq 0. \tag{3.1}$$

At this point, we think geometrically. The graph of $y = Ax^2 + Bx + C$ is a parabola. (We know that $A > 0$, or else all of the components of p would be zero, contradicting the hypothesis.) A well known fact from high school mathematics says that a parabola with $A > 0$, has its lowest point (its turning point) at $x = \frac{-B}{2A}$. Substituting this into (3.1), we get that

$$A\left(\frac{-B}{2A}\right)^2 + B\left(\frac{-B}{2A}\right) + C \geq 0.$$

Combining the first two terms of this display into a single fraction, this simplifies to

$$-\frac{B^2}{4A} + C \geq 0.$$

Adding $\frac{B^2}{4A}$ to both sides of the inequality and multiplying both sides by $4A$, which we know is positive by definition of A, we get that

$$B^2 \leq 4AC.$$

Now, substituting the values for A, B and C given above, we get that

$$\left(2\sum_{i=1}^{n} p_i q_i\right)^2 \leq 4\left(\sum_{i=1}^{n} p_i^2\right) \cdot \left(\sum_{i=1}^{n} q_i^2\right),$$

or just that

$$4\left(\sum_{i=1}^{n} p_i q_i\right)^2 \leq 4\left(\sum_{i=1}^{n} p_i^2\right) \cdot \left(\sum_{i=1}^{n} q_i^2\right).$$

We divide both sides by 4 to get

$$\left(\sum_{i=1}^{n} p_i q_i\right)^2 \leq \left(\sum_{i=1}^{n} p_i^2\right) \cdot \left(\sum_{i=1}^{n} q_i^2\right).$$

Now we take the square root of both sides, remembering that $\sqrt{m^2} = |m|$, and we get

$$\left|\sum_{i=1}^{n} p_i q_i\right| \leq \sqrt{\left(\sum_{i=1}^{n} p_i^2\right)} \cdot \sqrt{\left(\sum_{i=1}^{n} q_i^2\right)},$$

which is what we were trying to prove.

What if all the components of p are 0? Then the CBS inequality clearly holds. ∎

3.1. BASIC RESULTS AND EXAMPLES

Definition 3.9 *Suppose that $p, q \in R^n$ and that $p = (p_1, p_2, p_3, ..., p_n)$ and $q = (q_1, q_2, q_3, ..., q_n)$. Define the following function on $R^n \times R^n$, known as the **usual metric on** R^n: $d(p, q) = \sqrt{\sum_{i=1}^{n} (p_i - q_i)^2}$*
$= \sqrt{(p_1 - q_1)^2 + (p_2 - q_2)^2 + (p_3 - q_3)^2 + ... + (p_n - q_n)^2}$.

In R^2, $d(p, q) = \sqrt{(p_1 - q_1)^2 + (p_2 - q_2)^2}$ which is the usual distance between points in R^2, and in R^3, $d(p, q) = \sqrt{(p_1 - q_1)^2 + (p_2 - q_2)^2 + (p_3 - q_3)^2}$ which is the usual distance between points in R^3.

We now have

Theorem 3.10 *R^n with the usual metric, is a metric space.*

Proof. Condition (1) of a metric space clearly holds since $d(p, q) \geq 0$ and if $d(p, q) = 0$, then

$$\sqrt{(p_1 - q_1)^2 + (p_2 - q_2)^2 + (p_3 - q_3)^2 + ... + (p_n - q_n)^2} = 0,$$

which implies, by squaring both sides, that $(p_1 - q_1)^2 + (p_2 - q_2)^2 + (p_3 - q_3)^2 + ... + (p_n - q_n)^2 = 0$. We now have a sum of nonnegative terms equal to zero, which tells us that each term in this sum is 0. From this we get that $p_1 = q_1, p_2 = q_2, ..., p_n = q_n$, so that p and q are the same point.

Since $d(p, q) = d(q, p)$ is clear, we need only verify the triangle inequality. So let $r = (r_1, r_2, r_3, ..., r_n)$. We now show that $d(p, q) \leq d(p, r) + d(r, q)$.

Let $a_i = p_i - r_i$, and $b_i = r_i - q_i$. Observe that $a_i + b_i = p_i - q_i$. So we have the following:

$$d(p, q) = \sqrt{\sum_{i=1}^{n} (a_i + b_i)^2}, \tag{3.2}$$

$$d(p, r) = \sqrt{\sum_{i=1}^{n} a_i^2}, \quad \text{and} \tag{3.3}$$

$$d(r, q) = \sqrt{\sum_{i=1}^{n} b_i^2}. \tag{3.4}$$

Consider the following string of equalities and inequalities:

$$\sum_{i=1}^{n}(a_i+b_i)^2 = \sum_{i=1}^{n}(a_i)^2 + \sum_{i=1}^{n}(b_i)^2 + 2\sum_{i=1}^{n}(a_ib_i)$$

$$\leq \sum_{i=1}^{n}(a_i)^2 + \sum_{i=1}^{n}(b_i)^2 + 2\left|\sum_{i=1}^{n}(a_ib_i)\right|$$

(since a number is always $\leq$ its absolute value)

$$\leq \sum_{i=1}^{n}(a_i)^2 + \sum_{i=1}^{n}(b_i)^2 + 2\sqrt{\sum_{i=1}^{n}(a_i)^2} \cdot \sqrt{\sum_{i=1}^{n}(b_i)^2}$$

(by Theorem 3.8)

$$= \left(\sqrt{\sum_{i=1}^{n}(a_i)^2} + \sqrt{\sum_{i=1}^{n}(b_i)^2}\right)^2$$

$\left(\text{since } (x+y)^2 = x^2+y^2+2xy. \text{Here we are taking } x = \sqrt{\sum_{i=1}^{n}(a_i)^2} \text{ and } y = \sqrt{\sum_{i=1}^{n}(b_i)^2}\right)$.
This string of equalities and inequalities shows that

$$\sum_{i=1}^{n}(a_i+b_i)^2 \leq \left(\sqrt{\sum_{i=1}^{n}(a_i)^2} + \sqrt{\sum_{i=1}^{n}(b_i)^2}\right)^2.$$

Taking the square root of both sides we get that

$$\sqrt{\sum_{i=1}^{n}(a_i+b_i)^2} \leq \sqrt{\sum_{i=1}^{n}(a_i)^2} + \sqrt{\sum_{i=1}^{n}(b_i)^2}.$$

Using (3.2), (3.3) ,and (3.4) this last display can be rewritten as

$$d(p,q) \leq d(p,r) + d(r,q).$$

Since we have shown that all 3 properties of a metric space hold, R^n with the usual metric is a metric space. ∎

Definition 3.11 *Suppose that (M,d) is a metric space and that S is a subset of M. If we restrict d to $S \times S$ and call the restricted function d_S, conditions $(1) - (3)$ of the definition of metric space hold in S since they hold in M. Thus, (S, d_S) is a metric space. S with the metric d_S is called a **subspace** of (M,d).*

3.2. SEQUENCES IN METRIC SPACES

Remark 3.12 *A subset, S, of a metric space (M, d), is itself a metric space when we use the same metric on S that M carries. This provides us with infinitely many metric spaces. Also, S may carry many different metrics giving us several different metric spaces. It should be noted, however, that in order for S to be called a subpace of M it MUST carry the same metric as M.*

Example 3.13 *Consider the set of rational numbers, Q. It is a subset of R. We can define the trivial metric on Q, and with this metric, Q is not a subspace of R. But it is a subspace of R if we use the usual metric on S. That is, if we define $d(r, s) = |r - s|$ for any rational numbers r and s.*

EXERCISES

1. In what follows, $p = (x, y)$ and $q = (a, b)$ are arbitrary points in $R \times R$. Show that the functions defined in (a) and (b) are not metrics on $R \times R$, while that in (c) is.

   (a) $d(p, q) = x^2 + a^2$

   (b) $d(p, q) = e^{\sqrt{(x-a)^2 + (y-b)^2}}$

   (c) $d(p, q) = |x - a| + |y - b|$

2. Suppose that U is the unit circle in R^2, that is, $U = \{(x, y) | x^2 + y^2 = 1\}$. For any two points, P and Q, on this circle, define $d(P, Q)$ to be the length of smallest arc of the circle joining the points P and Q. Is this a metric on U?

3. Suppose that d is a metric on S. Define d_λ on $S \times S$ as follows: For $\lambda > 0$, $d_\lambda(x, y) = \lambda d(x, y)$. Is d_λ a metric on S?

4. Suppose that (M, d) is a metric space. Show that $|d(x, y) - d(x, y_0)| \leq d(y, y_0)$ for any x, y, and y_0 in M.

5. Define the following metric on R^2: $d((x_1, y_1), (x_2, y_2)) = \sup(|x_1 - x_2|, |y_1 - y_2|)$. Verify this is a metric.

3.2 Sequences in Metric Spaces

The same way that one can talk about sequences of real numbers, one can also talk about sequences in a metric space. We can also talk about notions of convergence, boundedness, etc., and we will do so now.

Intuitively, a sequence in a metric space is a listing, $x_1, x_2, x_3, \ldots$ of elements of the space. The formal definition of a sequence in a metric space (M, d) is that it is a function from Z^+ to M, where $f(1) = x_1$, $f(2) = x_2$, etc.

Definition 3.14 *We say that a sequence of points $\{a_n\}$ in a metric space (M,d) **converges to a point** a in the metric space, if for any $\varepsilon > 0$, one can find a positive N, such that $n > N$ implies that $d(a_n, a) < \varepsilon$. When $\{a_n\}$ converges to the point a, we say that a is the limit of the sequence $\{a_n\}$.*

Notice how similar this definition is to the definition of convergence of real numbers. All we did was replace $|a_n - a|$ in the definition of convergence of a sequence of real numbers by $d(a_n, a)$. Just as in the case of the real numbers with the usual metric, the N in this definition usually depends on ε. Let us look at some examples to see what this means in specific cases.

Example 3.15 *Take R with the usual metric. Saying that $\{a_n\}$ converges to the real number L means that for any $\varepsilon > 0$, there exists an N such that $n > N$ implies that $d(a_n, L) < \varepsilon$. But $d(a_n, L)$ in R with the usual metric is $|a_n - L|$. So, replacing $d(a_n, L)$ with $|a_n - L|$ in the previous sentence, we see that our definition of convergence in metric spaces agrees with our definition of convergence of sequences on the real line, when the real line carries the usual metric.*

Example 3.16 *Let (M, d) be R^2 with the usual metric. Since points in R^2 are ordered pairs, any sequence in R^2 consists of points which are ordered pairs. Thus, if $\{a_n\}$ is a sequence in R^2, $a_n = (x_n, y_n)$. Saying that a_n converges to $a = (x, y)$ means that for any $\varepsilon > 0$, there is an N such that $n > N$ implies that $d(a_n, a) = \sqrt{(x_n - x)^2 + (y_n - y)^2} < \varepsilon$. So, for example, suppose that we wanted to show that the sequence $a_n = (\frac{1}{n}, \frac{n}{n+1})$ converges to $(0,1)$. Then we would want to make $d(a_n, a)$ or $\sqrt{\left(\frac{1}{n} - 0\right)^2 + \left(\frac{n}{n+1} - 1\right)^2} < \varepsilon$ for $n > N$. This square root simplifies to $\sqrt{\frac{1}{n^2} + \frac{1}{(n+1)^2}}$, which is less than $\sqrt{\frac{1}{n^2} + \frac{1}{n^2}} = \sqrt{\frac{2}{n^2}} = \frac{\sqrt{2}}{n}$, and this can be made less than ε if $n > \frac{\sqrt{2}}{\varepsilon} = N$.*

Example 3.17 *Let $[0, 1]$ carry the discrete metric. With this metric, does the sequence $\{1/n\}$ converge? The answer is no. One might think it converges to 0. But if that were the case, then $d(1/n, 0)$ could be made less than any $\varepsilon > 0$. So let's take $\varepsilon = 1/2$. Now, in this metric, $d(1/n, 0) = 1$ for all n, which obviously can't be made less than ε. In fact, if a sequence in a discrete metric space converges, it must eventually be constant. Proving this is one of the exercises of this chapter. So if a sequence is eventually constant, like the sequence $1, 1/3, 1/5, 1/5, 1/5, 1/5....$, it will converge, otherwise it won't.*

3.2. SEQUENCES IN METRIC SPACES

Let us return to Example 3.7, where we discussed $C[0,1]$, the set of continuous real valued functions defined on the closed interval $[0,1]$. The distance between any two functions $f(x)$ and $g(x)$ was defined to be the maximum of $|f(x) - g(x)|$ where x runs through all values in $[0,1]$. Now, suppose a sequence, $f_n(x)$, of functions in $C[0,1]$ converges to $f(x)$ in $C[0,1]$ with this metric. This means that for each $\varepsilon > 0$, we can make $d(f_n, f) = \max|f_n(x) - f(x)| < \varepsilon$ for $n > N$, for some N. But $\max|f_n(x) - f(x)| < \varepsilon$ if an only if $|f_n(x) - f(x)| < \varepsilon$ for all x in $[0,1]$ which is equivalent to $f(x) - \varepsilon < f_n(x) < f(x) + \varepsilon$ (for $n > N$). The geometric interpretation of convergence of functions in this metric space is that for any $\varepsilon > 0$, the functions $f_n(x)$ eventually lie within a band of width ε around $f(x)$. (See Figure 3.1 for a picture of this band within which all the functions lie after a certain point.)

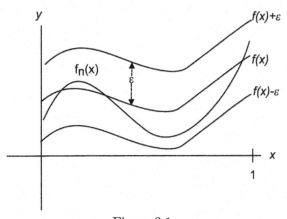

Figure 3.1.

Another way that we might say this is that the functions become uniformly close to $f(x)$ after a certain point, where by uniformly close we mean the $f_n(x)$ are within ε of $f(x)$ after a certain point, regardless of x. We will exploit this idea in Chapter 5.

The definition of convergence of $\{a_n\}$ to a in the abstract case, namely that $d(a_n, a)$ can be made less than any $\varepsilon > 0$ after a certain point, is very clean and generalizes the notion of convergence on the real line. As a result, we can often prove things in the abstract case about sequences and other things, by simply mimicking the proofs on the real line. Let us illustrate this by proving the next theorem first on the real line, and then in an arbitrary metric space.

Theorem 3.18 *A convergent sequence in a metric space has a unique limit.*

Proof. On the real line: Suppose that L and M are both limits of the sequence $\{a_n\}$. We will establish that $L = M$, thus establishing the uniqueness of the limit.

Suppose then that ε is any positive number. Since $\{a_n\}$ converges to both L and M we have, from the definition of a limit, that there is an N_1 such that

$$|a_n - L| < \frac{\varepsilon}{2} \text{ for } n > N_1 \qquad (3.5)$$

and there is an N_2 such that

$$|a_n - M| < \frac{\varepsilon}{2} \text{ for } n > N_2. \qquad (3.6)$$

Now, let N be any number bigger than both N_1 and N_2. Then we have

$$\begin{aligned} |L - M| &= |L - a_n + a_n - M| \\ &\leq |L - a_n| + |a_n - M| \\ &= |a_n - L| + |a_n - M| \\ &\leq \varepsilon/2 + \varepsilon/2 \text{ (by (3.5) and (3.6))} \\ &= \varepsilon \end{aligned}$$

This string of inequalities shows that we can make $|L - M| \leq \varepsilon$ for any $\varepsilon > 0$, and hence by Lemma 2.41, $L = M$. ∎

Now we turn to the proof in an abstract metric space.

Proof. In the abstract: Suppose that L and M are both limits of the sequence $\{a_n\}$. We will establish that $L = M$, thus establishing the uniqueness of the limit. Suppose then that ε is any positive number. Since $\{a_n\}$ converges to both L and M we have, from the definition of a limit, that there is an N_1 such that

$$d(a_n, L) < \frac{\varepsilon}{2} \text{ for } n > N_1 \qquad (3.7)$$

and there is an N_2 such that

$$d(a_n, M) < \frac{\varepsilon}{2} \text{ for } n > N_2. \qquad (3.8)$$

Now, let N be any number bigger than both N_1 and N_2. Then we have

$$\begin{aligned} &d(L, M) \\ &\leq d(L, a_n) + d(a_n, M) \text{ (by the triangle inequality)} \\ &= d(a_n, L) + d(a_n, M) \\ &\leq \varepsilon/2 + \varepsilon/2 \text{ (by (3.7) and (3.8))} \\ &= \varepsilon. \end{aligned}$$

3.2. SEQUENCES IN METRIC SPACES

This string of inequalities shows that we can make the fixed number $d(L, M) \leq \varepsilon$ for any $\varepsilon > 0$, and hence by Lemma 2.41, $d(L, M) = 0$ (using $a = d(L, M) = |d(L, M)|$). Therefore by condition (1) of the definition of a metric space, $L = M$. ∎

This illustrates the similarities of the proofs for metric spaces and those for the real line with the usual metric. So we will simply state the theorems we have already proven that are also true in general metric spaces. However, we have to define terms analogous to those we gave on the real line in general metric spaces.

Definition 3.19 *A set S in a metric space (M, d) is called **bounded** if there is some positive K such that $d(x, y) \leq K$ for all $x, y \in S$.*

Remark 3.20 *For the real line with the usual metric, we now have 3 definitions of bounded. A set, S, is bounded if it is bounded above and below, or, if there exists a $K > 0$ such that $|a| \leq K$, for all a in S, and our most recent definition, that $d(a, b) = |a - b| \leq K$ for all a and b contained in S. In the next section, we will give yet another definition of bounded set in a metric space. We will ask you to show that all four definitions of bounded set on the real line are equivalent, meaning, any one of them implies any other.*

Definition 3.21 *A sequence $\{a_n\}$ in a metric space (M, d) is called a Cauchy sequence if, for any $\varepsilon > 0$, there exists an N (which usually depends on ε), such that $m, n > N$ implies that*
$$d(a_n, a_m) < \varepsilon.$$

Definition 3.22 *A subsequence of a sequence $\{a_n\}$ in a metric space has the same meaning as we gave before, in Definition 2.61 which informally is a sequence obtained from the original sequence by knocking out certain terms (possibly knocking out no terms). More formally, it is a sequence $a_{n_1}, a_{n_2}, a_{n_3}, ...$, where $n_1 < n_2 < n_3 < ...$.*

Theorem 3.23 *If a sequence, $\{a_n\}$ in a metric space converges to L, then any subsequence of $\{a_n\}$ also converges to L.*

Theorem 3.24 *If a sequence $\{a_n\}$ in a metric space converges, then it is Cauchy.*

Theorem 3.25 *If $\{a_n\}$ is a sequence in a metric space which converges to L, then $\{a_n\}$ is bounded.*

Theorem 3.26 *A Cauchy sequence in a metric space is bounded.*

Theorem 3.27 *If a Cauchy sequence, $\{a_n\}$, in a metric space has a subsequence converging to L, then $\{a_n\}$ converges to L.*

We would like to talk a bit about sequences in R^n.

Remark 3.28 *If you are thinking that Theorem 3.26 might immediately from Theorem 3.25 because every Cauchy sequence converges, you are mistaken. On the real line every Cauchy sequence converges, but it is not true in a general metric space that every Cauchy sequence converges. For example, let $M = (0,1]$ carry the usual metric from R. The Cauchy sequence $\{1/n\}$ is contained in M, but doesn't converge in M. (Converging in M means its limit is in M.) It does however converge in R (which is one way to determine it is Cauchy.)*

Suppose that we have the sequence $\{a_n\}$ in R^3 where $a_n = \left(\frac{1}{n}, 3 + \frac{3}{n}, \frac{n-1}{2n+1}\right)$. Now, (in R with the usual metric) the sequence $\left\{\frac{1}{n}\right\}$, of first components, converges to 0, the sequence of second components, $\left\{3 + \frac{3}{n}\right\}$, converges to 3, and the sequence of third components, $\left\{\frac{n-1}{2n+1}\right\}$, converges to $\frac{1}{2}$. So to what point in R^3 (with the usual metric) does the sequence $\{a_n\}$ converge? It would be gratifying, and would make our life easier if it converged to $(0, 3, \frac{1}{2})$, but does it? The next theorem tells us yes, and more. It tells us that this generalizes to sequences in R^k. To discuss this further, we need some notation.

Every point, a, in R^k has k components. We will call the first component a^1, the second a^2, etc. Now, if $\{a_n\}$ is a sequence of points in R^k, each a_n has k components, so $a_1 = (a_1^1, a_1^2, a_1^3, ...a_1^k)$ and $a_2 = (a_2^1, a_2^2, a_2^3, ...a_2^k)$, and so on. The next theorem tells us that to determine if a sequence, $\{a_n\}$, in R^k with the usual metric converges, we need only look at the sequences of first components $\{a_n^1\}$, the sequence of second components $\{a_n^2\}$, and so on. If they all converge, then so does $\{a_n\}$. Furthermore, if they converge to $a_0^1, a_0^2, a_0^3, ..., a_0^k$, respectively, then $\{a_n\}$ converges to the point $a_0 = (a_0^1, a_0^2, a_0^3, ..., a_0^k)$.

Theorem 3.29 *If $\{a_n\}$ is a sequence in R^k, then $a_n \to a_0 = (a_0^1, a_0^2, a_0^3, ...a_0^k)$, in the usual metric on R^k, if and only if $a_n^1 \to a_0^1, a_n^2 \to a_0^2, ..., a_n^k \to a_0^k$ in R with the usual metric.*

Proof. ($\Longrightarrow$) Suppose that $a_n \to a_0$, then for any $\varepsilon > 0$, there is an N such that $n > N$ implies that $d(a_n, a_0) = \sqrt{(a_n^1 - a_0^1)^2 + (a_n^2 - a_0^2)^2 + (a_n^3 - a_0^3)^2 + ... + (a_n^k - a_0^k)^2} < \varepsilon$. But

$$\sqrt{(a_n^1 - a_0^1)^2} < \sqrt{(a_n^1 - a_0^1)^2 + (a_n^2 - a_0^2)^2 + (a_n^3 - a_0^3)^2 + ... + (a_n^k - a_0^k)^2} < \varepsilon, \quad (3.9)$$

and since $\sqrt{(a_n^1 - a_0^1)^2} = |a_n^1 - a_0^1|$. It follows from (3.9), that for $n > N$, $|a_n^1 - a_0^1| < \varepsilon$. This says that $a_n^1 \to a_0^1$. In a similar manner $a_n^2 \to a_0^2$ and so on.

($\Longleftarrow$) Suppose that $a_n^i \to a_0^i$ for $i = 1, 2, 3, .., k$. Then by definition of convergence, for any $\varepsilon > 0$ there are numbers $N_1, N_2, .., N_k$, such that when $n > N_1, |a_n^1 - a_0^1| <$

3.3. OPEN AND CLOSED SETS

$\varepsilon/\sqrt{k}$, when $n > N_2$, $|a_n^2 - a_0^2| < \varepsilon/\sqrt{k}$, when $n > N_3$ $|a_n^3 - a_0^3| < \varepsilon/\sqrt{k}$ and so on, to when $n > N_k$, $|a_n^k - a_0^k| < \varepsilon/\sqrt{k}$. Now, let N be any number bigger than $N_1, N_2, .., N_k$. Then when $n > N$, all these inequalities hold. Now, consider

$$\begin{aligned} d(a_n, a_0) &= \sqrt{(a_n^1 - a_0^1)^2 + (a_n^2 - a_0^2)^2 + (a_n^3 - a_0^3)^2 + ... + (a_n^k - a_0^k)^2} \\ &< \sqrt{\left(\frac{\varepsilon}{\sqrt{k}}\right)^2 + \left(\frac{\varepsilon}{\sqrt{k}}\right)^2 + \left(\frac{\varepsilon}{\sqrt{k}}\right)^2 + ... + \left(\frac{\varepsilon}{\sqrt{k}}\right)^2} \\ &= \varepsilon. \end{aligned}$$

We have shown that for any $\varepsilon > 0$, there is an N such that $n > N$ implies that $d(a_n, a_0) < \varepsilon$. This is what it means for $\{a_n\}$ to converge to a_0. ∎

EXERCISES

1. Show that if a sequence in a discrete metric space converges, it must eventually be constant.

2. Prove each of the Theorems 3.23 to 3.27.

3. In R^4, what is the limit of the sequence $\{p_n\}$ where
$$p_n = \left(\frac{1}{n}, \frac{3n}{n+1}, 2^{(2/n)}, \cos\left(\frac{1}{n^2}\right)\right)?$$

3.3 Open and Closed Sets

Definition 3.30 *Suppose that (M, d) is a metric space, that $a \in M$, and that the real number r is greater than 0. The set $B_r(a) = \{x \in M | \, d(x, a) < r\}$ is called the **open ball** with center a and radius r.*

Example 3.31 *Let our metric space be R with the usual metric. That is, for any real numbers a and b, $d(a, b) = |a - b|$. Then*

$$\begin{aligned} B_r(a) &= \{x \in R | \, d(x, a) < r\} \\ &= \{x \in R | \, |x - a| < r\} \\ &= \{x \in R | \, -r < x - a < r\} \\ &= \{x \in R | \, a - r < x < a + r\} \\ &= (a - r, a + r). \end{aligned}$$

Thus, open balls $B_r(a)$, in R with the usual metric, are open intervals whose centers are at a, and extend in each direction from the center a distance of r.

Example 3.32 *Let our metric space be R^2 with the usual metric. That is, if $p = (p_1, p_2)$ and $q = (q_1, q_2)$, then $d(p,q) = \sqrt{(p_1 - q_1)^2 + (p_2 - q_2)^2}$. What do our open balls look like here? Well, letting $P = (x, y)$ represent a point in R^2 and letting $Q = (a, b)$ we have*

$$\begin{aligned} B_r(Q) &= \{P \in R^2 |\ d(P, Q) < r\} \\ &= \{(x, y) \in R^2 |\ \sqrt{(x - a)^2 + (y - b)^2} < r\} \\ &= \{(x, y) \in R^2 |\ (x - a)^2 + (y - b)^2 < r^2\}, \end{aligned}$$

which we recognize as being the interior of a circle whose center is (a, b) and whose radius is r. Thus open balls in R^2 are interiors of circles.

Example 3.33 *Let our metric space be R^3 with the usual metric. That is, if $p = (p_1, p_2, p_3)$ and $q = (q_1, q_2, q_3)$ then $d(p, q) = \sqrt{(p_1 - q_1)^2 + (p_2 - q_2)^2 + (p_3 - q_3)^2}$. What do our open balls look like now? Well, letting $P = (x, y, z)$ represent a point in R^3 and letting $Q = (a, b, c)$ we have*

$$\begin{aligned} B_r(Q) &= \{P \in R^3 |\ d(P, Q) < r\} \\ &= \{(x, y, z) \in R^3 |\ \sqrt{(x - a)^2 + (y - b)^2 + (z - c)^2} < r\} \\ &= \{(x, y, z) \in R^3 |\ (x - a)^2 + (y - b)^2 + (z - c)^2 < r^2\}, \end{aligned}$$

which we recognize as the interior of a sphere whose center is at Q and whose radius is r.

Example 3.34 *Let M be any set, and let M carry the discrete metric, which we recall is defined by*

$$d(x, y) = \begin{cases} 0 & \text{if } x = y \\ 1 & \text{if } x \neq y \end{cases}.$$

What do the open balls look like now? Well, if $a \in M$, then $B_r(a) = \{x \in M | d(x, a) < r\}$. Now, since the maximum distance between any two points in this set is 1, if $r > 1$ we have $B_r(a) = M$, the entire metric space, while if $r \leq 1$, $B_r(a) = \{a\}$. So the only open balls are the whole metric space and sets consisting of single points.

Example 3.35 *Suppose S is a subspace of M. What do the open balls in S look like? It turns out that they are the open balls in M intersected with S. To see this, let $B_r^S(p)$ represent an open ball in S and let $B_r(p)$ represent an open ball in M. Then by definition,*

$$\begin{aligned} B_r^S(p) &= \{x \in S |\ d(x, p) < r\} \\ &= \{x \in M |\ d(x, p) < r\} \cap S \\ &= B_r(p) \cap S. \end{aligned}$$

3.3. OPEN AND CLOSED SETS

Thus, open balls in a metric space can have very strange shapes. A look at Figure 3.2 shows an open ball in a subspace, S, of R^2 (represented by a "blob"). The open ball is represented by the shaded area.

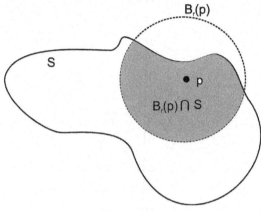

Figure 3.2.

As another example, let M be R with the usual metric and consider the set $S = [0, 1)$. Now consider the set $S \cap (-1, 1/2) = [0, 1/2)$. This set is not an open ball in M, but is an open ball in S, since it is the intersection of S with an open ball in R.

Remark 3.36 *The notion of convergence of a sequence in metric spaces can be cast in terms of open balls. The original definition, stated another way, says that a sequence $\{a_n\}$ converges to a in a metric space (M, d), if for each $\varepsilon > 0$, there is an N such that $n > N$ implies that $d(a_n, a) < \varepsilon$. But if $d(a_n, a) < \varepsilon$, then $a_n \in B_\varepsilon(a)$ and vice versa. Said more colloquially, a sequence $\{a_n\}$ converges to the point a if and only if the terms of the sequence are eventually in every open ball you choose centered at a.*

Remark 3.37 *Using balls, we can now give an alternate definition of bounded in a metric space. A set B in a metric space (M, d) is said to be bounded, if $B \subset B_r(p)$ for some p and some r. Thus a set is bounded if it can be enclosed by a ball. You will show in the exercises that this is equivalent to our other definition of bounded given on page 75.*

Definition 3.38 *If S is a subset of a metric space (M, d), then we say that S is* **open** *if for each point, $a \in S$, there is an open ball $B_r(a) \subset S$.*

Remark 3.39 *One should take special note of the words, "for **each** point" $a \in S$, there must be an open ball with center a contained in S for S to be open.*

Suppose that in R^2 with the usual metric we let $S = \{(x,y)|y > x^2\}$. The picture of $y = x^2$ is shown in Figure 3.3. S is the region inside but not on the parabola.

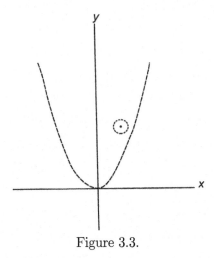

Figure 3.3.

We observe that for each point in S there is an open ball totally contained within S. But if we were to add the origin to S, this new set would not be open since no ball about the origin would be contained totally within S.

Example 3.40 *In R with the usual metric, the set $[0,1)$ is not open. There is no open ball with center 0 (open balls are open intervals in this metric space) that is totally contained within $[0,1)$. For a set to be open, there must be an open ball contained in the set for every point in the set. Notice, however, that all open intervals are open sets in R with the usual metric since if (a,b) is an open interval, and p is any point in (a,b), then the open ball with radius $r = \min\{p-a, b-p\}$ is contained in S. You can easily verify that even infinite intervals like $(-\infty, a)$ are open for any real number a.*

Example 3.41 *If M is any set with the discrete metric, then any subset, S, of M, is open. For if $p \in S$, then $B_{1/2}(p) = \{p\}$, which is totally contained in S since p is.*

Theorem 3.42 *In any metric space (M,d), $B_r(p)$ is open for all $p \in M$.*

Proof. We need to show that if we take any point $q \in B_r(p)$, there is an open ball $B_s(q)$ such that $B_s(q) \subset B_r(p)$. We are guided by Figure 3.4 (where we use R^2 as a model for our intuition, and hope that our ideas carry through to any metric space).

3.3. OPEN AND CLOSED SETS

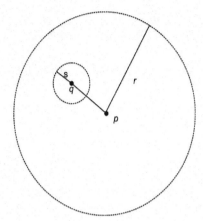

Figure 3.4a.

Let $s < r - d(q,p)$, then certainly in Figure 3.4a, $B_s(q) \subset B_r(p)$. To see that this works in general metric spaces we pick an arbitrary point $x \in B_s(q)$. (Use Figure 3.4b as a reference.)

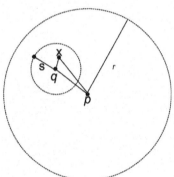

Figure 3.4b.

Then
$$d(q, x) < s. \tag{3.10}$$
To show that $x \in B_r(p)$, we need to show that $d(x, p) < r$. But

$$\begin{aligned} d(x,p) &\leq d(x,q) + d(q,p) \\ &< s + d(q,p) \quad \text{(by (3.10))} \\ &< r \quad \text{(using the definition of } s\text{).} \end{aligned}$$

From these relationships, we have that $d(x, p) < r$. We have taken an arbitrary point $x \in B_s(q)$ and have shown that it is in $B_r(p)$. Thus $B_s(q) \subset B_r(p)$. To summarize, we took an arbitrary point q in $B_r(p)$ and found an open ball centered at q, totally contained in $B_r(p)$, which shows that $B_r(p)$ is open. ∎

Remark 3.43 *It is clear that in any metric space (M, d), M is open, since no matter which point $p \in M$ you choose and no matter which radius you use, $B_r(p) \subset M$. By convention, we declare the null set to be open in any metric space (though one can prove that the null set is open using vacuous reasoning).*

Theorem 3.44 *In a metric space (M, d), the arbitrary union of open sets is open.*

Proof. Suppose that $\{O_\alpha\}$ is an arbitrary collection of open sets in (M, d). Then we wish to show that $\cup O_\alpha$ is open. So pick an arbitrary $p \in \cup O_\alpha$. If we can show that there is an open ball, $B_r(P) \subset \cup O_\alpha$, then we will be done. But this is easy to do since if $p \in \cup O_\alpha$, then $p \in O_{\alpha_0}$ for some α_0. But since O_{α_o} is open, there is an open ball $B_r(P) \subset O_{\alpha_o} \subset \cup O_\alpha$. So $B_r(p) \subset \cup O_\alpha$, and we are done. ∎

Theorem 3.45 *In a metric space, any finite intersection of open sets is open.*

Proof. Suppose that $O_{\alpha_1}, O_{\alpha_2}, O_{\alpha_3}, ..., O_{\alpha_n}$ is an arbitrary collection of open sets in (M, d). Suppose $p \in \bigcap_{i=1}^{n} O_{\alpha_i}$. Since p is in O_{α_1} and O_{α_1} is open, we can find an open ball of radius r_1 such that $p \in B_{r_1}(p) \subset O_{\alpha_1}$. Similarly since p is in all the other O_{α_i}, all of which are open, there is an open ball of radius r_2 such that $B_{r_2}(p) \subset O_{\alpha_2}$ and an open ball of radius r_3 such that $B_{r_3} \subset O_{\alpha_3}$ etc. All of these open balls have p as their centers. Now, let $r = \min\{r_1, r_2, r_3, ..., r_n\}$. Then $B_r(p) \subset B_{r_i}(p) \subset O_{\alpha_i}$ for each $i = 1, 2, 3..., n$, (verify!), so $B_r(p)$ must be in $\bigcap_{i=1}^{n} O_{\alpha_i}$. We have chosen an arbitrary $p \in \bigcap_{i=1}^{n} O_{\alpha_i}$ and constructed an open ball, $B_r(p)$, containing it that is contained totally within $\bigcap_{i=1}^{n} O_{\alpha_i}$. This shows that $\bigcap_{i=1}^{n} O_{\alpha_i}$ is open. ∎

Example 3.46 *To show that an arbitrary intersection of open sets need not be open, we take our metric space to be R with the usual metric. We let $O_n = (-1/n, 1/n)$. Now, $\bigcap_{1}^{n} O_n = \{0\}$. This is not open in R. (There clearly is no open ball containing 0 which is totally contained in $\{0\}$.)*

Definition 3.47 *Suppose that (M, d) is a metric space and that F is a subset of M. We say that F is **closed** if the complement of F is open.*

Example 3.48 *In R with the usual metric, $\{0\}$ is closed. The set $\{0\}^c = (-\infty, 0) \cup (0, \infty)$, and this is the union of open sets by Example 3.40. So by Theorem 3.44, this set is open, and hence $\{0\}$ is closed.*

3.3. OPEN AND CLOSED SETS

Example 3.49 *In R^2 with the usual metric, the set $S = \{(x,y)|x^2 + y^2 \leq 1\}$ is closed. The complement is $S^c = \{(x,y)|x^2 + y^2 > 1\}$, which is the exterior of a circle in R^2, a set which is open. Drawing a picture will help you to see this, since every point in S^c can be enclosed in an open ball totally contained within S^c.*

Example 3.50 *In R with the usual metric, the set $S = [0,1)$ is neither open nor closed.*

Before giving the next result, recall a version of Demorgan's Laws (Theorem 1.14), which states that for any collection $\{O_\alpha\}$ of sets, $(\cup O_\alpha)^c = \cap O_\alpha^c$ and $(\cap O_\alpha)^c = \cup O_\alpha^c$. These relationships are the natural driver for the next two results.

Theorem 3.51 *In a metric space,*
 (1) the arbitrary intersection of closed sets is closed, and
 (2) the finite union of closed sets is closed.

Proof. We will only prove (1), leaving (2) to you. Let $\{F_\alpha\}$ be an arbitrary collection of closed sets. To show that $\cap F_\alpha$ is closed, we need only show that $(\cap F_\alpha)^c$ is open. But $(\cap F_\alpha)^c = \cup F_\alpha^c$. And since each F_α^c is open, by virtue of F_α being closed, $\cup F_\alpha^c$ is open by Theorem 3.44. We have shown that $(\cap F_\alpha)^c$ is open, so, $(\cap F_\alpha)$ is closed. ∎

Example 3.52 *We have already pointed out that in any discrete metric space (M, d), every subset is open. (See Example 3.41.) As a result, every subset S of M is closed since the complement of S is a subset of M. So in this metric space, all sets are both open and closed. So we see the words "open" and "closed" in metric spaces have different meanings than they do in English. In the English language, something like a door cannot be both open and closed, but in mathematics, a set can be both open and closed.*

Definition 3.53 *If (M,d) is a metric space and S is a subset of M, we say that $p \in M$ is a **limit point** of S if every $B_r(p)$ contains points of S other than p. The collection of all limit points of S is denoted by S'.*

Example 3.54 *In R with the usual metric, the point 0 is a limit point of the set $S = (0,1]$ since every open ball (=open interval) with center 0 contains infinitely many points of S, and therefore a point of the set other than 0. Similarly, 1 is a limit point of S. In fact, every point in this set is a limit point of the set. Notice that the limit point 0 is not in S, while the limit point 1 is in S. So limit points of a set may or may not be in the set. If we were to let $T = (0,1] \cup \{2\}$, then 2 would not be a limit point of T, since the open ball of radius $1/2$ with center at 2 (which is the open interval $(1.5, 2.5)$) contains no points of T other than 2.*

Example 3.55 *The sequence $\{1/n\}$ in R, with the usual metric, converges to zero. Thus every open interval with center 0 contains infinitely many points of the set $\{1/n\}$ (by Remark 2.40), hence contains a point of the set other than 0. So zero is a limit point of the set of numbers of the form $1/n$ where $n = 1, 2, 3, \ldots$.*

Example 3.56 *If (M, d) is a discrete metric space and S is a subset of M, S has no limit points in M. For if p is any point of M, $B_{1/2}(p) = \{p\}$. So this ball can't contain points of S other than p.*

Example 3.57 *In R^2 with the usual metric, every point in the set $S = \{(x, y) | x^2 + y^2 \leq 1\}$ (which is a circle together with its interior), is a limit point of S. Draw a picture to help you see why.*

Remark 3.58 *It is useful to ask what it means for a point p in a metric space not to be a limit point of a set S. This means it is not true that every open ball centered at p contains a point of S other than p. Thus, **some** open ball centered at p will have either no intersection with S, or will have only $\{p\}$ as its intersection with S.*

Theorem 3.59 *If p is a limit point of a set S in a metric space, then every open ball centered at p contains infinitely many points of S.*

Proof. Suppose it is not true that every open ball centered at p contains infinitely many points of S. Then there is an open ball $B_s(p)$ which contains only finitely many points of S in common with it. Call these points $p_1, p_2, p_3, \ldots p_n$. So

$$B_s(p) \cap S = \{p_1, p_2, p_3, \ldots p_n\}. \tag{3.11}$$

Let $r = \min\{d(p, p_1), d(p, p_2), d(p, p_3), \ldots d(p, p_n)\}$. Then it seems intuitively clear (at least in R^2 and R^3) that $B_r(p)$ has no points in common with S. To prove it, suppose that $q \in B_r(p) \cap S$. Since $B_r(p) \subset B_s(p)$, (verify!), $B_r(p) \cap S \subset B_s(p) \cap S = \{p_1, p_2, p_3, \ldots p_n\}$. So if $q \in B_r(p) \cap S$, q must be in $B_r(p)$ and q must equal some p_i. But since q is in $B_r(p)$, $d(p, q) < r$, and since $q = p_i$, this yields $d(p, p_i) < r$. We have our contradiction. This says that the distance between p and p_i is less than r, the minimum of all such distances, which is impossible. Since our assumption that some open ball had only finitely many points in common with S led to a contradiction, every open ball with center at p must contain infinitely many points of S. ∎

Corollary 3.60 *If p is a limit point of a set S in a metric space, then there is a sequence of distinct points from S different from p, which converges to p. Said another way, a limit point of a set is the limit of a sequence of distinct points coming from the set (which is why we call it a limit point!).*

3.3. OPEN AND CLOSED SETS

Proof. Pick a point p_1 in S which lies in $B_1(p)$. (We know such a p_1 exists since $B_1(p)$ contains infinitely many points of S.) Pick a point p_2 in S, which lies in $B_{1/2}(p)$ where $p_2 \neq p_1$. (Again, we know such a p_2 exists since $B_{1/2}(p)$ contains infinitely many points of S, which means there certainly is a point other than p_1 in this ball which lies in S). Continue in this manner picking p_3 in S not equal to p_1 or p_2 which is in $B_{1/3}(p)$, and p_n not equal to any of the p_i which we picked before and such that $p_n \in B_{1/n}(p)$. Now, let $\varepsilon > 0$ and choose N so that $n > N$ implies $1/n < \varepsilon$. Because $p_n \in B_{1/n}(p)$, $d(p_n, p) < 1/n < \varepsilon$, for $n > N$. This is what it means for $\{p_n\}$ to converge to p. ∎

Remark 3.61 *The converse of this theorem is also true, namely if a point p is the limit of a sequence of distinct points from a set S (in a metric space) different from p, then p must be a limit point of S. You are asked to prove this in the exercises.*

Example 3.62 *The word '"distinct" is important in the theorem. For example consider the set $S = [0, 1] \cup \{2\}$. The point 2 is not a limit point of the set, but is the limit of a sequence of points from the set, namely, the sequence $2, 2, 2, \ldots$ No sequence coming from the set which converges to 2 can have all distinct points. So 2 cannot be a limit point of the set. The point 1 is the limit of the sequence $1, 1, 1, \ldots$, but is also the limit of the sequence $\{1 - 1/n\}$ where the points are distinct. As long as there is SOME sequence of distinct points from S that converges to p, p is a limit point of S. So 1 is a limit point of the set.*

Theorem 3.63 *A set S in a metric space is closed if and only if $S' \subset S$. That is, if and only if S contains all its limit points.*

Proof. ($\Longrightarrow$) Here we are assuming that S is closed. We want to show that $S' \subset S$. Now, if $S' = \emptyset$, then of course $S' \subset S$. So let's assume that $S' \neq \emptyset$, and pick any point p in S'. Now, if p is not in S, then $p \in S^c$. But S^c is open by virtue of S being closed. That implies that there is an open ball, $B_r(p)$, totally contained in S^c. Thus, $B_r(p) \cap S = \emptyset$. But this contradicts that p is a limit point of S since every open ball centered at p must have a point in common with S other than p. This contradiction, which arose from assuming that the limit point p was not in S, tells us that p must be in S. Since p was an arbitrary point in S', $S' \subset S$.

($\Longleftarrow$) We are now assuming that $S' \subset S$, and we want to show that S is closed. We will do this by showing that S^c is open. So, pick a point $p \in S^c$. (If $S^c = \emptyset$, then $S = M$ which we know is closed, so let's assume $S^c \neq \emptyset$). Then p can't be in S, and hence can't be in S' since $S' \subset S$. So p is not a limit point of S. Thus, there is an open ball $B_r(p)$ whose intersection with S is either $\{p\}$ or $\emptyset$ by Remark 3.58. The intersection with S can't be p or else p would be in S and we chose p from S^c. So $B_r(p) \cap S = \emptyset$. Since $B_r(p)$ has no intersection with S, $B_r(p) \subset S^c$. To recap, we chose an arbitrary point in S^c and showed that there was an open ball containing p, $B_r(p)$, where $B_r(p) \subset S^c$. This shows that S^c is open. Hence S is closed. ∎

Example 3.64 *In R with the usual metric, the set [0,1) is not closed since it doesn't contain its limit point 1. In R^2 with the usual metric, $S = \{(x,y) | x^2 + y^2 = 1\}$ is closed since $S' = S$, hence S' is a subset of S. In R^3 with the usual metric, the set $\{(0,0,0)\}$ has no limit points, so $S' = \varnothing$. Since $S' \subset S$, S is closed.*

The following is a corollary of the previous theorem.

Corollary 3.65 *A set S in a metric space is closed, if and only if whenever $p_n \to p$ (where $p_n \in S$), it follows that p is in S. That is, a set S in a metric space is closed, if and only if limits of convergent sequences coming from S are in S.*

Proof. ($\Longrightarrow$) Suppose not. Then there is a convergent sequence $\{p_n\}$ of points coming from S, which converges to a point p in S^c. Since S^c is open, there is an open ball $B_r(p)$, whose intersection with S is empty. Since the sequence converges to p, the terms of the sequence are eventually in $B_r(p)$, by Remark 3.36, and this contradicts the hypothesis that the points come from S (since $B_r(p)$ has nothing in common with S). So, every sequence convergent sequence $\{p_n\}$ of points coming from S, converges to a point in S.

($\Longleftarrow$) We want to prove S is closed. If S is not closed, then by the theorem, it is not true that S contains all its limit points. So there is a limit point p of S which is not in S. But we know from Corollary 3.60 that there is a sequence of distinct points from S converging to p. But then p must be in S by our assumption. Here is our contradiction, since we said p didn't come from S. This contradiction which arose from assuming that S is not closed, shows us that S must be closed. ∎

A very useful notion that occurs over and over in different analysis related courses is the notion of completeness.

Definition 3.66 *A metric space (M, d) is called **complete**, if every Cauchy sequence in M converges to a point in M.*

Example 3.67 *R with the usual metric is complete. This follows from Theorem 2.89.*

Example 3.68 *Every discrete metric space is complete because any Cauchy sequence is eventually constant, by Example 3.17, and every constant sequence converges.*

Example 3.69 *It is more difficult to show that $C[0,1]$ (where $d(f,g)$ is defined to be $\max |f(x) - g(x)|$) is complete. We don't yet have the machinery to show this, but will eventually show it in Chapter 5.*

3.3. OPEN AND CLOSED SETS

Example 3.70 *We will soon show that any closed subset of a complete metric space is complete. This affords us with infinitely many complete metric spaces, by just looking at closed subsets of complete metric spaces.*

Example 3.71 *While R with the usual metric is complete, there are subsets of R which when considered as subspaces, are not complete. For instance, in Remark 3.28, we saw that the subspace $M = (0,1]$ was not complete, since the Cauchy sequence $\{1/n\}$ contained in M, does not converge to a point in M.*

Theorem 3.72 R^k *is complete.*

Proof. We will use the notation preceding Theorem 3.29. Suppose that $\{a_n\}$ is Cauchy, then $\{a_n^i\}$ is Cauchy for $i = 1, 2, ..., k$. (The proof is almost identical to the proof of the first part of Theorem 3.29.) Since $\{a_n^i\}$ is Cauchy for each i, $\{a_n^i\}$ converges to some number a_0^i for each i, since R is complete. Since $\{a_n^i\}$ converges to some number a_0^i for each $i = 1, 2.3, ..., k$, $\{a_n\}$ converges to $a_0 = (a_0^1, a_0^2, ...a_0^k)$ by Theorem 3.29. ∎

Theorem 3.73 *A closed subspace, S, of a complete metric space (M, d) is complete.*

Proof. Suppose $\{p_n\}$ is a Cauchy sequence in S. Then it is Cauchy in M since S carries the same metric as M. Since M is complete, this sequence converges to a point, p, in M. But by Corollary 3.65, this sequence converges to a point in S. So, p is in S since the limit of a sequence is unique. By the definition of complete (every Cauchy sequence in S converges to a point in S), S is complete. ∎

EXERCISES

1. Give an example of a set A on the real line such that $\sup A \in A$, and $\inf A \in A$, but A is not closed.

2. Show that any finite set in a metric space is closed.

3. Define the following metric on R^2: $d((x_1, y_1), (x_2, y_2)) = |x_1 - x_2| + |y_1 - y_2|$. What does the open ball with radius 1 centered at the origin look like? What do open balls with center (a, b) generally look like?

4. Define the following metric on R^2: $d((x_1, y_1), (x_2, y_2)) = \sup\{|x_1 - x_2|, |y_1 - y_2|\}$. What does $B_1(0,0)$ look like? What does $B_r(a,b)$ look like?

5. Define the following metric on Z^+: $d(m,n) = \left|\frac{1}{n} - \frac{1}{m}\right|$.

    (a) Verify that this is a metric.

(b) Show that the sequence $\{n\}$ is Cauchy in this metric.

(c) Is (Z^+, d) complete?

6. Show that when Z^+ carries the usual metric from R, Z^+ is complete.

7. Suppose that p is a point in a metric space (M, d). Prove that $S = \{q \in M$ such that $d(p, q) \leq r\}$ is a closed set.

8. Two metrics defined on a set S are said to be equivalent, if each open ball either metric contains an open ball from the other metric. Show, by drawing pictures, that the usual metric on R^2 is equivalent to either of the metrics in questions 3 and 4.

9. The set of points on the x-axis strictly between 0 and 1 is an open set in R when R carries the usual metric. It is not an open set in R^2 when R^2 carries the usual metric. Is this a contradiction of anything?

10. For the real line with the usual metric, we now have 4 definitions of bounded: (1) A set S is bounded, if it is bounded above and below (which means it can be contained in an interval), or, (2) if there exists an $M > 0$ such that $|a| \leq M$, for all a in S, and our most recent definition, (3) that $d(a, b) = |a - b| \leq M$ for all a and b in S and some M, and finally, (4) $S \subset B_r(p)$ for some $p \in S$ and some $r > 0$. Show that these definitions are all equivalent.

11. Determine which of the following sets are open, closed or neither. Answer the question in R with the usual metric for those sets that are subsets of R, and answer the question in R^2 with the usual metric for those that are subsets of R^2.

    (a) $\{x \in R \mid x \geq 2\}$
    (b) $\{x \in R \mid x > 2 \text{ or } x < 1\}$
    (c) $\{(x, y) \in R^2 \mid y < 2x\}$
    (d) $\{(x, y) \in R^2 \mid y \leq 2x\}$
    (e) $\{(x, y) \in R^2 \mid x \in Q \text{ and } y \in Q\}$

12. Find the set of limit points in R with the usual metric of each of the following sets:

    (a) $(1, \infty)$
    (b) $(-3, 2) \cup \{2^{-n}\}$ where n is a positive integer

3.3. OPEN AND CLOSED SETS

  (c) the set of numbers of the form $(-1)^n + 2^{-n}$
  (d) the set of rational numbers
  (e) the set of irrational numbers
  (f) the set of numbers of the form $3^{-n} + 5^{-m}$ where m and n are positive integers

13. Determine whether each of the following subsets is open, closed, or neither, in R with the usual metric. Justify your answer.

    (a) $(0,1) \cup (1,2)$
    (b) $\{2^{-n}\} \cup \{0\}$ where n is a positive integer
    (c) Z^+
    (d) Q
    (e) $R - Q$

14. We showed that open balls in a subspace S of a metric space M, were the intersections of open balls in M with S. Show that every open *set* in M is the union of open balls, and then show that every open set in S is the intersection of an open set in M with S.

15. Let $E = (-\infty, 2] \cup [3, \infty)$ carry the usual metric on R. Show that $[3,4)$ is an open ball in E. Then show that $[3, \infty)$ is both open and closed in E.

16. Suppose that if S is a subset of a metric space M and $p \in M$. If there is a sequence of distinct points of S, converging to p then p is a limit point of S.

17. Suppose that (M, d) is a metric space. Let $A = M \times M$. We will define a metric, d^* on A as follows: If $p = (a, b)$ and $q = (c, d)$, then $d^*(p, q) = d(a, c) + d(b, d)$.

    (a) Verify that d^* is a metric on A. This metric is called the product metric.
    (b) Show that if $p_n = (a_n, b_n)$ is a sequence in A, then $p_n \to p = (a, b)$ in (A, d^*) if an only if $a_n \to a$ and $b_n \to b$ in (M, d), and that p_n is Cauchy in (A, d^*) if and only if $\{a_n\}$ and $\{b_n\}$ are Cauchy in M.
    (c) Show that if C is closed in M, then $C \times C$ is closed in the product metric space.
    (d) Show that if K is complete in M, then $K \times K$ is complete in (A, d^*).

18. If S is a subset of a metric space, we say that $p \in S$ is an interior point of S if there is an open ball of radius r with center p totally contained within S. The interior of a set, S, denoted by $int(S)$ is defined to be the set of all its interior points.

    (a) Find $int(R), int(Q)$.
    (b) Find $int([a, b])$.
    (c) Find $int(S)$ where $S = \{(x, y) | \ y \leq x\} \cup (2, 3)$.
    (d) Show that if U is open, $U = int(U)$.
    (e) Show that if A and B are subsets of a metric space, then $int(A) \cup int(B) \subset int(A \cup B)$, but in general, they are not equal. What can you say about $int(A \cap B)$ versus $int(A) \cap int(B)$? Are they equal?

19. The closure of a set E in a metric space is defined to be $E \cup E'$. We denote the closure of E by $\overline{E}$.

    (a) Show that if $A \subset B$, then $\overline{A} \subset \overline{B}$.
    (b) True or False: In any metric space, $\overline{A \cup B} = \overline{A} \cup \overline{B}$. Prove your answer.

20. Show that if E is a subset of a metric space, $\overline{E}$ is closed.

21. Show that a set E in a metric space is closed if and only if equals its closure.

22. Show that the closure in a metric space of a set, A, is the intersection of all the closed sets containing A.

Chapter 4

Limits, Compactness and Uniform Continuity

4.1 Limits

In calculus, one studies the limit of a function $f(x)$, as x approaches a. At first glance, this notion seems to have little connection to the limit of sequences, but they are in fact very much tied together. Before making the connection, let us review a little of that material from calculus. Intuitively, we say that the limit of $f(x)$ as x approaches a is L, and write $\lim_{x \to a} f(x) = L$, if as x gets closer and closer to a, the height of the graph of $f(x)$ gets closer and closer to L. For example, looking at Figure 4.1 below,

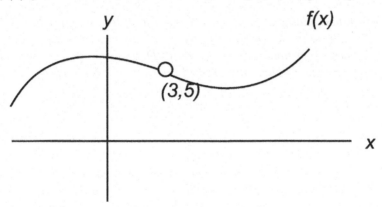

Figure 4.1.

the limit of the function as x approaches 3 is 5, since as x gets closer and closer to 3, the height of the graph gets closer and closer to 5. Notice that the function does not have to be defined at 3 in order for the limit to exist at $x = 3$. However, before we can talk about the limit of $f(x)$ as x approaches 3, there must exist points in the domain of $f(x)$ approaching 3. Fortunately, Corollary 3.60 guarantees this once x is a limit point of the domain. Let us develop a formal definition of $\lim_{x \to a} f(x) = L$ for real valued functions.

Our goal is to assure that $f(x)$, the height of the curve, gets close to L as x gets close to a. But how close is close? What is considered close to one person is not necessarily close to another. So, an acceptable definition will have to satisfy everyone's definition of "$f(x)$ is close to L." Closeness between real numbers is measured on the real line by the distance between them, and distance between any two points, a and b (again, on the real line), is $|a - b|$. Thus, saying that $f(x)$ is close to L means that the distance between them, $|f(x) - L|$, is small. We let ε represent this measure of closeness to L. We want to make $|f(x) - L| < \varepsilon$ (as x gets close to a). Of course to satisfy everyone, we have to be able to make $|f(x) - L| < \varepsilon$ for **any** $\varepsilon > 0$, by making x sufficiently close to a; that is, by making $|x - a|$, the distance between x and a, less than some number δ. How close x has to be to a (that is, the value of δ needed to make $|f(x) - L| < \varepsilon$) will usually depend on how small ε is, and what a is. That brings us to the formal definition of the limit of $f(x)$ as $x \to a$, for real valued functions $f(x)$ with domain E.

Definition 4.1 *Suppose that $f : E \subset R \to R$, where E is a subset of the real line, and that a is a limit point of E. Then we say that the **limit of** $f(x)$ **as** x **approaches** a **is** L, and write $\lim_{x \to a} f(x) = L$, if for every $\varepsilon > 0$, there is a $\delta > 0$, such that if x is in E and $0 < |x - a| < \delta$, then we have $|f(x) - L| < \varepsilon$.*

Remark 4.2 *When finding a limit, our goal is to find, for any given $\varepsilon > 0$, the $\delta > 0$, that "works." That is, the δ that makes $|f(x) - L| < \varepsilon$ when $0 < |x - a| < \delta$. Notice the part of the definition that says $0 < |x - a|$. This means that we are only looking at points x not equal to a. Thus, the limit of the function as x approaches a does not have anything to do with the function value at a. In fact, $f(x)$ may not be defined at a. (So the limit point a need not be in E.) When finding the δ such that $0 < |x - a| < \delta$ implies that $|f(x) - L| < \varepsilon$, we may assume that $\delta < 1$. To see that, suppose that for a given ε the δ needed is found to be 2. That is, if $0 < |x - a| < 2$, it will be true that $|f(x) - L| < \varepsilon$. Well, if $|x - a| < 1$, or if $|x - a| < 1/2$, or if $|x - a| < 1/3$ (where $x \neq a$), then it will still be true that $|x - a| < 2$, and hence it will be true that $|f(x) - L| < \varepsilon$. The device of assuming that $\delta < 1$ (or any fixed positive number) is useful in showing limits exist, as is illustrated in the following example.*

Example 4.3 *Show that $\lim_{x \to 2} x^2 = 4$. We need to show that we can make $f(x) = x^2$ close to 4, by making x sufficiently close to 2 (but not equal to 2). In more formal terms, for any given $\varepsilon > 0$, we want to be able to make $|x^2 - 4| < \varepsilon$ by making*

4.1. LIMITS

$0 < |x - 2| < \delta$. Our job is to find δ. By the above remark, if there is a δ, we may assume δ to be less than 1. Now, if

$$|x - 2| < \delta < 1, \tag{4.1}$$

we get that $-1 < x - 2 < 1$, or that $1 < x < 3$, and this in turn implies that

$$|x| < 3. \tag{4.2}$$

We are now ready to proceed. We want to make $|x^2 - 4| < \varepsilon$. Now, assuming that (4.1) holds, we examine $|x^2 - 4|$. We have

$$\begin{aligned}
|x^2 - 4| &= |x - 2||x + 2| \\
&< \delta \cdot |x + 2| \quad \text{(by (4.1))} \\
&\leq \delta(|x| + |2|) \quad \text{(triangle inequality)} \\
&< 5\delta \quad \text{(by (4.2))}.
\end{aligned}$$

We want to make this less than ε. Our task is now easy if we take δ to be smaller than $\varepsilon/5$ and also smaller than 1 (so that (4.2) holds), we are assured that $|x^2 - 4|$ will be less than ε. Of course, this is true for each ε we choose. So we have found the δ that "works" and have shown that $\lim_{x \to 2} x^2 = 4$. In the exercises we will ask you to show, in a similar manner, that $\lim_{x \to 3} x^2 = 9$. You will find that δ will be different, which highlights the fact that δ depends on both ε and a.

Even for this simple example, proving that $\lim_{x \to 2} x^2 = 4$ required some work. Since we don't want to do this $\delta - \varepsilon$ process each time we deal with a limit, we need some theorems to help us along. The following theorem has a proof that is almost identical to the proof of Theorem 2.45, and includes some of the limit theorems one learns in calculus. We state it and prove only one part. We will revisit this theorem soon, from a different and more general point of view, and the proof of all parts of it, will be very quick.

Theorem 4.4 *Suppose that $f(x)$ and $g(x)$ are real valued functions defined on a set $E \subset \mathbb{R}$, and suppose that a is a limit point of E. Then if $\lim_{x \to a} f(x) = L$ and $\lim_{x \to a} g(x) = M$, it follows that*

(1) $\lim_{x \to a} (f(x) + g(x)) = L + M$,

(2) $\lim_{x \to a} (f(x) - g(x)) = L - M$,

(3) $\lim_{x \to a} (f(x)g(x)) = LM$,

(4) $\lim_{x \to a} \left(\dfrac{1}{g(x)} \right) = \dfrac{1}{M}$ provided $M \neq 0$, and

(5) $\lim_{x \to a} \dfrac{f(x)}{g(x)} = \dfrac{L}{M}$ provided $M \neq 0$.

Proof. (of (1)) Pick an $\varepsilon > 0$. Since $\lim_{x \to a} f(x) = L$, there is a δ_1 such that if $x \in E$ and $0 < |x - a| < \delta_1$, we have

$$|f(x) - L| < \varepsilon/2. \tag{4.3}$$

Similarly, since $\lim_{x \to a} g(x) = M$, there exists a δ_2 such that if $x \in E$ and $0 < |x - a| < \delta_2$, we have

$$|g(x) - M| < \varepsilon/2. \tag{4.4}$$

Now we let δ be any positive number smaller than δ_1 and δ_2. Then both (4.3) and (4.4) are true. So if $0 < |x - a| < \delta$, we have

$$\begin{aligned} & |f(x) + g(x) - (L + M)| \\ = & |f(x) + g(x) - L - M| \\ = & |f(x) - L + g(x) - M| \\ \leq & |f(x) - L| + |g(x) - M| \quad \text{(triangle inequality)} \\ < & \varepsilon/2 + \varepsilon/2 \quad \text{((4.3) and (4.4))} \\ = & \varepsilon. \end{aligned}$$

We have found a δ such that if $0 < |x - a| < \delta$, then $|f(x) + g(x) - (L + M)| < \varepsilon$, and this is the definition of $\lim_{x \to a}(f(x) + g(x)) = L + M$. ∎

Remark 4.5 *The notion of a limit that we have presented is a very general one, and includes the concept of $\lim_{x \to a^+} f(x)$ (the limit as x approaches a from the right) and $\lim_{x \to a^-} f(x)$ (the limit as x approaches a from the left), that were studied in calculus. Thus, if $E = [a, b)$ and $f : [a, b) \subset R \to R$, and we wish to talk about the limit of $f(x)$ as $x \to a$, a look at Definition 4.1 tells us that when we approach a it must be through points $p \in E$. So if $p \in E$, then $p \in [a, b)$, which means that p is to the right of a. Similarly, if we want to talk about the limit of $f(x)$ as $x \to b$, a look at Definition 4.1 tells us that when we approach b it must be through points $p \in E = [a, b)$. So p must be to the left of b. The words "$p \in E$" in Definition 4.1, are critical in understanding the nature of the limit. This discussion also tells us that the previous theorem remains valid when we are taking limits from the right or limits from the left.*

EXERCISES

1. Show that $\lim_{x \to 2}(2x + 1) = 5$ using the definition of limit.

2. Show, using the definition of limit, that $\lim_{x \to 3} x^2 = 9$.

3. Show, using the definition of limit that $\lim_{x \to a} \sqrt{x} = \sqrt{a}$.

4. A common question teachers ask is, "When finding $\lim_{x \to 1} \dfrac{x^2 - 1}{x - 1}$, we do the following: $\lim_{x \to 1} \dfrac{x^2 - 1}{x - 1} = \lim_{x \to 1} \dfrac{(x - 1)(x + 1)}{(x - 1)} = \lim_{x \to 1} x + 1 = 2$. But when $x = 1$, we are cancelling 0s, and that is illegal." How would you answer that question?

5. Using the theorems of the chapter, show that $\lim_{x \to 1} \dfrac{x^2 + 2x + 1}{x - 2} = -4$.

6. Prove that if $\lim_{x \to a} f(x) = L$, then $\lim_{x \to a} |f(x)| = |L|$.

7. Suppose that a is a limit of a set A of real numbers. We say that the limit of $f(x)$ as $x \to a$ is ∞, and write $\lim_{x \to a} f(x) = \infty$, if for any N there is a $\delta > 0$, such that $0 < |x - a| < \delta$ implies that $f(x) > N$. (Intuitively, $f(x)$ gets large as we get close to a). Using this definition, show that $\lim_{x \to 0} \dfrac{1}{x^2} = \infty$. But it is not true $\lim_{x \to 0} \dfrac{1}{x} = \infty$.

4.2 Limits of Functions Defined On Metric Spaces

We now move away from real valued functions and discuss the notion of a limit for a function between two metric spaces, (M_1, d_1) and (M_2, d_2). Our definition will be the natural one. We will no longer use the letter x as the independent variable of $f(x)$ in this section unless we are talking about functions from subsets of R to R. We instead use the letter "p" to indicate a point in a set, since p can be anything and not necessarily a numerical value.

Definition 4.6 *Suppose that $f : E \subset M_1 \to M_2$, where (M_1, d_1) and (M_2, d_2) are metric spaces. (That is, we have a function defined only on E, a subset of a metric space.) Suppose that a is a limit point of E (not necessarily contained in E). Then we say that the **limit of** $f(p)$ **as** $p \to a$ is L, and write $\lim_{p \to a} f(p) = L$, if for each $\varepsilon > 0$, there exists a $\delta > 0$ such that if $p \in E$ and $0 < d_1(p, a) < \delta$, it follows that $d_2(f(p), L) < \varepsilon$.*

Remark 4.7 *If we have a function from R^2 to some other metric space, then p and a in the definition of a limit are both ordered pairs of numbers. If we have a function from R^3, then p and a are ordered triples of numbers. Note also, that in the definition above, L need not be a real number, rather it is an element of M_2.*

Remark 4.8 *Let f be any function from R with the usual metric to R with the usual metric. So, $d(x, a) = |x - a|$ and $d(f(x), L) = |f(x) - L|$. Our definition of $\lim_{p \to a} f(p) = L$ becomes, in this context, (since p is now x): For any fixed $\varepsilon > 0$, there exists a $\delta > 0$ such that if $0 < |x - a| < \delta$, then $|f(x) - L| < \varepsilon$. This is the usual definition of $\lim_{x \to a} f(x) = L$ for real valued functions.*

Example 4.9 *Let M_1 be R^2 with d_1 as the usual metric, and let M_2 be R with d_2 as the usual metric. Define the function f from $E = R^2 - (0,0) \subset M_1 \to M_2$, by $f(x, y) = \dfrac{x^2 y^2}{x^2 + y^2}$. We claim that $\lim_{(x,y) \to (0,0)} f(x) = 0$. Notice that a in this case is $(0,0)$, and what we called p in the definition of a limit is now our ordered pair (x, y). Now, $d_1(p, a) = \sqrt{x^2 + y^2}$, while $d_2(f(p), 0) = \left| \dfrac{x^2 y^2}{x^2 + y^2} - 0 \right|$ or just $\dfrac{x^2 y^2}{x^2 + y^2}$, since the expression inside the absolute value is nonnegative. Pick an $\varepsilon > 0$. We want to find a δ such that if $0 < d_1(p, a) < \delta$, then $d_2(f(p), 0) < \varepsilon$. That is, we want to find a δ such that if $0 < \sqrt{x^2 + y^2} < \delta$, then $\dfrac{x^2 y^2}{x^2 + y^2} < \varepsilon$. But since $x^2 \leq x^2 + y^2$ and $y^2 \leq x^2 + y^2$, we have*

$$\begin{aligned}
\frac{x^2 y^2}{x^2 + y^2} &\leq \frac{(x^2 + y^2)(x^2 + y^2)}{x^2 + y^2} \\
&= x^2 + y^2 \\
&= \left(\sqrt{x^2 + y^2} \right)^2 \\
&< \delta^2 \\
&< \varepsilon,
\end{aligned}$$

when we take $\delta < \sqrt{\varepsilon}$. So we have found our δ.

The following theorem links the limit of a function as $p \to a$, to limits of the function values for sequences converging to a. The theorem is telling us that we can reduce all results about limits of functions to limits of sequences. The motivation for this theorem comes from studying a function, $f(x)$, from R to R. If the limit of $f(x)$ as $x \to a$ is L, then the height of the graph of $f(x)$ gets closer and closer to L as x gets closer and closer to a. So of course, for sequences converging to a, the height of the

4.2. LIMITS OF FUNCTIONS DEFINED ON METRIC SPACES

curve at these points of the sequence will also approach L. Conversely, if the height of the curve approaches L for each sequence converging to a, it seems (at least visually) that the height of the curve approaches L as $x \to a$. Although the next theorem is motivated by the picture in R^2, because we are dealing with metric spaces, we don't have pictures. So now, to prove the analogous result we have to appeal to definitions.

Theorem 4.10 *Suppose that $f : E \subset M_1 \to M_2$, where (M_1, d_1) and (M_2, d_2) are metric spaces and a is a limit point of E. Then $\lim_{p \to a} f(p) = L$ if and only if for every sequence of points $\{p_n\}$ from E converging to a, where $p_n \neq a$ for all n, we have that $f(p_n) \to L$.*

Proof. ($\Longrightarrow$) Since $\lim_{p \to a} f(p) = L$, we know that for each $\varepsilon > 0$ there exists a $\delta > 0$, such that for all $p \in E$ with

$$0 < d_1(p, a) < \delta, \text{ we have } d_2(f(p), L) < \varepsilon. \tag{4.5}$$

Now, pick an $\varepsilon > 0$ and let $\{p_n\}$ be **any** sequence of points converging to a, where $p_n \neq a$ for all n. Since $p_n \neq a$, $d(p_n, p) > 0$. Since $p_n \to a$, there is an $N > 0$ such that $n > N$ implies $d_1(p_n, a) < \delta$. By (4.5),

$$d_2(f(p_n), L) < \varepsilon \text{ for all } n > N. \tag{4.6}$$

But (4.6) is precisely the definition of $\{f(p_n)\}$ converging to L (in M_2), which is what we wanted to prove. So we are done.

($\Longleftarrow$) Now we are assuming that for every sequence $\{p_n\}$ converging to a, where $p_n \neq a$ for all n, it follows that $f(p_n)$ converges to L. To show that $\lim_{p \to a} f(p) = L$, we use proof by contradiction. That is, we suppose that $\lim_{p \to a} f(p) \neq L$. What does this mean? It means that there exists an $\varepsilon > 0$, such that for each $\delta > 0$ there is a point $p \in E$ with $0 < d_1(p, a) < \delta$, where $d_2(f(p), L) \geq \varepsilon$. So, we can find a point $p_1 \neq a$, such that $d_1(p_1, a) < 1$ and $d_2(f(p_1), L) \geq \varepsilon$; we can find a point $p_2 \neq a$, such that $d_1(p_2, a) < 1/2$ and $d_2(f(p_2), L) \geq \varepsilon$; we can find a point $p_3 \neq a$, such that $d_1(p_3, a) < 1/3$ and $d_2(f(p_3), L) \geq \varepsilon$; etc. Since $d_1(p_n, a) < 1/n$ for $n = 1, 2, 3, ...$, the sequence $\{p_n\}$ converges to a, (and none of the terms are equal to a by choice). Furthermore, since the distance between $f(p_n)$ and L is always $\geq \varepsilon$, the sequence $\{f(p_n)\}$ can't converge to L since the terms can't get very close to L. We have constructed a sequence, $\{p_n\}$, converging to a, $p_n \neq a$, for which $f(p_n)$ doesn't converge to L. This contradicts our hypothesis. Since this contradiction arose from assuming that $\lim_{p \to a} f(p) \neq L$, we must conclude that $\lim_{p \to a} f(p) = L$. ∎

Since the previous theorem reduced the study of limits to sequences, we can now easily prove the limit theorems for functions one usually sees in calculus using what we know about sequences. One advantage of this theorem is that it is in the more general context of real valued functions defined on metric spaces.

Theorem 4.11 *Suppose that $f, g : E \subset M_1 \to R$, where (M_1, d) is a metric space, and R carries the usual metric. Suppose also, that a is a limit point of E. Then, if $\lim_{p \to a} f(p) = L$ and $\lim_{p \to a} g(p) = M$, we have*

(1) $\lim_{p \to a}(f + g)(p) = L + M$,

(2) $\lim_{p \to a}(f - g)(p) = L - M$,

(3) $\lim_{p \to a}(fg)(p) = LM$, and

(4) $\lim_{p \to a}(f/g)(p) = L/M$ provided $M \neq 0$

Proof. (1) Since $\lim_{p \to a} f(p) = L$ and $\lim_{p \to a} g(p) = M$, we have by the previous theorem, that for any sequence,$\{p_n\}$,of points converging to a, $\lim f(p_n) = L$ and $\lim g(p_n) = M$. To show that $\lim_{p \to a}(f(p) + g(p)) = L + M$, we have to show that for any sequence $\{p_n\}$ of points converging to a, where $p_n \neq a$, $\lim(f+g)(p_n) = L+M$. But this follows immediately from Theorem 2.45 for sequences since $(f + g)(p_n) = f(p_n) + g(p_n)$. So, $\lim(f + g)(p_n) = \lim f(p_n) + \lim g(p_n) = L + M$. ∎

The proofs of (2), (3) and (4) are similar, and are left to you as good practice.

We have seen (Theorem 4.10) that $\lim_{p \to a} f(p) = L$ if and only if for every sequence of points $\{p_n\}$ converging to a (and $p_n \neq a$), $f(p_n) \to L$. Thus, to show that $\lim_{p \to a} f(p) \neq L$, we need only produce a sequence $\{p_n\}$ converging to a ($p_n \neq a$) for which $f(p_n)$ doesn't converge to L.

Example 4.12 *Let $f : R - \{0\} \to R$ be defined by $f(x) = \sin(1/x)$. $x = 0$ is a limit point of the domain. We want to show that $\lim_{x \to 0} f(x)$ does not exist. Suppose it does, and is equal to L. Then for any sequence $\{x_n\}$ of points converging to 0, $\{f(x_n)\} = \{\sin(1/x_n)\}$ must approach L, and only L. Consider the sequence of points $\{x_n\}$ where $x_n = \dfrac{2}{(2n+1)\pi}$, which converges to 0. Now look at the sequence $f(x_n) = \sin(1/x_n) = \sin((2n+1)\frac{\pi}{2})$. When n is even, say $n = 2k$, then $f(x_n) = \sin((4k+1)\frac{\pi}{2}) = 1$. When n is odd, say $n = 2k+1$ $f(x_n) = \sin((4k+3)\frac{\pi}{2}) = -1$. Since we are getting two different limits for $f(x_n)$ using these subsequences of $\{x_n\}$, $\lim_{x \to 0} f(x) = \lim_{x \to 0} \sin(1/x)$ doesn't exist. Our computations show that the graph of $f(x)$ oscillates between -1 and 1 infinitely often as x goes to zero; a fact which is probably already familiar to some.*

4.2. LIMITS OF FUNCTIONS DEFINED ON METRIC SPACES

Example 4.13 *A related example is the following. Let $f : R - \{0\} \to R$ be defined by $f(x) = x\sin(1/x)$, and let R carry the usual metric. Again, $x = 0$ is a limit point of the domain. Only this time, $\lim_{x \to 0} f(x)$ does exist and is zero. You might think this is obvious. However, we cannot use part (3) of the previous theorem to prove this since $\lim_{x \to 0} \sin(1/x)$ does not exist. We need to appeal to the definition of a limit. So, suppose that $\varepsilon > 0$ is given. To show that $\lim_{x \to 0} x\sin(1/x) = 0$, we need to find a $\delta > 0$ such that if $0 < d(x, 0) < \delta$, then $d(f(x), 0) < \varepsilon$. This means that if $0 < |x| < \delta$, then $|f(x)| < \varepsilon$ (since $d(x, 0) = |x|$ in the usual metric, with a similar statement for $d(f(x), 0)$). But*

$$\begin{aligned}
|f(x)| &= |x\sin(1/x)| \\
&= |x||\sin(1/x)| \\
&\leq |x| \quad \text{(since the maximum value of the sine function is 1)} \\
&< \delta.
\end{aligned}$$

So we need only take $\delta < \varepsilon$ to guarantee that $|f(x)| < \varepsilon$.

EXERCISES

1. Show that $\lim_{x \to 0} \dfrac{1}{x^2}$ is not finite by showing there is a sequence $\{x_n\}$ converging to 0 where $f(x_n)$ does not converge to a finite number.

2. Let $f : R^3 - \{(0, 0, 00\} \to R$ be defined by $f(x, y, z) = \dfrac{x^2}{\sqrt{x^2 + y^2 + z^2}}$. Show that $\lim_{(x,y,z) \to (0,0,0)} f(x, y, z) = 0$.

3. Let $f : R^3 - \{(0, 0, 0)\} \to R$ be defined by $f(x, y, z) = \dfrac{xy}{x^2+y^2+z^2}$. Show that $\lim_{(x,y,z) \to (0,0,0)} f(x, y, z)$ does not exist.

4. Show that if $\lim_{(x,y,z) \to (a,b,c)} f(x, y, z) = p > 0$, then $\lim_{(x,y,z) \to (a,b,c)} \sqrt{f(x, y, z)} = \sqrt{p}$.

5. Suppose that $f(x)$ is defined on some set A and that c is a limit point of A. Suppose further, that for each $x \in A$, $a \leq f(x) \leq b$. Show that if $\lim_{x \to c} f(x) = L$, then $a \leq L \leq b$.

6. Prove the Squeeze Theorem from calculus: If $f, g,$ and h are defined in some open interval I containing c, but not necessarily at c, and if $f(x) \leq g(x) \leq h(x)$ in this interval, then if $\lim_{x \to c} f(x) = \lim_{x \to c} h(x) = L$, we have that $\lim_{x \to c} g(x) = L$.

100   CHAPTER 4. LIMITS, COMPACTNESS AND UNIFORM CONTINUITY

4.3 Continuity

In calculus, the notion of continuity is very intuitive. A function $f(x)$ is said to be continuous at a point a in the domain of f, if the graph of $f(x)$ has no holes, no breaks, and no jumps at $x = a$. So, if one looks at Figure 4.1, that function is not continuous at $x = 3$ because there is a hole in the graph at $x = 3$.

Similarly, the function whose graph is shown in Figure 4.2 is not continuous at $x = 3$ (even though $f(x)$ is defined at $x = 3$), because there is a break at $x = 3$.

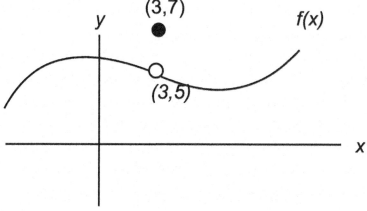

Figure 4.2.

The same is true for the function whose graph is shown in Figure 4.3.

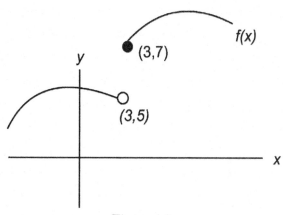

Figure 4.3.

But the function whose graph is shown in Figure 4.4 is continuous at $x = 3$.

4.3. CONTINUITY

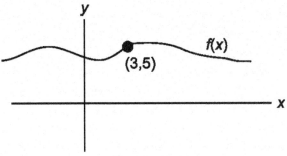

Figure 4.4.

Saying that there are no breaks in a curve at $x = a$, intuitively means that the height of the curve at a coincides with what the height of the curve approaches as x approaches a, or in more formal terms, that $\lim_{x \to a} f(x) = f(a)$. In terms of δ and ε, continuity at $x = a$ means that for each $\varepsilon > 0$ there is a $\delta > 0$ such that if $0 < |x - a| < \delta$, we have that $|f(x) - f(a)| < \varepsilon$. Now, however, since $f(a)$ is defined, the condition that $0 < |x - a| < \delta$, can be strengthened to $|x - a| < \delta$. So we allow x to be equal to a. This leads us to a similar definition of continuity for functions between metric spaces.

Definition 4.14 *Suppose that (M_1, d_1) and (M_2, d_2) are metric spaces, and that $f : M_1 \to M_2$. We say that f is continuous at the point a of the domain, if for each $\varepsilon > 0$ there is a $\delta > 0$ such that if $d_1(p, a) < \delta$, it follows that $d_2(f(p), f(a)) < \varepsilon$.*

Notice that the function must be defined at a for it to be continuous at a. Because of this, we can dispense of statements like $f : E \subset M_1 \to M_2$, where a might be a limit point of E but is not contained in E. For continuity at limit points, the function must be defined at the limit point, so a must be in E. Therefore, we will forego the E in the above definition and write instead, $f : M_1 \to M_2$, and then talk about continuity at a point $a \in M_1$.

Theorem 4.15 *Suppose that (M_1, d_1) and (M_2, d_2) are metric spaces. The function $f : M_1 \to M_2$ is continuous at a, where a is a limit point of the domain, if and only if $\lim_{p \to a} f(p) = f(a)$.*

Proof. ($\Longrightarrow$) Here is the definition of f being continuous at a : (1) For each $\varepsilon > 0$, there is a $\delta > 0$ such that if $d_1(p, a) < \delta$, it follows that $d_2(f(p), f(a)) < \varepsilon$. Here is the definition of $\lim_{p \to a} f(x) = f(a)$: (2) For each $\varepsilon > 0$, there is a $\delta > 0$ such that if $0 <$

102 CHAPTER 4. LIMITS, COMPACTNESS AND UNIFORM CONTINUITY

$d_1(p,a) < \delta$, it follows that $d_2(f(p), f(a)) < \varepsilon$. Clearly, (1) $\Longrightarrow$ (2). That is, if $f(x)$ is continuous at a, then $\lim_{p \to a} f(x) = f(a)$.

($\Longleftarrow$) Now, if (2) holds, $d_2(f(p), f(a)) < \varepsilon$ for the given δ and $p \neq a$. But we know that $d_2(f(p), f(a)) < \varepsilon$ even when $p = a$, since when $p = a$, $d_2(f(p), f(a)) = 0$. So we can relax the condition that $0 < d_1(p, a) < \delta$ and replace it with $d_1(p,a) < \delta$. So (2) $\Longrightarrow$ (1). That is, if $\lim_{p \to a} f(p) = f(a)$, then f is continuous at a. ∎

Definition 4.16 *Suppose that (M,d) is a metric space, and $S \subset M$. A point, $p \in S$, which is not a limit point of S is called an **isolated point** of S. Another way of saying this is that $p \in S$ is an isolated point of S if there is an open ball $B_\delta(p)$ whose intersection with S is just p, or equivalently, $p \in S$ is an isolated point of S, if there is a $\delta > 0$ such that if $a \in S$ and $d(a,p) < \delta$, then $a = p$. Notice that an isolated point of a set is a point in the set.*

Remark 4.17 *A look at Figure 4.5 shows the complete graph of a function, and one can see that $x = 5$ is an isolated point of the domain. We have drawn an open ball of radius δ and having 5 as a center, whose intersection with the domain is just $\{5\}$.*

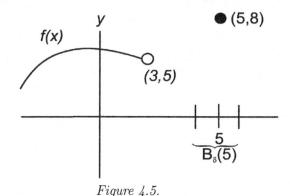

Figure 4.5.

*How should we treat continuity at isolated points? Should we say that f is continuous or discontinuous at an isolated point a? Because there is nothing in the domain close to a, there really isn't a break in the curve at a. But visually, to some, it might look like there is. Mathematicians have decided that we should consider the graph continuous at isolated points, because it is convenient to do so. And in fact, it follows from the $\varepsilon - \delta$ definition of continuity between metric spaces, that f is **automatically continuous at isolated points of the domain**. For if a is an isolated point in M_1, and $f : M_1 \to M_2$, then there is a δ such that $d_1(a,p) < \delta$, implies that $a = p$. So it immediately follows that $d_2(f(a), f(p)) < \varepsilon$, since $f(a) = f(p)$.*

4.3. CONTINUITY

Theorem 4.18 *Suppose that (M_1, d_1) and (M_2, d_2) are metric spaces, and that $f : M_1 \to M_2$. Then f is continuous at a if and only if for each $\varepsilon > 0$ there is an open ball $B_\delta(a)$, such that $f(B_\delta(a)) \subset B_\varepsilon(f(a))$.*

Proof. ($\Longrightarrow$) Saying that f is continuous at $p = a$ means that for each $\varepsilon > 0$, there is a $\delta > 0$ such that if $d_1(p, a) < \delta$, it follows that $d_2(f(p), f(a)) < \varepsilon$. Let us concentrate on the first statement. Saying that $d_1(p, a) < \delta$ means that

$$p \in B_\delta(a). \tag{4.7}$$

Looking at the second inequality, saying that $d_2(f(p), f(a)) < \varepsilon$ means that

$$f(p) \in B_\varepsilon(f(a)). \tag{4.8}$$

So, from (4.7) and (4.8) we have $f(B_\delta(a)) \subset B_\varepsilon(f(a))$.

($\Longleftarrow$) The condition that (for any $\varepsilon > 0$, there is a $\delta > 0$ such that) $f(B_\delta(a)) \subset B_\varepsilon(f(a))$, tells us that if $p \in B_\delta(a)$, $f(p) \in B_\varepsilon(f(a))$. Another way of saying this is, if $d_1(p, a) < \delta$, then $d_2(f(p), f(a)) < \varepsilon$; and this is the definition of f being continuous at $p = a$. ∎

A look at Figure 4.6 illustrates this theorem.

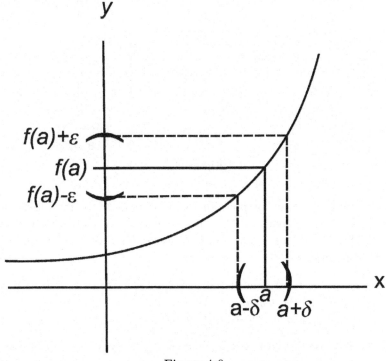

Figure 4.6.

104   CHAPTER 4. LIMITS, COMPACTNESS AND UNIFORM CONTINUITY

The function represented by the graph in Figure 4.6 is continuous at $x = a$. Recall that open balls on the real line with the usual metric are open intervals. We have picked an open ball of radius ε with center $f(a)$ on the y-axis, as shown. On the x-axis, we see an open ball of radius δ which maps into it as guaranteed by the previous theorem. Figure 4.7 illustrates a function that is continuous at both a_1 and a_2. We have taken open balls of radius ε about both $f(a_1)$ and $f(a_2)$, and have shown open balls on the x-axis, with radii δ_1 and δ_2, respectively, which map into them. Of course, the size of the largest open ball on the x-axis that maps into the open ball $B_\varepsilon(f(a))$, will depend not only on ε but on the point a.

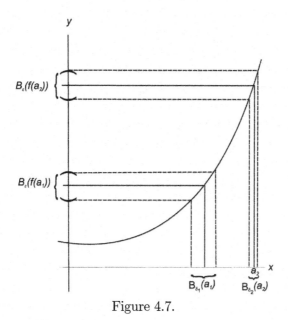

Figure 4.7.

Since the study of limits of functions can be reduced to the study of limits of sequences, as we saw in Theorem 4.10, it is no surprise that the continuity of functions can be reduced to statements about limits of sequences. We have the following result for real valued functions.

Theorem 4.19 *Let (M_1, d) be a metric space and $f : M_1 \to R$ (where R carries the usual metric), and let a be a point in M_1. Then f is continuous at a if and only if for any sequence $\{p_n\}$ which converges to a, $\{f(p_n)\}$ converges to $f(a)$.*

Proof. ($\Longrightarrow$) If a is an isolated point of the domain, then the only sequences in M_1 that converge to a are those whose terms are eventually all equal to a. So after a

4.3. CONTINUITY

certain point, all the terms, $f(p_n)$, will be equal to $f(a)$, and of course this means that the sequence $\{f(p_n)\}$ converges to $f(a)$. If a is a limit point of M_1, then since $\lim_{p \to a} f(p) = f(a)$ (Theorem 4.15), we have that for any sequence $\{p_n\}$ converging to a, that $f(p_n) \to f(a)$ (by Theorem 4.10).

($\Longleftarrow$) If a is an isolated point, then there is nothing to prove since a function is always continuous at an isolated point. If a is a limit point of M_1, and if for any sequence $\{p_n\}$ converging to a it follows that $f(p_n) \to f(a)$, we have by Theorem 4.10, that $\lim_{p \to a} f(p) = f(a)$; and this, by Theorem 4.15, implies that $f(p)$ is continuous at a. ∎

It is now easy to prove a variety of theorems about continuity that you learned in calculus.

Theorem 4.20 *Suppose $f, g : M \to R$ where (M, d) is a metric space and R carries the usual metric. If f and g are both continuous at a, then*
(1) $f + g$ is continuous at a,
(2) $f - g$ is continuous at a,
(3) fg is continuous at a, and
(4) f/g is continuous at a, provided $f(a)$ is not 0.
(5) If $f : M \to R$, $g : f(M) \to R$, and if g is continuous at $f(a)$, then $(g \circ f)$ is continuous at a. (Loosely speaking, this says that a continuous function "of" a continuous function is continuous.)

Proof. (1) By the previous theorem, we need only show that for any sequence $\{p_n\}$ converging to a, $(f+g)(p_n) \to (f+g)(a)$. So, suppose that $\{p_n\}$ is any sequence converging to a. By the continuity of f and g at a we know that $f(p_n) \to f(a)$, and $g(p_n) \to g(a)$ (Theorem 4.19). Thus, $(f+g)(p_n) = f(p_n) + g(p_n) \to f(a) + g(a)$ by Theorem 2.45, and we are done. All the other parts are proved in the same way. ∎

Definition 4.21 *Suppose that (M_1, d_1) and (M_2, d_2) are metric spaces, and that $f : M_1 \to M_2$. Then we say that f **is continuous on** M_1 if f is continuous at each point of M_1.*

Example 4.22 *Consider the function defined from $E = [-1, 1]$ to R, with the usual metric, where*
$$f(x) = \begin{cases} -1 & \text{if } -1 \leq x \leq 0 \\ 1 & \text{if } 0 < x \leq 1 \end{cases}.$$

This function is not continuous on E since it has a discontinuity at 0. However, it is continuous on $E - \{0\}$, since if we pick any point in this set, the function is continuous at that point.

106   CHAPTER 4. LIMITS, COMPACTNESS AND UNIFORM CONTINUITY

A natural question to ask is "How do we apply these theorems when computing limits?" The next example illustrates how we use some of the rules that allow us to compute limits.

Example 4.23 *Let $f, g : R \to R$, where R carries the usual metric. Define f by the rule $f(x) = x$, and g by the rule $g(x) = c$, where c is constant. Both of these are continuous everywhere. To see that $f(x)$ is continuous, suppose that $\{p_n\}$ is any sequence converging to p, where p is any real number. Then $f(p_n)$ converges to $f(p)$ (since $f(p_n) = p_n$, and $f(p) = p$). Similarly, $g(p_n) \to g(p)$ since all the terms here are c. So by Theorem 4.19, both of these functions are continuous everywhere. Therefore, by the previous theorem, $f \cdot f = x^2$ is continuous everywhere, $f \cdot f \cdot f = x^3$ is continuous everywhere, etc. Since we can multiply these by constant functions and still get continuous functions, we have that $h(x) = cx^n$ is continuous everywhere, and since sums and differences of these are continuous everywhere, we have that polynomials are continuous everywhere. Once we have the continuity of polynomials, we find that quotients of polynomials are continuous everywhere except where the denominator is 0. We can then go on, using part (5) of the theorem, to prove that things like $\sin(x^2 + 3x - 1)$ is continuous everywhere (assuming $\sin x$ is), since compositions of continuous functions are continuous. In this way we can build an empire of continuous functions.*

How do we find the limits of these functions? Well, at points of continuity, $\lim_{x \to a} f(x) = f(a)$. What this says, is that to find the limit of $f(x)$ as $x \to a$, you just substitute in a for x. So $\lim_{x \to 2}(3x^2 - 4x + 1) = 3(2)^2 - 4(2) + 1$. In the same way, we can find limits of quotients of polynomials, as long as the denominator isn't zero at a. Similarly, since $\sin(x^2 + 3x - 1)$ is continuous everywhere, to find the limit as $x \to a$, we replace x by a. At points of discontinuity, there are all kinds of tricks used to find limits, like factoring, rationalizing, and using the squeeze theorem (some of which occur in the exercises and which you have undoubtedly seen in calculus).

This same type of thing can be done with functions of several variables. Let us prove that $f, g, h, i : R^3 \to R$, defined by $f(x, y, z) = x$, $g(x, y, z) = y$, $h(x, y, z) = z$ and $i(x, y, z) = c$, where c is a constant, are continuous everywhere. Once proven it will follow that all polynomials in three variables are continuous everywhere, and therefore quotients of polynomials are also continuous everywhere, except where the denominator is 0. The continuity for f, g, and h, is easy. We show it for f. Suppose that $p_n = (x_n, y_n, z_n)$ and $p = (a, b, c)$. If $p_n \to p$, then $(x_n, y_n, z_n) \to (a, b, c)$. It follows by Theorem 3.29 that $x_n \to a$. But $f(p_n) = x_n$ and $f(p) = a$. So saying that $x_n \to a$, means that $f(p_n) \to f(p)$, which by Theorem 4.19 establishes the continuity of f at any point (a, b, c).

Lemma 4.24 *If x is a real number, then there is a sequence of rational numbers*

4.3. CONTINUITY

converging to x. There is also a sequence of irrational numbers converging to x.

Proof. According to Theorem 1.47, we can find a rational number r_1 in the interval $(x-1, x+1)$. Also, we can pick a rational number r_2 in the interval $(x-1/2, x+1/2)$, and then a rational number r_3 in the interval $(x - 1/3,\ x + 1/3)$, and so on. Then $|r_n - x| < 1/n$, which can be made less than any $\varepsilon > 0$. So $\{r_n\}$ converges to x. The same proof with the word "irrational" substituted for "rational" works to prove there is a sequence, $\{i_n\}$, of irrational numbers converging to x. Notice the existence of these irrational numbers that occur in our sequences is guaranteed by Theorem 1.49. ∎

Now we give two very interesting examples which are often used to illustrate concepts in continuity as well as in integration.

Example 4.25 *Let*

$$f(x) = \begin{cases} 0 & \text{if } x \text{ is rational} \\ 1 & \text{if } x \text{ is irrational} \end{cases}.$$

This function is discontinuous everywhere. For if x is rational, we can find, by the previous lemma, a sequence, $\{i_n\}$, of irrational numbers converging to i. But $\{f(i_n)\}$ is 1 for all n, and $f(x) = 0$, so $f(i_n)$ does not converge to $f(x)$. Therefore, by Theorem 4.19 this can't be continuous at x. If x is irrational we take a sequence, $\{r_n\}$, of rational numbers converging to x, and similar reasoning yields the same conclusion.

Example 4.26 *Let*

$$f(x) = \begin{cases} \dfrac{1}{q} & \text{if } x = \dfrac{p}{q} \text{in lowest terms} \\ 0 & \text{if } x \text{ is irrational.} \end{cases}$$

Surprisingly, f is continuous at every irrational number, but discontinuous at every rational number. Here is an outline of the proof. The discontinuity at any rational number r, is clear, since we can take a sequence, $\{i_n\}$, of irrational numbers converging to r, and $f(i_n)$ does not converge to $f(r)$, since $f(i_n) = 0$ and $f(r)$ is not zero. For the continuity at every irrational number, we have to be more careful. Suppose that i is irrational and that $\{r_n\}$ is a sequence of rational numbers converging to i. Then, by Theorem 2.44, this sequence is bounded, and all the r_n are in some interval, $[-M, M]$. Now observe that in this finite interval, there are only a finite number of fractions with denominators of 1 (the integers in the interval!). Similarly, there are only a finite number of fractions with denominators of 2 in the interval $(\pm 1/2, \pm 1, \pm 3/2, ...$ until we go outside $[-M, M]$, and a finite number of fractions with denominator 3 in the interval, and so on. So if we pick a fixed integer K, there are only a finite number of fractions with denominators $\leq K$. All the rest have

108   CHAPTER 4. LIMITS, COMPACTNESS AND UNIFORM CONTINUITY

denominators $> K$. Now pick $\varepsilon > 0$, and pick K so that $1/K < \varepsilon$. Pick a sequence of rational numbers $\{r_n | r_n = \dfrac{a_n}{b_n}$ in lowest terms$\}$ converging to our irrational number i. By what we have just shown, all terms of this sequence, $\dfrac{a_n}{b_n}$, have denominators $b_n > K$ after a certain point (for $n > N$). So, $1/b_n < 1/K < \varepsilon$ for $n > N$. Now, $f(r_n) = 1/b_n < 1/K < \varepsilon$ for $n > N$. Since we can make $f(r_n) < \varepsilon$ for any $\varepsilon > 0$, this sequence converges to 0, and so is equal to $f(i)$. In summary, we have shown that for any sequence $\{r_n\}$ of rational numbers converging to an irrational number i, the sequence $f(r_n)$ converges to $f(i)$. But to show continuity at a point x, we have to show that for any sequence $\{x_n\}$ of real numbers converging to x, $f(x_n)$ converges to $f(x)$. So, we haven't finished the proof. The reader should try to finish the proof.

The following characterizes continuous functions in terms of open sets. This is used quite a bit in higher analysis, and is a fundamental result.

Theorem 4.27 *If $f : M_1 \to M_2$ where (M_1, d_1) and (M_2, d_2) are metric spaces, then f is continuous if and only if for any open set O in M_2, $f^{-1}(O)$ is open in M_1.*

Proof. ($\Longrightarrow$) Suppose that O is open in M_2, and x is any point in $f^{-1}(O)$. To show that $f^{-1}(O)$ is open, we need to show that there is an open ball containing x, totally contained in $f^{-1}(O)$. But x being in $f^{-1}(O)$ implies that $f(x) \in O$, and since O is open, there is an open ball $B_\varepsilon(f(x)) \subset O$. But since f is continuous, at x, there is an open ball $B_\delta(x)$ such that

$$f(B_\delta(x)) \subset B_\varepsilon(f(x)) \subset O,$$

by Theorem 4.18. Applying f^{-1} to all parts of this set inclusion chain, we get

$$f^{-1}\left(f(B_\delta(x))\right) \subset f^{-1}(B_\varepsilon(f(x))) \subset f^{-1}(O). \tag{4.9}$$

From (4.9) it follows that

$$f^{-1}\left(f(B_\delta(x))\right) \subset f^{-1}(O). \tag{4.10}$$

But we know from Theorem 1.37, that

$$B_\delta(x) \subset f^{-1}\left(f(B_\delta(x))\right). \tag{4.11}$$

From (4.11) and (4.10), (together with the fact that $x \in B_\delta(x)$), we have that

$$x \in B_\delta(x) \subset f^{-1}(O).$$

4.3. CONTINUITY

To summarize, we started with an arbitrary $x \in f^{-1}(O)$ and showed there was an open ball containing x totally contained within $f^{-1}(O)$. So, $f^{-1}(O)$ is open.

($\Longleftarrow$) To show that f is continuous if $f^{-1}(O)$ is open, we will use Theorem 4.18 and show that for each $\varepsilon > 0$, there is an open ball $B_\delta(a)$, such that $f(B_\delta(a)) \subset B_\varepsilon(f(a))$. So begin with $B_\varepsilon(f(a))$. Since $B_\varepsilon(f(a))$ is open by Theorem 3.42, we have by hypothesis that $f^{-1}(B_\varepsilon(f(a)))$ is open. Since a is contained in $f^{-1}(B_\varepsilon(f(a)))$, and $f^{-1}(B_\varepsilon(f(a)))$ is open, there is an open ball with center a, $B_\delta(a)$, such that $B_\delta(a) \subset f^{-1}(B_\varepsilon(f(a)))$. Applying f to this set inclusion, we get that

$$f(B_\delta(a)) \subset f(f^{-1}(B_\varepsilon(f(a)))). \qquad (4.12)$$

But by part (2) of Theorem 1.37 we know that

$$f(f^{-1}(B_\varepsilon(f(a)))) \subset B_\varepsilon(f(a)). \qquad (4.13)$$

From (4.12) and (4.13), we have that $f(B_\delta(a)) \subset B_\varepsilon(f(a))$, which is what we wanted to show. This, together with Theorem 4.18, establishes the continuity of f. ∎

A useful corollary, in the study of the Lebesgue Integral and something called Borel sets, is:

Corollary 4.28 *Suppose that $f : M_1 \to R$, where (M_1, d_1) is a metric space and R carries the usual metric. Then if f is continuous, we have that for all a and b that, $f^{-1}(a, \infty)$, $f^{-1}(-\infty, a)$, and $f^{-1}(a, b)$, are open sets in M_1.*

Proof. Since each of the sets (a, b), (a, ∞) and $(-\infty, a)$, are open in R with the usual metric, their pre-images must be open by the theorem. ∎

Example 4.29 *To illustrate the theorem, Figure 4.8 shows a portion of a continuous function on R (with the usual metric). O is an open ball on the y-axis, and its pre-image is the union of two open intervals, (a, b) and (c, d), which is open in R with the usual metric.*

110   CHAPTER 4. LIMITS, COMPACTNESS AND UNIFORM CONTINUITY

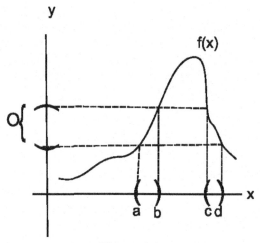

Figure 4.8.

Contrast this with Figure 4.9 which is not continuous at b. In Figure 4.9, the pre-image of O is $(a, b]$, which is not open in R.

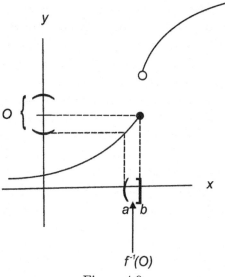

Figure 4.9.

Theorem 4.30 *Suppose that (M_1, d_1) and (M_2, d_2) are metric spaces, and that $f : M_1 \to M_2$. Then f is continuous if and only if for any closed set F in M_2, $f^{-1}(F)$ is closed.*

4.3. CONTINUITY

Proof. ($\Longrightarrow$) If F is closed, then F^c is open, and so by the previous theorem we have that
$$f^{-1}(F^c) \text{ is open.} \tag{4.14}$$
But $f^{-1}(F^c) = (f^{-1}(F))^c$ (Theorem 1.37) so this, with (4.14), tells us that $(f^{-1}(F))^c$ is open. And of course, that means that $(f^{-1}(F))$ is closed.

($\Longleftarrow$) Pick any open set O in M_2. Then $(O)^c$ is closed. So by our assumption, $f^{-1}(O^c)$ is closed. But $f^{-1}(O^c) = (f^{-1}(O))^c$ (Theorem 1.37). So since $f^{-1}(O^c) = (f^{-1}(O))^c$ is closed, $f^{-1}(O)$ is open. Now, by Theorem 4.27, f is continuous. ∎

Corollary 4.31 *If $f : R \to R$, is continuous when R carries the usual metric, then for any real number a, $f^{-1}(\{a\})$, $f^{-1}[a, \infty)$, $f^{-1}(-\infty, a]$ and $f^{-1}[a, b]$, are closed.*

Proof. All of the sets we are taking the pre-images of are closed. So by the theorem, their pre-images are closed. ∎

EXERCISES

1. Is $f(x) = \begin{cases} \dfrac{\sqrt{1+x}-1}{x} & \text{if } x \neq 0 \\ \dfrac{1}{2} & \text{if } x = 0 \end{cases}$ continuous at $a = 0$?

2. Given the function $f(x, y) = \begin{cases} \dfrac{x^2 - y^2}{x^2 - xy} & \text{if } x \neq y \\ 3 & \text{if } x = y \end{cases}$. Is this function continuous at $a = (2, 2)$? Why?

3. Let $f(x) = \begin{cases} x & \text{if } x \text{ is rational} \\ 0 & \text{otherwise} \end{cases}$. True or false: $f(x)$ is only continuous at $a = 0$. Justify your answer.

4. If $f(x) = \begin{cases} 2x & \text{if } x \text{ is rational} \\ x+1 & \text{if } x \text{ is irrational} \end{cases}$ where is $f(x)$ continuous?

5. Show that if $f(x)$ is continuous at $x = a$, then so is $|f(x)|$.

6. Prove that the only continuous function defined on $[a, b]$ which takes on the value 0 on the rationals is $f(x) = 0$.

7. Suppose that $f : R \to R$ is a real valued function that satisfies $f(x + y) = f(x) + f(y)$ for all x and y. If $f(1) = 1$ and $f(x)$ is continuous on R, show that

    (a) $f(0) = 0$

(b) $f(n) = n$ for each integer n

(c) $f(p/q) = p/q$ where p and q are integers and q is not zero

(d) $f(x) = x$ for all real numbers x

8. Show that if $f(x)$ is continuous at c, and $f(c) > 0$, then there is an open interval containing c on which $f(x) > 0$.

9. Suppose that (M, d) is a metric space.

    (a) Show that $|d(x, y) - d(x, y_0)| \leq d(y, y_0)$ for any x, and y_0 in M.
    (b) Show that function defined by $f_x(y) = d(x, y)$ (where $y \in M$) is continuous.
    (c) Show that if $x_n \to x$ and $y_n \to y$ in a metric space (M, d), then $d(x_n, y_n) \to d(x, y)$.

10. Suppose $M_1 = R$ with the usual metric and $M_2 = R$ with the discrete metric. Let $f : M_1 \to M_2$ be defined by $f(x) = x$, and let $g : M_2 \to M_1$ be defined by $g(x) = x$. Is $f(x)$ continuous? What about $g(x)$?

11. How do you know that $f(x, y, z) = 3x^2y + y^3 + z + 1$ is continuous on R^3 when R^3 carries the usual metric?

12. Explain how you know that $\sin(x^2 y)$ is continuous on R^2 when R^2 carries the usual metric.

13. Show that $f : (M_1, d_1) \to (M_2, d_2)$ is continuous, if and only if the preimage of every open *ball* in (M_2, d_2) is open in (M_1, d_1). Is the preimage of an open ball necessarily an open ball?

14. Suppose that $f : A \cup B \to R$ where A and B are open R. If f restricted to A is continuous on A and if f restricted to B is continuous, show that f is continuous on $A \cup B$. What if A and B are closed? Can we make the same conclusion?

4.4 Compactness

The notion of compactness is an important one in analysis.

Definition 4.32 *Suppose that $S \subset M$, where (M, d) is a metric space. By an **open cover** of S, we mean a collection, $\{O_\alpha\}$, of open sets from M, such that $S \subset \cup O_\alpha$. Any collection of sets from $\{O_\alpha\}$, which also covers S, is called a **subcover** of S. A finite collection of sets from $\{O_\alpha\}$ which also covers S is called a **finite subcover** of S.*

4.4. COMPACTNESS

Example 4.33 *In R with the usual metric, the sets $O_n = (0, n)$, $n = 1, 2, 3, ...$, is a collection of open sets, and $(0, 5)$ is covered by these sets since $(0, 5) \subset \cup O_n$. In fact, any open interval (a, b) with a and b positive, is covered by this collection since $\cup O_n = R^+$. One possible finite subcover of $(0, 5)$ would be the set $\{O_5\} = \{(0, 5)\}$. Another finite subcover would be the collection of intervals $\{(0, 1), (0, 2), (0, 3), (0, 4), (0, 5), (0, 6)\}$. There are infinitely many finite subcovers from $\{O_n\}$ covering $(0, 5)$. Observe, for instance, that any finite collection of sets from $\{O_n\}$ which contains $(0, 5)$ is a finite subcover of $(0, 5)$.*

Example 4.34 *The collection of open sets $G_n = (0, 1 - 1/n)$, $n = 2, 3, 4...$ covers $(0, 1)$ since $(0, 1) \subset \cup G_n$. In fact, in this case, $\cup G_n = (0, 1)$. However, there is no finite subcover of $\{G_n\}$ which covers $(0, 1)$ (Verify!).*

Definition 4.35 *A set S in a metric space is called **compact** if every open cover of S has a finite subcover of S.*

Remark 4.36 *Notice the word "every." One might think that $(0, 5)$ is compact, because the open cover we presented above, $\{O_n\}$, in Example 4.33, has a finite subcover. That is not the case. That was only one possible open cover of $(0, 5)$, with many finite subcovers. There are other possible open covers of $(0, 5)$. For example, the collection of sets $G = \{(0, i_x) | i_x$ is an irrational number between 0 and $5\} \cup (1, 2)$ is an open cover of $(0, 5)$ since the union of this collection of sets covers $(0, 5)$. However, no finite subcover of G will cover $(0, 5)$ (Why?). So, $(0, 5)$ is not compact.*

Example 4.37 *Any finite set of points $F = \{x_1, x_2, x_3, ..., x_k\}$ in a metric space is compact. For if $\{O_\alpha\}$ covers the finite set F, we just pick one set from $\{O_\alpha\}$ containing x_1, and one set from $\{O_\alpha\}$ containing x_2, etc. This collection of k sets that we have picked is a finite subcover of F.*

It is not clear that anything other than finite sets are compact. After all, sets can have infinitely many open covers. How can we guarantee that for a particular set, *all* the open covers of the set have a finite subcover? Certainly any theorem that can establish a set as compact, would have to be a powerful one. The following result is the key to proving such a theorem on the real line. Its proof is a bit subtle and clever, and the conclusion is unexpected.

Theorem 4.38 *Let I be a collection of open intervals which cover $[a, b]$. Then, there is a finite subcollection of I which also covers $[a, b]$.*

Proof. Let $C = \{x \in (a, b] | [a, x]$ is covered by a finite number of intervals from $I\}$. First we show that C is not empty. Since the intervals cover $[a, b]$, a belongs to some interval $I_1 = (s_1, t_1)$ from I. See Figure 4.10(a).

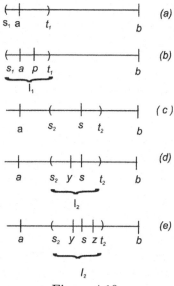

Figure 4.10.

Choose a point p between a and t_1, as shown in Figure 4.10(b). Then $[a, p] \subset I_1$, so $[a, p]$ is covered by a finite collection of intervals from I, namely, the single interval I_1. So $p \in C$, and C is not empty.

C is clearly bounded since it consists of points from $[a, b]$, which is bounded. So by the Completeness Theorem (Theorem 2.16) the set C has a least upper bound, which we call s. Since every element of C is $\leq b$, $s = \sup C \leq b$ and hence $s \in [a, b]$.

Claim : $s = b$

Suppose not. Then $s < b$. Since $s \in [a, b]$, and I covers $[a, b]$, s is in some $I_2 = (s_2, t_2)$ coming from I. See Figure 4.10(c). Since $s_2 < s = \sup C$, by Theorem 2.20, there is a $y \in C$, such that $s_2 < y \leq s$. And since $y \in C$, $y > a$. See Figure 4.10(d). Also, since $y \in C$, we have, by the definition of C, that $[a, y]$ is covered by a finite number of sets from I. Furthermore, if you choose a z between s and t_2 (see Figure 4.10(e)), then $[y,z]$ is also $\subset I_2 = (s_2, t_2)$. So $[y, z]$ is covered by one interval from I, namely I_2. Now to the final step. Since $[a, y]$ and $[y, z]$ are both covered by a finite number of sets from I, their union, $[a, z]$, is covered by a finite collection of sets from I. We now have a contradiction: s was the supremum of all sets $[a, x]$ such that $[a, x]$ is covered by a finite subcollection of intervals from I. We found something bigger than s, namely z, such that $z \in C$. Our contradiction arose from assuming that s, which was in $[a, b]$, was less than b. Thus $s = b$, and therefore b is the supremum of all $x \in (a, b]$ such that $[a, x]$ can be covered by a finite collection of sets from I. Said another way, $[a, b]$ can be covered by a finite subcollection of sets from I. ∎

4.4. COMPACTNESS

Theorem 4.39 *(Heine-Borel)* $[a,b]$ *is a compact subset of R, when R carries the usual metric.*

Proof. Suppose that $\{O_\alpha\}$ is *any* collection of open sets in R which covers $[a,b]$. Then by definition of open cover, $[a,b] \subset \cup O_\alpha$, and each x in $[a,b]$ is contained in an open set O_{α_x} from our collection. Since O_{α_x} is open, for each $x \in [a,b]$, there is an open ball I_x such that
$$x \in I_x \subset O_{\alpha_x}. \tag{4.15}$$
Furthermore, the open balls in R with the usual metric are open intervals. So our collection $\{I_x\}$ is really a collection of open intervals. Since x runs through $[a,b]$, and $x \in I_x$, this collection $\{I_x\}$ covers $[a,b]$. By the previous theorem, this collection has a finite subcover, $\{I_{x_i}\}$, $i = 1, 2, ..., n$, of $[a,b]$. So by (4.15), $[a,b] \subset \bigcup_{i=1}^{n} I_{x_i} \subset \bigcup_{i=1}^{n} O_{\alpha_{x_i}}$, and we see that $\{O_{\alpha_{x_i}}\}$, where $i = 1, 2, 3, ..., n$, is our finite subcollection of open sets which covers $[a,b]$. ∎

Theorem 4.40 *A closed subset F of a compact set K in a metric space (M, d) is compact.*

Proof. Suppose that $\{O_\alpha\}$ is any open cover for F. We want to show that there is a finite subcover of F. By definition of an open cover for F, $F \subset \cup O_\alpha$. Now, the string of set inclusions, $M = F \cup F^c \subset (\cup O_\alpha) \cup F^c \subset M$, shows that $(\cup O_\alpha) \cup F^c$ is sandwiched between M and M, and so $M = (\cup O_\alpha) \cup F^c$, and since $K \subset M$,
$$K \subset (\cup O_\alpha) \cup F^c.$$
Since F is closed, F^c is open, so we have an open cover, $\{O_\alpha\} \cup \{F^c\}$, of K. But since K is compact, this open cover has a finite subcover. So,
$$K \subset (\bigcup_{i=1}^{n} O_{\alpha_i}) \cup F^c.$$
But $F \subset K$, so
$$F \subset (\bigcup_{i=1}^{n} O_{\alpha_i}) \cup F^c.$$
And since no points of F are contained in F^c, we can write the last set inclusion as
$$F \subset (\bigcup_{i=1}^{n} O_{\alpha_i}).$$
To recap, we started with any open cover $\{O_\alpha\}$ for F and found a finite subcover of F. So F is compact.

116 CHAPTER 4. LIMITS, COMPACTNESS AND UNIFORM CONTINUITY

One can, if one wishes, visualize this proof using the figure below, where one sees M, F, K, and the part of the open cover for F (represented by open balls). We notice that not all of K is covered by these open balls. The shaded portion represents F^c, which is open. From the figure we see that $M = \cup O_\alpha \cup F^c$, that $K \subset (\cup O_\alpha) \cup F^c$, and that no points of F are in F^c. So all the pieces of this proof are indicated in Figure 4.11.

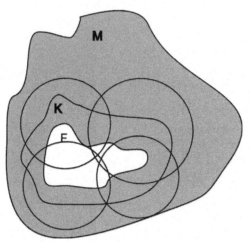

Figure 4.11.

■

Corollary 4.41 *In R with the usual metric, any closed and bounded set S (not necessarily an interval), is compact.*

Proof. Since S is bounded, S is contained in some closed interval $[-M, M]$. Since $[-M, M]$ is compact by the Heine-Borel theorem, S is compact by this theorem since it is a closed subset of a compact set. ■

Example 4.42 *Consider the set $\{1/n\} \cup \{0\}$, where n runs through the positive integers. The only limit point of this set is 0. Since 0 is contained in the set, this set is closed by Theorem 3.63. This set is also bounded since it is contained, for example, in the set $[-1, 1]$. So by the corollary to the previous theorem, this set has to be compact.*

We have seen that every closed and bounded set in R is compact. We will soon prove that in R with the usual metric, every compact set is closed and bounded. From this it will follow that,a set in R is closed and bounded if and only if it is compact. This is also true in R^n, though we won't prove it. This result, however, is not true in general metric spaces.

4.4. COMPACTNESS

Example 4.43 *Consider Z^+, the set of positive integers, with the discrete metric. We saw in Example 3.41 that when a set carries the discrete metric, every set is both open and closed. So in particular, the set, E, of even numbers, is closed in this metric space. It is also bounded, since the distance between any two points is less than or equal to 1. (See Remark 3.20.) So E is closed and bounded. But E is not compact. For $E \subset \{2\} \cup \{4\} \cup \{6\} \cup ...$, (in fact, E is equal to that union, but that is irrelevant), and the sets $\{2\}, \{4\}, \{6\}$, etc., are all open since every subset of a discrete metric space is open. But no finite subcollection of this collection of sets will cover E, since E is infinite.*

We now begin the process of showing that every compact set in R (and more generally in any metric space) is closed and bounded.

Lemma 4.44 *If p and q are distinct points in a metric space (M, d), then there are open balls with center p and q, respectively, that are disjoint.*

Proof. The proof in general metric spaces is suggested by Figure 4.12, which shows a picture in R^2.

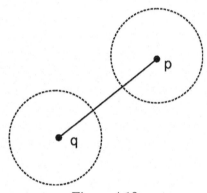

Figure 4.12.

Let $r = \dfrac{d(p,q)}{3}$. Then $B_r(p)$ and $B_r(q)$ are disjoint. To see that, suppose they weren't. Then there would be a point z in both $B_r(p)$ and $B_r(q)$. So $d(z,p) < \dfrac{d(p,q)}{3}$, and $d(z,q) < \dfrac{d(p,q)}{3}$. By the triangle inequality, $d(p,q) \leq d(p,z) + d(z,q) < \dfrac{d(p,q)}{3} + \dfrac{d(p,q)}{3} = \dfrac{2d(p,q)}{3}$, and this implies that $d(p,q) < \dfrac{2d(p,q)}{3}$, which is impossible. So there is no point in $B_r(p) \cap B_r(q)$, which means this intersection is empty. ∎

118   CHAPTER 4. LIMITS, COMPACTNESS AND UNIFORM CONTINUITY

Theorem 4.45 *Let K be a compact subset of a metric space (M,d), and suppose that there is a point $q \in M$ which is not in K. Then there are open sets U and V containing K and q respectively, such that $U \cap V = \varnothing$.*

Proof. By the previous lemma, for each point $p_i \in K$, there is an open ball, U_{p_i} with center p_i, and a corresponding open ball V_{q_i} with center q and radius r_i such that

$$U_{p_i} \cap V_{q_i} = \varnothing. \tag{4.16}$$

Since $K \subset \bigcup_{p_i \in K} U_{p_i}$, and K is compact, there is a finite subcollection of the U_{p_i} that cover K, say $U_{p_1}, U_{p_2}, U_{p_3}, ..., U_{p_k}$. If we let

$$U = \bigcup_{k=1}^{n} U_{p_k}, \tag{4.17}$$

then U is open, being the finite union of open sets, and since these sets cover K, we have $K \subset U$. Now, let

$$V = \bigcap_{k=1}^{n} V_{q_k}, \tag{4.18}$$

the intersection of the V sets that correspond to the U sets. V is open since it is the finite intersection of open sets and, in fact, is an open ball with radius $\min\{r_i\}$ and center q. And since each V_{q_k} contains q, so does V.

We claim that $U \cap V = \varnothing$. To see this, suppose that $z \in U \cap V$. Then since $z \in U$, by (4.17), $z \in U_{p_i}$ for some i. Also, by (4.18), since $z \in V$, $z \in V_{q_i}$ for *each* i. But this contradicts (4.16). This contradiction tells us that there is no $z \in U \cap V$, so $U \cap V = \varnothing$.

Refer to Figure 4.13 to see a picture demonstrating the flow of this proof. In part(a) we show the open balls U_{p_i} and V_{q_i} and in part(b) we illustrate the intersection of the open balls V_{qi} and the case when K is covered with three open balls (our Up_i).

4.4. COMPACTNESS

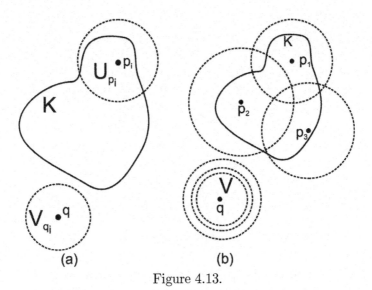

Figure 4.13.

■

Theorem 4.46 *Compact subsets of metric spaces are closed.*

Proof. Suppose that K is a compact set in a metric space (M, d). We will show that K^c is open, so that K is closed. Pick a point $q \in K^c$. Since q is not in K, the previous theorem provides us with open sets U and V such that $K \subset U$, $q \in V$, and $U \cap V = \varnothing$. Since $U \cap V = \varnothing$ and U contains K, $K \cap V = \varnothing$. So $V \subset K^c$. We found, for any point q in K^c, an open ball V with center q, which is totally contained in K^c. Thus K^c is open, and hence K is closed. ■

Theorem 4.47 *Compact subsets of metric spaces are bounded.*

Proof. If we examine Figure 4.14, we get an idea of how to proceed. That figure shows a compact set K, covered by three open balls. If we take any two points, p and q in the set, then the distance from p to q is less than or equal to the distance from p to A, plus the distance from A to B, plus the distance from B to q.

CHAPTER 4. LIMITS, COMPACTNESS AND UNIFORM CONTINUITY

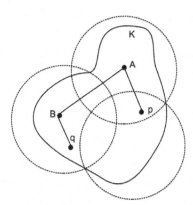

Figure 4.14.

Now, to the proof in metric spaces. Drawing a picture as you read, will help. Suppose that K is a compact subset of a metric space M. Enclose each point $x \in K$ by an open ball $B_1(x)$ of radius 1 centered at x. Since $K \subset \bigcup_{x \in K} B_1(x)$ and K is compact, this collection of open balls covering K has a finite subcover. So,

$$K \subset \bigcup_{i=1}^{n} B_1(x_i). \tag{4.19}$$

Let L' be larger than all of the distances between any two centers of these open balls. So, $d(x_i, x_j) < L'$, for all $i, j = 1, 2, ..., n$. Now, pick any points, x and y in K. By (4.19), $x \in B_1(x_i)$ for some i, and $y \in B_1(x_j)$ for some j. So, $d(x, x_i) < 1$, and $d(x_j, y) < 1$. Now, $d(x, y) \leq d(x, x_i) + d(x_i, x_j) + d(x_j, y) \leq 1 + L' + 1 = 2 + L' = L$. Since the distance between any two points is less than or equal to the fixed number L, K is bounded. (This is one of the equivalent definitions of a set being bounded in a metric space.) ∎

We have seen that closed and bounded sets in R with the usual metric are compact, and this theorem, coupled with the previous one says that compact subsets of R are closed and bounded. So we have already proved the following:

Theorem 4.48 *In R with the usual metric, a set K is compact if and only if it is closed and bounded.*

Although we won't prove it, as it is a bit tedious, the same theorem is true in R^n. So we have:

Theorem 4.49 *In R^n with the usual metric, a set K is compact if and only if is closed and bounded.*

4.4. COMPACTNESS

As we pointed out, it is not true that closed and bounded implies compact in an arbitrary metric space.

Compactness is a notion that is useful in proving some theorems in calculus. You may recall finding the maxima and minima of continuous functions on closed intervals in beginning calculus, and then finding the absolute maxima and minima of a continuous function defined on a closed and bounded set in multivariable calculus. The question that one needs to answer before one even begins to solve those problems is, "How do you know that the maximum or minimum exists in the first place?" The next two theorems will answer that.

Theorem 4.50 *Suppose that $f : M_1 \to M_2$, where M_1 and M_2 are metric spaces, and that f is continuous. Then, if K is a compact subset of M_1, $f(K)$ is a compact subset of M_2. (This is often stated as "The continuous image of a compact set is compact.")*

Proof. To show that $f(K)$ is compact, we have to start with any open cover of $f(K)$ and show that there is a finite subcover. So let $\{O_\alpha\}$ be a collection of open sets which cover $f(K)$. Then by definition of open cover,

$$f(K) \subset \cup O_\alpha. \tag{4.20}$$

Apply the pre-image to both sides of (4.20) to get

$$f^{-1}(f(K)) \subset f^{-1}(\cup O_\alpha). \tag{4.21}$$

Since (by Theorem 1.37) $K \subset f^{-1}(f(K))$, and $f^{-1}(\cup O_\alpha) = \cup f^{-1}(O_\alpha)$, (Theorem 1.34), (4.21) tells us that

$$K \subset f^{-1}(f(K)) \subset f^{-1}(\cup O_\alpha) = \cup f^{-1}(O_\alpha), \tag{4.22}$$

from which we can extract that

$$K \subset \cup f^{-1}(O_\alpha). \tag{4.23}$$

Since f is continuous, Theorem 4.27 tells us that $f^{-1}(O_\alpha)$ is open for each α. So (4.23) shows that $\{f^{-1}(O_\alpha)\}$ is an open cover of K. Since K is compact, there exists a finite subcover of K. That is, there are sets, $f^{-1}(O_{\alpha_1}), f^{-1}(O_{\alpha_2}), f^{-1}(O_{\alpha_3}), ..., f^{-1}(O_{\alpha_n})$, such that

$$K \subset \bigcup_{i=1}^{n} f^{-1}(O_{\alpha_i}). \tag{4.24}$$

Applying f to both sides of 4.24, we get that

$$f(K) \subset f\left(\bigcup_{i=1}^{n} f^{-1}(O_{\alpha_i})\right) \qquad (4.25)$$

$$= \bigcup_{i=1}^{n} f\left(f^{-1}(O_{\alpha_i})\right) \text{ (by Theorem 1.35)} \qquad (4.26)$$

$$\subset \bigcup_{i=1}^{n} O_{\alpha_i} \qquad \text{(Theorem 1.37).} \qquad (4.27)$$

(4.25 – 4.27) yield

$$f(K) \subset \bigcup_{i=1}^{n} O_{\alpha_i}. \qquad (4.28)$$

We started with an open covering $\{O_\alpha\}$ of $f(K)$, and showed there was a finite subcover, the one exhibited in (4.28). Thus, $f(K)$ is compact. ∎

Corollary 4.51 *(Extreme Value Theorem from elementary calculus):* If $f : [a,b] \to R$, and f is continuous, then f has an absolute maximum and minimum on $[a,b]$.

Proof. Since f is continuous and $[a,b]$ is a compact subset of R (with the usual metric), the range of the function, $f([a,b])$, is compact. By Theorem 4.48, this means that $f([a,b])$ is closed and bounded. Since $f([a,b])$ is bounded, we can talk about M, the supremum of $f([a,b])$. Since the range is closed, M is part of the range. (See Theorem 2.54 and Corollary 3.65.) So $M = f(x)$ for some $x \in [a,b]$. But when the supremum of a set is in the set, it is the largest element of the set. So M is the absolute maximum of the range, or said another way, M is the absolute maximum of the function on $[a,b]$. There is a similar proof to show that the absolute minimum of the range also exists. ∎

Corollary 4.52 *If $f : K \to R$, is continuous, where K is a compact metric space, then f has an absolute maximum and minimum on K.*

Proof. The proof is identical to the proof of previous corollary. Just replace $[a,b]$ by K. ∎

Corollary 4.53 *If $f : R^n \to R$, and K is any closed and bounded set in R^n, then there is an absolute maximum and minimum of $f(x)$ on K.*

Proof. We need only point out that K is compact by Theorem 4.49, and the result follows from the previous corollary. ∎

EXERCISES

1. In Example 4.34 we asked you to show that the cover G_n had no finite subcover. Do that.

2. Show that each of the following are open covers of $[2, \infty)$:

    (a) $\{(-3, \infty)\}$
    (b) $\{(-n, n)|\ n \text{ is a positive integer}\}$
    (c) $\{(r-1, r+1)|\ r \text{ is a rational number}\}$

3. In R with the usual metric, give an open cover of the subset Z^+ which has no finite subcover. Then give an open cover of Z^+ which does have a finite subcover.

4. In R with the usual metric, give an open cover of the subset $\left\{\dfrac{1}{n}|n \in Z^+\right\}$ with no finite subcover.

5. Give an example of a complete space which is not compact.

6. Give an example (if possible) of a compact subset which is not complete. If it is not possible, justify your answer.

7. Prove that there is no continuous function defined on $[a, b]$ with the usual metric whose range is an interval $[c, d)$.

8. We proved, using compactness, that any continuous function $f : [a, b] \to R$ is bounded. Develop the following alternate proof: If f is not bounded, then there is a sequence $\{p_n\} \subset [a, b]$ where $f(p_n) > n$. By the Bolzano-Weierstrass Theorem, $\{p_n\}$ has a convergent subsequence, $\{p_{n_k}\}$ which converges to some number v. Finish the proof.

9. Develop the following proof of the fact that a continuous function on a closed and bounded interval actually takes on its maximum and minimum values. Since f is bounded, the supremum, s, of the range exists. Thus, there exists a sequence $\{x_n\} \subset [a, b]$ such that $f(x_n) \to s$. Extract a subsequence of $\{x_n\}$ which converges to d, and use the continuity to show that $f(d) = s$. So the function takes on its maximum value.

10. Suppose that $f : R \to R$ is continuous where R carries the usual metric. Show that $f^{-1}(K)$ where K is compact, is closed.

4.5 Some Useful Theorems Related to Compactness

In this section we give some theorems that, to a beginner, may seem to be a bit off the beaten track, but turn out to be very useful in higher analysis. We will also use them here. First, we need a definition.

Definition 4.54 *We say that a collection of sets, $\{S_\alpha\}$, has the **finite intersection property**, if the intersection of any finite subcollection of sets is non empty.*

Example 4.55 *Let $S_x = (-x, x)$ on the real line, where x is not zero. Then any finite intersection of these sets is non-empty, containing, at the very least, 0.*

Example 4.56 *Let C be the collection of all subsets of Z^+ whose complements in Z^+ are finite. So, $C_1 = \{3, 4, 5, ...\}$ is a set in C. Suppose that $C_1, C_2, ..., C_n$ are sets in C. If $\bigcap_{k=1}^{n} C_k$ were empty, then $(\bigcap_{k=1}^{n} C_k)^c = \bigcup_{k=1}^{n} C_k^c$ would be Z^+. But since each C_k^c is finite, so is $\bigcup_{k=1}^{n} C_k^c$, and thus can't equal Z^+ since Z^+ is infinite. So this contradiction (which arose from assuming that the finite intersection of sets from C is empty), shows that the finite intersection of any of these sets is non-empty, and therefore, this collection of sets has the finite intersection property.*

Remark 4.57 *In the first example, $\cap S_x = \{0\}$. So the intersection is not empty. In the second example $\cap C_\alpha$, where $C_\alpha \in C$, is empty as one can see by examining the sets $C_1 = \{1\}^c$, $C_2 = \{2\}^c$, etc. So, given a collection of sets with the finite intersection property, the intersection of the entire collection may or may not be empty. The following result gives a condition on the sets that will guarantee a non empty intersection.*

Theorem 4.58 *If $\{K_\alpha\}$ is a collection of compact sets with the finite intersection property in a metric space (M, d), then $\cap K_\alpha$ is not empty.*

Proof. Pick from among these sets one set K_{α_0}, and keep it fixed. We will show that there is a point p in K_{α_0} that is in all the other K_α, and this will prove the theorem since p will then be in the intersection of all the K_α. We prove this by contradiction. Suppose $\cap K_\alpha = \emptyset$. Then, no matter which p we pick in K_{α_0}, $p \notin \cap K_\alpha$. Thus, for each $p \in K_{\alpha_0}$, there is a K_α^p such that $p \notin K_\alpha^p$. So, $p \in (K_\alpha^p)^c$. Since each p in K_{α_0} is contained in some $(K_\alpha^p)^c$, as we have just shown, we have that

$$K_{\alpha_0} \subset \cup (K_\alpha^p)^c.$$

4.5. SOME USEFUL THEOREMS RELATED TO COMPACTNESS

where p, in the union, runs through points in K_{α_0}. Notice that this is a union of open sets since each K_{α_i} is closed (Theorem 4.46), making each complement open. Since K_{α_0} is compact, there is a finite subcover of K_{α_0}. So,

$$K_{\alpha_0} \subset \bigcup_{i=1}^{n} (K_{\alpha}^{p_i})^c = (\cap K_{\alpha}^{p_i})^c$$

(by Demorgan's Law). But, $K_{\alpha_0} \subset \left(\bigcap_{i=1}^{n} K_{\alpha}^{p_i}\right)^c$ means that $K_{\alpha_0} \cap \left(\bigcap_{i=1}^{n} K_{\alpha}^{p_i}\right) = \varnothing$. We have our contradiction. This is a finite collection of sets from the K_{α} whose intersection is empty, and this contradicts the hypothesis that $\{K_{\alpha}\}$ has the finite intersection property. This contradiction arose from assuming that no point of K_{α_0} was in all the other sets. So there must be some point $p \in K_{\alpha_0}$, which is in all the other sets in $\{K_{\alpha}\}$. This means that $\cap K_{\alpha}$ is not empty. ∎

Definition 4.59 *A collection of sets $S_1, S_2, S_3, ...$, is called a **nested sequence** of sets, if S_1 contains S_2 and S_2 contains S_3 and S_3 contains S_4, and so on.*

Example 4.60 *Let $S_n = (2 - 1/n, 2 + 1/n)$. Then, $S_1 = (1, 3)$, $S_2 = (1\frac{1}{2}, 2\frac{1}{2})$, $S_3 = (1\frac{2}{3}, 2\frac{1}{3})$, and so on. This collection of sets is nested since S_1 contains S_2, which contains S_3, and so on.*

Corollary 4.61 *If $\{K_n\}$ is a nested sequence of non-empty, compact sets in a metric space, then $\bigcap_{i=1}^{\infty} K_i \neq \varnothing$.*

Proof. Because this sequence is nested, it has the finite intersection property. (The intersection of any finite subcollection is always the smallest set in the subcollection.) So by the previous theorem, $\bigcap_{i=1}^{\infty} K_i \neq \varnothing$. ∎

Corollary 4.62 *If $\{K_n\}$ is a nested sequence of compact sets in a metric space such that $\bigcap_{i=1}^{\infty} K_i = \varnothing$, then one of the sets in this subcollection must be empty.*

Proof. If all the sets were non-empty, then by the theorem, the intersection would be non-empty and it isn't. So, one of the sets must be empty. ∎

Corollary 4.63 *(Nested Interval Property) If $I_n = [a_n, b_n]$ is a nested sequence of closed intervals on the real line with the usual metric, then $\cap I_n \neq \varnothing$.*

Proof. The intervals are compact by Theorem 4.39. So, by the previous corollary, the intersection must be non-empty. ∎

There is another characterization of compactness which we will give shortly. Compactness, can be expressed in terms of sequences. This is quite an important characterization. First, we need a definition.

126   CHAPTER 4. LIMITS, COMPACTNESS AND UNIFORM CONTINUITY

Definition 4.64 *Let S be a set in the metric space (M,d). Then, we say that S **sequentially compact** if every sequence coming from S has a subsequence converging to a point in S.*

Theorem 4.65 *A set S in a metric space (M,d) is compact if and only if it is sequentially compact.*

The complete proof of the theorem is rather intricate and is omitted, although we ask you to prove the easier part of it in the exercises.

EXERCISES

1. If $A_1, A_2, A_3, ..., A_n$ are compact subsets of a metric space, show that $\bigcup_{i=1}^{n} A_i$ and $\bigcap_{i=1}^{n} A_i$ are compact. Show that the first statement is not true for an infinite number of compact sets.

2. Explain why each of the following sets in R^2 with the usual metric is not compact.

    (a) $\{(x,y)|y = x \text{ or } y = -x\}$
    (b) $\{(x,y)||x|+|y| < 1\}$
    (c) $\{(x,y)| y < 1\}$

3. Give an example of an open cover for $\{(x,y)|x^2+y^2 < 1\}$ with no finite subcover.

4. Show that every sequence coming from a compact set K in a metric space, has a convergent sequence which converges to a point in K.

5. One can use the Nested Interval Property to prove the following theorem from calculus: If $f(x)$ is continuous on $[a,b]$ and $f(a)$ and $f(b)$ have opposite signs, then there is a p in (a,b) such that $f(p) = 0$. Develop the following proof: We will suppose that $f(a)$ is positive and that $f(b) < 0$. If it is the reverse, a similar proof will work. Call $[a,b]$, I_1. Divide I_1 into two intervals $[a,p_1]$ and $[p_1,b]$ where p_1 is the midpoint of the interval. If $f(p_1) = 0$, we are done. If $f(p_1) > 0$, then let $I_2 = [a_2, b_2] = [p_1, b]$. If $f(p_1) < 0$. let $I_2 = [a, p_1]$. Either way, $I_2 \subset I_1$ and the value of the function at the endpoints have opposite signs. Moreover, and this is the important part, the value of f at the left endpoint is positive and at the right endpoint, negative. Now, divide I_2 in half. Call the midpoint p_2. If $f(p_2) = 0$, we are done. If $f(p_3) < 0$, let $I_3 = [a_2, p_3]$, otherwise, let $I_3 = [p_3, b_2]$. Call $I_3 = [a_3, b_3]$. Continue in this manner generating nested

sequence of intervals $I_n = [a_n, b_n]$ with lengths $\frac{b-a}{2^n}$ and where $f(a_n) > 0$ and $f(b_n) < 0$ (if we don't reach a point where $f(p_n) = 0$, in which case we would be done). The lengths of these intervals get smaller and smaller, and can be made less that any specified ε. By the Nested Interval Property, there must be a point, p in the intersection of these intervals. Show $f(p) = 0$.

6. A corollary of the previous exercise is the famous, Intermediate Value Theorem which has numerous applications. If $f : [a, b] \to R$ is continuous, then for any c where $f(a) < c < f(b)$ there is a $d \in (a, b)$ such that $f(d) = c$. [Hint; Consider $h(x) = f(x) - c$.]

7. Prove that if $f : [a, b] \to R$, and f is continous on $[a, b]$, $f(x)$ takes on all values between the maximum and minimum value of $f(x)$ on $[a, b]$.

8. Let c_0 be the set of sequences that converge to 0. Let $\{a_n\}$ and $\{b_n\}$ be sequences in c_0. Show that the function, d defined on $c_0 \times c_0$ by $d(\{a_n\}, \{b_n\}) = \sup |a_n - b_n|$ is a metric. Show that (c_0, d) is not a compact metric space.

9. Suppose that $f : K \to R$ is a $1-1$ real valued function defined on the compact set K. ($1-1$ means that, every value in the range arises from a single value in the domain.) Suppose also, that f is continuous. Then the function f^{-1} : Range of $f \to K$, is also continuous.

10. Use Theorem 4.65 to show that if $f(x)$ is continuous on $[a, b]$, then $K = \{(x, f(x) | x \in [a, b]\}$ (the graph of $f(x)$) is a compact subset of R^2 with the usual metric.

4.6 Uniform Continuity

Recall the definition of continuity: Suppose that (M_1, d_1) and (M_2, d_2) are metric spaces and that $f : M_1 \to M_2$. We say that f is continuous at $p = a$ if for each $\varepsilon > 0$ there is a $\delta > 0$, such that if $d_1(p, a) < \delta$, then $d_2(f(p), f(a)) < \varepsilon$. In general, the δ one finds depends on both ε and a, but not always. (Consider, for example, constant functions.)

A stronger form of continuity is called uniform continuity.

Definition 4.66 *Suppose that (M_1, d_1) and (M_2, d_2) are metric spaces, and that $f : M_1 \to M_2$. We say that f is **uniformly continuous** on M_1 if for each $\varepsilon > 0$ there is a $\delta > 0$, such that if $d_1(p, q) < \delta$, then $d_2(f(p), f(q)) < \varepsilon$, for all p and q in M_1.*

Remark 4.67 *There are a few differences between the notion of continuity and uniform continuity. First, uniform continuity is defined on a set. You never talk about uniform continuity at a point. In the definition of continuity at a, the δ depends on both ε and the point a. In the definition of uniform continuity, the same δ works for all points p and q in the set whose distance from one another is $< \delta$.*

Let us give some examples on the real line.

Example 4.68 *Let $f : R \to R$ be defined by $f(x) = 3x$, and let R carry the usual metric. So, $d_1(p,q) = |p - q|$ and $d_2(f(p), f(q)) = |f(p) - f(q)|$. Notice that, f is uniformly continuous, for if we take any $\varepsilon > 0$, then we need only take $d_1(p,q) = |p - q| < \delta < \varepsilon/3$. It follows that, $d_2(f(p), f(q)) = |3p - 3q| = 3|p - q| < 3\delta < 3(\varepsilon/3) = \varepsilon$.*

Example 4.69 *The function from $[0, 1]$ to R with the usual metric where $f(p) = p^2$, is uniformly continuous on $[0, 1]$. Pick an $\varepsilon > 0$. We claim that any $\delta < \varepsilon/2$ will work. For if $d_1(p, q) = |p - q| < \delta$, then*

$$\begin{aligned} d_2(f(p), f(q)) &= |p^2 - q^2| \\ &= |p + q||p - q| \\ &\leq (|p| + |q|)|p - q| \text{ (triangle inequality)} \\ &\leq (1 + 1)|p - q| \text{ (since } p, q \text{ are from } [0, 1]) \\ &= 2|p - q| \leq 2\delta < 2(\varepsilon/2) = \varepsilon. \end{aligned}$$

Example 4.70 *Although the function $f(x) = x^2$ is uniformly continuous on $[0, 1]$, it is not uniformly continuous on R^+. If it were uniformly continuous on R^+, then for $\varepsilon = 1$, we would be able to find a $\delta > 0$ such that if $|p - q| < \delta$, then $|f(p) - f(q)| < \varepsilon = 1$. Now, choose an n large enough so that $\frac{1}{n} < \delta$. If we let $q = n$ and $p = n + \frac{1}{2n}$, then $|p - q| = \left|\frac{1}{2n}\right| < \frac{1}{n} < \delta$, but $|f(p) - f(q)| = \left|(n + \frac{1}{2n})^2 - n^2\right| = \left|1 + \frac{1}{4n^2}\right| = 1 + \frac{1}{4n^2} > 1$. Since we have found points p and q such that $|p - q| < \delta$, and for which it is not true that $|f(p) - f(q)| < \varepsilon$, this function cannot be uniformly continuous on R^+.*

Let us examine this notion geometrically. Think of $f(p)$ and $f(q)$ as numbers on the y-axis, and p and q as numbers on the x-axis. Saying that $|f(p) - f(q)| < 1$ is equivalent to $-1 < f(p) - f(q) < 1$, or that $f(q) - 1 < f(p) < f(q) + 1$. Thus, we may think of the condition $|f(p) - f(q)| < 1$ as describing roaming open intervals (open balls) of radius 1 on the y-axis with center $f(q)$, where q varies. The set of points for which $|p - q|$ is less than δ can also be considered roaming intervals (open balls) on

4.6. UNIFORM CONTINUITY

the x-axis with center q and radius δ. We are looking for intervals of this type on the x-axis such that they map into these intervals of radius 1 on the y-axis. Consider the interval $(0, 2)$ on the y-axis with center 1 and radius 1. Then the largest interval on the x-axis that maps into this is $(0, \sqrt{2})$, which has length ≈ 1.4 and radius .7. Now consider the interval $(9, 11)$ on the y-axis, which has radius 1 but has center 10. The largest interval on the x-axis that maps into this is $(3, \sqrt{11})$, whose length is $\approx .3$ and whose radius (δ) is $\approx .15$. Next, consider the interval $(9999, 10001)$ on the y-axis. This also has radius 1. The largest interval that maps into this is $(\sqrt{9999}, \sqrt{100001})$, whose length is approximately .01 and whose radius is .005. Our point, is that the further up on the y-axis we take our intervals of radius 1, the smaller the largest intervals on the x-axis that can map into them. Thus, there is no interval of *fixed* size on the x-axis that will map into an open interval of radius 1 on the y-axis. Said another way, there is no unique δ that will take points with $|p - q| < \delta$, into intervals of length 1 on the y-axis. Turn to Figure 4.15 to see these various intervals.

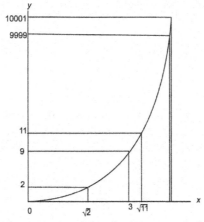

Figure 4.15. (Not drawn to scale.)

Let us contrast this last example to the function $f(x) = 3x$. If we take *any* open interval on the y-axis with radius 1, say $(a - 1, a + 1)$, then the interval on the x-axis that maps into this is the interval $(\frac{a-1}{3}, \frac{a+1}{3})$, whose length is *fixed* at $\frac{2}{3}$ and whose radius is $\delta = \frac{1}{3}$. Thus, you can always find an interval on the x-axis whose length is less than $2/3$, which will map into the interval on the y-axis we chose. Similarly, if we took an interval on the y-axis with radius ε, say $(a - \varepsilon, a + \varepsilon)$, then the corresponding interval on the x-axis that maps into this is $(\frac{a-\varepsilon}{3}, \frac{a+\varepsilon}{3})$. Its length is $\frac{2\varepsilon}{3}$ and its radius is fixed at $\delta = \frac{\varepsilon}{3}$, and so any interval with radius less than this will map into an interval on the y-axis whose radius is less than ε. That is what we found earlier.

As a final example for comparison, let us consider $f(x) = x^2$ on $[0,1]$ and draw a picture similar to Figure 4.15. See Figure 4.16.

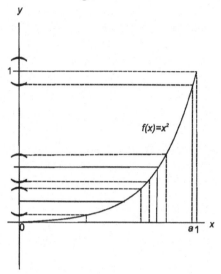

Figure 4.16.

We show open balls of radius ε on the y-axis and the open balls on the x-axis that map into them. As the open balls of radius ε on the y-axis move up, the open balls on the x-axis that map into them get smaller and smaller. But there is a limit to how small they need to get for them to map into open balls of radius ε on the y-axis. In fact, any open ball with the same length as $(a, 1)$ or smaller, shown in that figure, will map into an open ball of radius ε on the y-axis. So we expect (at least from the picture) that $f(x) = x^2$ is uniformly continuous on $[0, 1]$. We will formalize this soon.

Theorem 4.71 *Suppose that $f : M_1 \to M_2$, where M_1 and M_2 are metric spaces and f is uniformly continuous on M_1. Then f is continuous.*

Proof. To show that f is continuous we need only show that f is continuous at each point $q \in M_1$. Since f is uniformly continuous, for each $\varepsilon > 0$ there is a $\delta > 0$, such that if $d_1(p, q) < \delta$, then $d_2(f(p), f(q)) < \varepsilon$, for all p and q in M_1. Now fix any point q in M_1, and this reduces to the definition of $f(x)$ is continuous at q. ∎

Theorem 4.72 *Suppose that $f : M_1 \to M_2$, where (M_1, d_1) and (M_2, d_2) are metric spaces, f is continuous, and (M_1, d_1) is compact. Then f is uniformly continuous.*

Proof. Choose an $\varepsilon > 0$. Since f is continuous at each point a in M_1, we have, for each point $a \in M_1$ a $\delta_a > 0$, such that

$$\text{whenever} \quad d_1(p, a) < \delta_a, \text{then } d_2(f(p), f(a)) < \varepsilon/2. \tag{4.29}$$

4.6. UNIFORM CONTINUITY

Consider $\left\{ B_{\frac{\delta_p}{2}}(p) \right\}$ where p runs through all points in M_1. Certainly $M_1 \subset \bigcup_{p \in M_1} B_{\frac{\delta_p}{2}}(p)$, and since M_1 is compact, this open cover for M_1 has a finite subcover of M_1. So there exist points $p_1, p_2, p_3, ..., p_n$, such that $M_1 \subset \bigcup_{i=1}^{n} B_{\frac{\delta_{p_i}}{2}}(p_i)$. Now let δ be the minimum of the radii of these open balls. So $\delta = \min\left\{\frac{\delta_{p_i}}{2}\right\}$ for $i = 1, 2, ..., n$. Moreover, δ is certainly $\leq$ each $\frac{\delta_{p_i}}{2}$ in this set and δ being the minimum of a finite set of positive numbers is positive.

Claim: This is the δ that "works" in the definition of uniform continuity.

Proof of Claim: Suppose that x and y are any two points in M_1 such that $d_1(x,y) < \delta$. Since $x \in M_1$ and $M_1 \subset \bigcup_{i=1}^{n} B_{\frac{\delta_{p_i}}{2}}(p_i)$, x is in some fixed $B_{\frac{\delta_{p_k}}{2}}(p_k)$, which means that $d_1(x, p_k) < \frac{\delta_{p_k}}{2} < \delta_{p_k}$. Then, by (4.29)

$$d_2(f(x), f(p_k)) < \varepsilon/2. \tag{4.30}$$

Also, by the triangle inequality, since we are taking $d_1(x,y) < \delta$,

$$d_1(y, p_k) \leq d(y, x) + d(x, p_k)$$
$$< \delta + \frac{\delta_{p_k}}{2}$$
$$\leq \frac{\delta_{p_k}}{2} + \frac{\delta_{pk}}{2} = \delta_{p_k}.$$

So again, by (4.29), with $a = p_k$, we have

$$d_2(f(y), f(p_k)) < \varepsilon/2. \tag{4.31}$$

Using the triangle inequality once again, we get

$$d_2(f(x), f(y)) \leq d_2(f(x), f(p_k)) + d_2(f(p_k), f(y))$$
$$< \varepsilon/2 + \varepsilon/2 = \varepsilon \quad \text{(by (4.30) and (4.31))}.$$

Since we have shown that when $d_1(x,y) < \delta$ we have that $d_2(f(x), f(y)) < \varepsilon$, it follows that f is uniformly continuous. ∎

Remark 4.73 *Here is an intuitive way of thinking about the proof. By the continuity, each point p has an open ball of radius δ_p that "works". (That is, that maps into an* open ball of radius ε about $f(p)$.) However, the δ_p can all be different. We seek a "one size fits all δ." The compactness guarantees that we can find a finite set of δ_p that "work" and they cover M. One might think the smallest one would be the one that does the job. Well, not quite. If two points are in an open ball of radius δ, then their distance from each other can be as much as 2δ. So, we cut down the sizes of the δ_p to half of what they were originally. This guarantees that any two points whose distance is less than $\min\{\delta_p/2\}$ lie in the same open ball with radius δ_p, and so they will map into intervals of length less than ε. This is the idea. The rest is just putting it all together formally.

Corollary 4.74 *If $f : [a,b] \to R$ is continuous, then f is uniformly continuous on $[a.b]$.*

Proof. Since $[a,b]$ is compact by Theorem 4.39, the corollary follows immediately from the theorem. ∎

Determining whether or not a function is uniformly continuous is not always easy, but there are some theorems that can help us with this.

Theorem 4.75 *Suppose that $f : E \subset R \to R$ where R carries the usual metric, and suppose further, that f satisfies what is called the Lipschitz condition, namely,*

$$|f(x) - f(y)| \leq K|x - y| \text{ for all } x \text{ and } y \text{ in } E,$$

where K is some positive constant. Then f is uniformly continuous on E.

Proof. Pick $\varepsilon > 0$. Then if $|x - y| < \varepsilon/K$, it follows that

$$\begin{aligned} |f(x) - f(y)| &\leq K|x - y| \\ &\leq K(\varepsilon/K) = \varepsilon, \text{ for all } x \text{ and } y \text{ in } E. \end{aligned}$$

Therefore, we need only take δ to be any number $< \varepsilon/K$ to see that $|f(x) - f(y)| < \varepsilon$. So, f is uniformly continuous. ∎

At this point we recall a theorem from calculus whose validity we accept since we will not be talking about derivatives in this primer.

Theorem 4.76 (The Mean Value Theorem) *Suppose that $f : [a, b] \to R$ (where R carries the usual metric) is continuous on $[a, b]$, and has a derivative at each point of (a, b). Then*

$$\frac{f(b) - f(a)}{b - a} = f'(c), \text{ for some } c \in (a, b).$$

4.6. UNIFORM CONTINUITY

The geometric interpretation of this is shown in Figure 4.17.

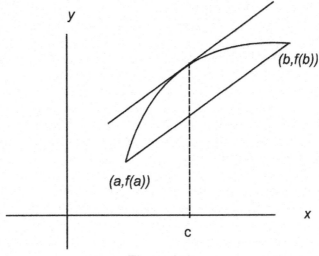

Figure 4.17.

The theorem says that under these conditions, the slope of the secant line joining $(a, f(a))$ to $(b, f(b))$ is parallel to the tangent line at some point c in (a, b). (This is because the secant line has slope $\frac{f(b)-f(a)}{b-a}$ while the tangent line at c has slope $f'(c)$, and two lines are parallel if and only if they have the same slope.) We won't be needing the geometric interpretation of this.

We also need to recall the following from calculus.

Theorem 4.77 *If a function $f : R \to R$ has a derivative at a point p, it is continuous at that point. (Again, R carries the usual metric.)*

A corollary of Theorem 4.76 which we will need in Chapter 5 is:

Corollary 4.78 *If $F(x)$ is an antiderivative of $f(x)$ on $[a, b]$, where $f : [a, b] \to R$, then for any $x, y \in [a, b]$, where $x < y$, we have $F(y) - F(x) = f(t)(y - x)$, for some t between x and y.*

Proof. Saying that F is an antiderivative of $f(x)$ means that $F'(x)$ exists and $= f(x)$. But since F' exists, F is continuous on $[x, y]$. Now, we apply the Mean Value Theorem to F on $[x, y]$ to get

$$\frac{F(y) - F(x)}{y - x} = F'(t), \tag{4.32}$$

134   CHAPTER 4.   LIMITS, COMPACTNESS AND UNIFORM CONTINUITY

for some $t \in (a,b)$. But since $F'(t) = f(t)$, equation (4.32) becomes,

$$\frac{F(y) - F(x)}{y - x} = f(t),$$

and we need only multiply both sides by $y - x$ to get the result stated in the theorem. ∎

Returning to our discussion on uniform continuity, we have the following result.

Theorem 4.79 *Let $f : E \subset R \to R$ (where R carries the usual metric) and E be any open interval, finite or infinite. If $f(x)$ has a bounded derivative at each point of E, then f is uniformly continuous on E.*

Proof. Saying that f has a bounded derivative on E means that $|f'(x)| \leq K$ for some positive constant, K. Now suppose that x, y are any two points in E, and suppose that $x < y$. Since the function has a derivative on E, f is continuous on E, by Theorem 4.77. Let us apply the Mean Value Theorem to the function on the interval $[x, y]$. We get that

$$\frac{f(y) - f(x)}{y - x} = f'(c), \qquad \text{for some } c \in (x, y).$$

It follows that

$$\left| \frac{f(y) - f(x)}{y - x} \right| = |f'(c)| \leq K,$$

and from this it follows that

$$\frac{|f(y) - f(x)|}{|y - x|} \leq K,$$

or that

$$|f(y) - f(x)| \leq K |y - x|.$$

The result now follows from Theorem 4.75. ∎

Example 4.80 *The function $f(x) = x^2$ is uniformly continuous on any finite open interval since the derivative. $f'(x) = 2x$. is always bounded on any finite interval.*

Example 4.81 *A natural question to ask is if there are uniformly continuous functions with unbounded derivatives. The function $f(x) = \sqrt{x}$ defined on $[0, 1]$ is uniformly continuous, by Corollary 4.74, so its restriction to $(0, 1)$ is also uniformly continuous. But its derivative is not bounded on $(0, 1)$.*

4.6. UNIFORM CONTINUITY

Definition 4.82 *The **closure** of a set E, in a metric space, is defined to be $E \cup E'$ and is abbreviated $\overline{E}$.*

Example 4.83 *The closure of the set $E = [0, 1)$ in R with the usual metric, is $\overline{E} = [0, 1]$. The closure of the set $F = \{(x, y)| x^2 + y^2 < 1\}$ in R^2 with the usual metric is $\overline{F} = \{(x, y)| x^2 + y^2 \leq 1\}$. The closure of any set in the discrete metric space is the set itself because no set in the discrete metric space has limit points.(Recall that in the discrete metric space $B_{1/2}(p) = \{p\}$. This open ball doesn't contain points of any set other than p. So p cannot be a limit point of any set.)*

The proof of the following is a bit elaborate, and so we will not prove it.

Theorem 4.84 *$f : E \subset R \to R$ is uniformly continuous on E if and only if f can be extended to a uniformly continuous function $g : \overline{E} \to R$.(What this means is that there is a uniformly continuous function g defined on $\overline{E}$, which equals f on E.)*

Example 4.85 *Prove that $f(x) = \dfrac{1}{x}$ is not uniformly continuous on $(0, 1]$.*

If $f(x)$ were uniformly continuous on $(0, 1]$, then we could extend it to a uniformly continuous function g on the closure of $(0, 1]$ (in R with the usual metric), which is $[0, 1]$. But if g is continuous on $[0, 1]$, the range of g is compact since $[0, 1]$ is, by Theorem 4.50, and hence is bounded, by Theorem 4.48. But g equals f on $(0, 1]$ and that implies that f is bounded on $(0, 1]$, which we know is not the case since the graph of $f(x) = \dfrac{1}{x}$ has a vertical asymptote at $x = 0$. This contradiction tells us that f cannot be uniformly continuous on $(0, 1]$.

EXERCISES

1. Show that if $f : (M_1, d) \to (M_2, d_2)$ is uniformly continuous, then f restricted to a subset S of the metric space (M_1, d) is also uniformly continuous.

2. True or False: If $\{a_n\}$ is a Cauchy sequence in a metric space (M, d) and f is real valued continuous on M, $f(a_n)$ is Cauchy. Prove your answer.

3. True or False: If $\{a_n\}$ is a Cauchy sequence in a metric space (M, d_1) and $f : (M_1, d_1) \to (M_2, d_2)$ is uniformly continuous, $f(a_n)$ is Cauchy. Prove your answer.

4. Show $\cos x$ uniformly continuous on $(-\infty, \infty)$.

5. A function $f : R \to R$ is called periodic, if there is some number positive number p, called the period, such that $f(x) = f(x + p)$ for all x. (So $\sin x$ is periodic with period 2π.) Show that a periodic function is uniformly continuous.

6. Show that $f(x) = \sqrt{x}$ is uniformly continuous on $[1, \infty)$.

7. Show that $f(x) = \sqrt{x}$ is uniformly continuous on $[0, \infty)$. [Hint: You can break the interval up in $[0, 2] \cup [1, \infty)$ and take δ small enough so that any two points whose distance is less than δ, forces them both to be in either $[0, 2]$ or $[1, \infty)$. See the next example for a different approach.]

8. We know that $|\sqrt{x} - \sqrt{y}|^2 = |\sqrt{x} - \sqrt{y}| \cdot |\sqrt{x} - \sqrt{y}| \leq |\sqrt{x} - \sqrt{y}| \cdot (\sqrt{x} + \sqrt{y}) = |\sqrt{x} - \sqrt{y}| \cdot |\sqrt{x} + \sqrt{y})| = |x - y|$. Use this to show that $f(x) = \sqrt{x}$ is uniformly continuous on $[0, \infty)$.

9. Suppose the domain of a uniformly continuous real valued function, f, is bounded. Show the range is also bounded. [Hint: If not, we can find a sequence such that $f(p_n) > n$ for all n. Use the Bolzano Weierstrass Theorem.]

10. Without using the definition, show the function $\sin(1/x)$ not uniformly continuous on $(0, 1]$, but $x \sin(1/x)$ is.

Chapter 5

Sequences of Functions

Sequences of functions are extremely important in higher analysis and its applications. Results that one would hope would be true for sequences of functions, turn out not to be true, and because of these issues, new developments in analysis arose. The notions of uniform convergence and the Lebesgue integral are some of these developments.

Definition 5.1 *Suppose that we have a sequence $\{f_n\}$ of functions from a set X to R, with the usual metric. We say that f_n converges to f **pointwise** on X, if for each fixed $x \in X$, the sequence $f_n(x)$ converges to $f(x)$. When f_n converges to f pointwise, we write $f_n \to f$.*

Example 5.2 *Let $X = [0, 1]$, and let $f_n(x) = \dfrac{n}{n+1}x$. So, $f_1(x) = \dfrac{1}{2}x$, $f_2(x) = \dfrac{2}{3}x$ and so on. For each fixed $x \in [0, 1]$, $f_n(x)$ converges to $f(x) = x$ since as $n \to \infty$, $\dfrac{n}{n+1} \to 1$. Furthermore, for each fixed $x \in [0, 1]$, $f_n(x)$ increases to $f(x)$. Suppose that we wanted to find an N for which $|f_n(1) - f(1)| < .1$, for $n > N$. How far out in the sequence $\{f_n(1)\}$ must we go before this happens? The answer is we take $N = 9$. For $f_9(1) = \dfrac{9}{10}$, and $|f_9(1) - f(1)| = \left|\dfrac{9}{10} - 1\right| = .1$. Also note that for $n > N$, $|f_n(1) - f(1)| < .1$, since the sequence increases to $f(1)$. Suppose we wanted to find an N such that $|f_n(\frac{1}{2}) - f(\frac{1}{2})| < .1$, when $n > N$. N now turns out to be 4, since $f_4\left(\frac{1}{2}\right) = \dfrac{2}{5}$ and $\left|f_4\left(\frac{1}{2}\right) - f\left(\frac{1}{2}\right)\right| = \left|\dfrac{2}{5} - \dfrac{1}{2}\right| = .1$. Also, for $n > 4$, $\left|f_n(\frac{1}{2}) - f(\frac{1}{2})\right| < .1$. So we see that the N needed to get a certain distance ε from the limit, varies with each x.*

Remark 5.3 *This last example shows us that an intuitive way of thinking about the idea of pointwise convergence is that for each x, the sequence $\{f_n(x)\}$ converges to $f(x)$. Note, however, that the speed at which $f_n(x)$ converges to $f(x)$ may differ from one x to another. When the speed of convergence is "essentially" the same for all x, we have what is called uniform convergence. We will define that later.*

Example 5.4 Let X be R. The sequence of functions $f_n(x) = \dfrac{x}{n}$ converges pointwise to $f(x) = 0$ on R This happens since for each fixed x, the denominators of the fractions $f_n(x)$ get larger and larger, making $\dfrac{x}{n}$ smaller and smaller. Therefore this makes the sequence $f_n(x) = \dfrac{x}{n}$ converge to 0 on R.

We will see other examples as we progress.

A natural question is,"Given a sequence of functions $f_n : X \to R$, which converges pointwise to $f(x)$, say on $[a, b]$, which of the following are true? (1) If each $f_n(x)$ is continuous on $[a, b]$, then $f(x)$ is continuous on $[a, b]$; (2) $\int_a^b f_n(x)dx \to \int_a^b f(x)dx$; (3) $f_n'(x) \to f'(x)$."

The answer, surprisingly, is, "None of them are true!" It should also be pointed out that historically, mathematicians made use of these false "facts" at times to demonstrate things that they assumed were true, and sometimes got themselves into a lot of trouble.

We will give several examples to illustrate that these statements are not true.

Example 5.5 Let $f_n(x) = x^n$, restricted to $[0, 1]$. Figure 5.1 shows a few of the graphs of $f_n(x)$.

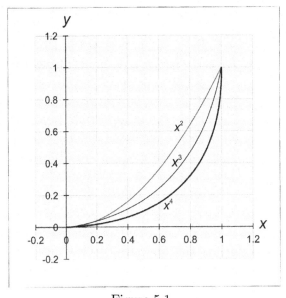

Figure 5.1.

They are all continuous on $[0,1]$, and for each $0 \le x < 1$, $f_n(x) \to 0$. However, $f_n(1) = 1$ for each positive integer n. Thus, $f_n(x) \to f(x) = \begin{cases} 0 & \text{if } 0 \le x < 1 \\ 1 & \text{if } x = 1 \end{cases}$.
This function is not continuous on $[0,1]$. So the pointwise convergence of a sequence of continuous functions, f_n, to $f(x)$, on a set does not guarantee that $f(x)$ is continuous.

Example 5.6 Let $f_n(x) = \dfrac{1}{\sqrt{n}} \sin nx$, defined only on $[0,1]$. Then using basic knowledge of derivatives from calculus, $f_n'(x) = \dfrac{1}{\sqrt{n}}(n \cos nx) = \sqrt{n} \cos nx$. Now, for each fixed x, $\sin nx$ is between -1 and 1, while $\dfrac{1}{\sqrt{n}}$ gets small as n gets large. So $\dfrac{1}{\sqrt{n}} \sin nx = f_n(x)$ approaches $f(x) = 0$. (A more formal way of seeing this is to realize that for a fixed x, $\sin nx$ is a sequence, and $-1 \le \sin nx \le 1$. From this, it follows that $\dfrac{-1}{\sqrt{n}} \le \dfrac{\sin nx}{\sqrt{n}} \le \dfrac{1}{\sqrt{n}}$. Taking the limit as $n \to \infty$ and realizing that as n goes to ∞, both $\dfrac{1}{\sqrt{n}}$ and $-\dfrac{1}{\sqrt{n}}$ converge to 0, we get, using the squeeze theorem (Corollary 2.51 with $L = 0$), that for each fixed x the sequence $\dfrac{\sin nx}{\sqrt{n}} \to 0$.) Now, $f_n'(0) = \sqrt{n} \cos 0 = \sqrt{n}$ and $f'(0) = 0$. So, $\lim f_n'(0) \ne f'(0)$.

For the next example, we need a lemma.

Lemma 5.7 If $p > 1$, then $\lim \dfrac{n^2}{p^n}$, where p is more than 1, approaches 0.

Proof. Since the numerator and denominator approach ∞ as n gets large, we can use L'Hospital's rule from calculus to find the limit. So $\lim \dfrac{n^2}{p^n} = \lim \dfrac{2n}{p^n \ln p} = \lim \dfrac{2}{p^n (\ln p)^2} = 0$ since as $n \to \infty$, the denominator gets very large and the numerator is fixed. ∎

Example 5.8 Let $f_n(x) = n^2 x (1-x)^n$, restricted to $[0,1]$. We first show that $f_n(x) \to 0$ for all $x \in [0,1]$. If $x = 0$ or 1, $f_n(x) = 0$, and so $\lim f_n(x) = 0$. If $0 < x < 1$, then $0 < 1-x < 1$, and $p = \dfrac{1}{1-x} > 1$. So for any fixed $x_0 \in (0,1)$, we have, letting

$p_0 = \dfrac{1}{1-x_0}$, that

$$\begin{aligned} f_n(x_0) &= n^2 x_0 (1-x_0)^n \\ &= x_0 \dfrac{n^2}{\left(\dfrac{1}{1-x_0}\right)^n} \\ &= x_0 \dfrac{n^2}{(p_0)^n}, \text{ where } p_0 > 1. \end{aligned}$$

So by the previous lemma, $f_n(x_0) \to 0$. Since x_0 was arbitrary in $(0,1)$ and $f_n(x_0) \to 0$ for these x, as well as for $x = 0$ and 1, we have that $f_n(x) \to f(x) = 0$ on $[0,1]$. Now, consider $\int_0^1 f_n(x) dx$ and $\int_0^1 f(x) dx$. We will show that $\int_0^1 f_n(x) dx \not\to \int_0^1 f(x) dx$. First, we compute $\int_0^1 f_n(x) dx = \int_0^1 n^2 x(1-x)^n dx$. We do this by the standard method of substitution from calculus. We let $u = 1 - x$. So, $du = -1 dx$, and $x = 1 - u$. While we are at it, we also change our limits. When $x = 0$, $u = 1$, and when $x = 1$, $u = 0$. So, we get

$$\begin{aligned} \int_0^1 n^2 x(1-x)^n dx &= -\int_1^0 n^2 (1-u) u^n du \\ &= \int_0^1 n^2 (1-u) u^n du \ \ \left(\text{since } -\int_a^b g(x) dx = \int_b^a g(x) dx\right) \\ &= n^2 \int_0^1 (u^n - u^{n+1}) du \\ &= n^2 \left[\dfrac{u^{n+1}}{n+1} - \dfrac{u^{n+2}}{n+2} \Big|_0^1 \right] \\ &= n^2 \left[\dfrac{1}{n+1} - \dfrac{1}{n+2} \right] \\ &= n^2 \left[\dfrac{1}{(n+1)(n+2)} \right] \\ &= \dfrac{n^2}{n^2 + 3n + 2}. \end{aligned}$$

We have shown that $\int_0^1 f_n(x) dx = \dfrac{n^2}{n^2+3n+2}$, so $\lim \int_0^1 f_n(x) dx = \lim \dfrac{n^2}{n^2+3n+2} = 1$. But the sequence $\{f_n(x)\}$ converges pointwise to $f(x) = 0$ on $[0,1]$, and since $\int_0^1 f(x) dx = 0$, we see that $\int_0^1 f_n(x) dx \to 1 \neq \int_0^1 f(x) dx$.

These last examples show that things we would like to be true about pointwise convergence, in terms of continuity, derivatives and integrals, aren't. So what additional conditions can we give that might give us the theorems we like? The next definition helps.

Definition 5.9 *Suppose that $\{f_n\}$ is a sequence of functions from a set X to R. We say that $\{f_n\}$ **converges uniformly** to $f(x)$, on X if for each $\varepsilon > 0$, there is an $N > 0$, such that $n > N$ implies that $|f_n(x) - f(x)| < \varepsilon$, for all $x \in X$.*

Notice the difference between uniform convergence and pointwise convergence. In pointwise convergence, the N that one must take varies depending on x, as we saw in Example 5.2. For uniform convergence, the same N works for each x.

The inequality $|f_n(x) - f(x)| < \varepsilon$ for all $x \in X$, is equivalent to the inequalities $f(x) - \varepsilon < f_n(x) < f(x) + \varepsilon$ for all $x \in X$. Geometrically, this means that if $\{f_n\}$ converges uniformly to $f(x)$ on X, all the terms $f_n(x)$ lie in a band of width ε about $f(x)$ after a certain point. (See Figure 3.1.)

Example 5.10 *Consider the sequence of functions defined by $f_n(x) = \dfrac{\sin nx}{n}$. This sequence converges pointwise to 0 on R We show that this sequence converges uniformly to the function $f(x) = 0$ on R To see that, pick any $\varepsilon > 0$, and let N be large enough so that $\frac{1}{N}$ is less than ε. Since the maximum value of $|\sin nx|$ is 1, we have that for $n > N, |f_n(x) - f(x)| = \left|\dfrac{\sin nx}{n}\right| = \dfrac{|\sin nx|}{n} \leq \dfrac{1}{n} < \dfrac{1}{N} < \varepsilon$. And this is true for all x. So, this sequence converges uniformly to the function $f(x) = 0$ on R.*

We defined pointwise convergence and uniform convergence for real valued functions defined on a set X. There is nothing to stop us from defining pointwise and uniform convergence for sequences of functions that go from a set X, to a metric space (Y, d). Here are the definitions.

Definition 5.11 *Suppose that $f_n, f : X \to Y$, where X is a set and (Y, d) is a metric space. We say that $f_n(x)$ converges to $f(x)$ pointwise on X, if for each fixed $x \in X$ and for each $\varepsilon > 0$, there exists an N such that $d(f_n(x), f(x)) < \varepsilon$, for $n > N$.*

Definition 5.12 *Suppose that $f_n, f : X \to Y$ where X is a set and (Y, d) is a metric space. We say that $f_n(x)$ converges to $f(x)$ uniformly on X, if for each $\varepsilon > 0$, there exists an N, such that $d(f_n(x), f(x)) < \varepsilon$ for $n > N$ and for all $x \in X$.*

Our first theorem is encouraging.

Theorem 5.13 *Suppose that $\{f_n(x)\}$ is a sequence of real valued functions with domain X, where (X, d) is a metric space, and suppose that $f_n(x)$ converges uniformly to $f(x)$ on X. If each $f_n(x)$ is continuous on X, then $f(x)$ is continuous on X.*

Proof. Pick an $\varepsilon > 0$. Then, since $f_n \to f$ uniformly on X, there is an integer $N > 0$ such that $|f_n(x) - f(x)| < \varepsilon/3$, for all $x \in X$ and all $n > N$. In particular, we can make

$$|f_{N+1}(x) - f(x)| < \varepsilon/3, \text{ for all } x \in X. \tag{5.1}$$

Since $f_{N+1}(x)$ is continuous, we know that we can find a $\delta > 0$ such that

$$|f_{N+1}(x) - f_{N+1}(x_0)| < \varepsilon/3, \text{ when } d(x, x_0) < \delta. \tag{5.2}$$

Now, let $d(x, x_0)$ be less than δ, and consider $|f(x) - f(x_0)|$. We have

$$
\begin{aligned}
&|f(x) - f(x_0)| \\
={}& |f(x) - f_{N+1}(x) + f_{N+1}(x) - f_{N+1}(x_0) + f_{N+1}(x_0) - f(x_0)| \\
\leq{}& |f(x) - f_{N+1}(x)| + |f_{N+1}(x) - f_{N+1}(x_0)| + |f_{N+1}(x_0) - f(x_0)| \quad \text{(triangle inequality)} \\
\leq{}& \varepsilon/3 + \varepsilon/3 + \varepsilon/3 = \varepsilon \quad \text{(by (5.1), (5.2) and (5.1), respectively).}
\end{aligned}
$$

We have shown that we can make $|f(x) - f(x_0)| < \varepsilon$ when $d(x, x_0) < \delta$, and this establishes the continuity of $f(x)$ at any point $x_0 \in X$ (using Definition 4.14). Since $f(x)$ is continuous at every point of X, f is continuous on X.

Notice how we jumped from f to f_{N+1}, and then back to f, to prove continuity. It is a clever proof. Figure 5.2 illustrates the proof. We want to show that the heights of the points A and B (which are $f(x_0)$ and $f(x)$, respectively), are close to each other when x and x_0 are sufficiently close. But the heights of A and B are close to the heights of C and D (which are $f_{N+1}(x_0)$ and $f_{N+1}(x)$, respectively), by the uniform convergence of the sequence $\{f_n\}$. The heights of C and D are close to each other when x and x_0 are sufficiently close, by the continuity of f_{N+1}. So, the heights of A and B are close when x and x_0 are sufficiently close.

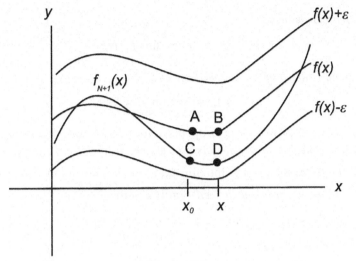

Figure 5.2.

∎

Remark 5.14 *It is interesting to note that at first, Cauchy, who in many senses is the father of analysis, thought that the pointwise limit of continuous functions was continuous, and even provided what he thought was a proof of it. Cauchy is not alone among the great mathematicians who believed (and thought they had proved) erroneous results.*

Remark 5.15 *Theorem 5.13 is still true if all the functions go from a metric space (X, d_1) to a metric space (Y, d_2). We do the exact same proof, simply replacing absolute values by distance functions. So, any term with an absolute value, for example, $|f(x) - f(y)|$, will be replaced by $d_2(f(x), f(y))$. (See Exercise 8.)*

Remark 5.16 *Suppose that $f_n \to f$, where all are real valued, defined on some metric space (X, d). Saying that f_n is continuous at x_0 means that*

$$\lim_{x \to x_0} f_n(x) = f_n(x_0). \tag{5.3}$$

Saying that f is continuous at x_0 means that

$$\lim_{x \to x_0} f(x) = f(x_0). \tag{5.4}$$

We can write (5.4) as

$$\lim_{x \to x_0} \lim_{n \to \infty} f_n(x) = f(x_0),$$

and this in turn can be written as
$$\lim_{x\to x_0}\lim_{n\to\infty} f_n(x) = \lim_{n\to\infty} f_n(x_0).$$

Finally, using (5.3), this can be written as
$$\lim_{x\to x_0}\lim_{n\to\infty} f_n(x) = \lim_{n\to\infty}\lim_{x\to x_0} f_n(x). \tag{5.5}$$

So, saying that a sequence $\{f_n(x)\}$ of continuous real valued functions converges to a continuous function $f(x)$ at x_0, means that we can interchange the order of $\lim_{x\to x_0}\lim_{n\to\infty}$. This gives us the following result: $\lim_{x\to x_0}\lim_{n\to\infty} f_n(x) = \lim_{n\to\infty}\lim_{x\to x_0} f_n(x)$. We cannot always do this as the following, previously seen example shows.

Example 5.17 *We saw in Example 5.5 that the sequence $\{f_n(x)\}$, where $f_n(x) = x^n$ on $[0,1]$, converged to the function $f(x) = \begin{cases} 0 & \text{if } 0 \le x < 1 \\ 1 & \text{if } x = 1 \end{cases}$.*

This function has a discontinuity at 1, and $\lim_{x\to 1}\lim_{n\to\infty} f_n(x) = \lim_{x\to 1} f(x) = 0$, while $\lim_{n\to\infty}\lim_{x\to 1} f_n(x) = \lim_{n\to\infty} 1 = 1$. So, (5.5) is not true here.

Remark 5.18 *It probably pays to show, geometrically, what (5.5) is saying. Let us consider the functions from Example 5.2, where the $f_n(x)$ converge uniformly to $f(x)$ on $[0,1]$.*

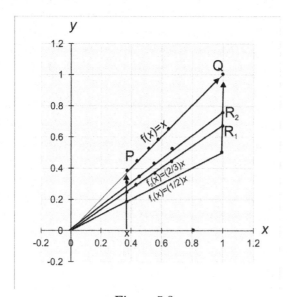

Figure 5.3.

Figure 5.3 shows a few of the terms of the sequence $f_n(x) = \frac{n}{n+1}x$, and the limit function $f(x)$. Suppose we pick an arbitrary, but fixed, x, and take $\lim f_n(x)$ as $n \to \infty$. Then these numbers, $f_n(x)$, can be visualized as a sequence of points moving up approaching P, the limit, as shown in the figure. Now, take the limit as $x \to 1$. P moves along $f(x)$ to the point Q. Now, let us go back and, instead of fixing x first, let us fix n and consider the limit of $f_n(x)$ as $x \to 1$. Then the points represented by $\lim_{x \to 1} f_n(x)$ move along the curve $f_n(x)$. (We have shown the case when $n = 2$ and $n = 3$. The points move along f_2 and f_3 towards points R_1, and R_2). Thus, we generate a sequence of points $\{R_n\}$. Take the limit of this sequence as $n \to \infty$. The points we get can be visualized as moving up along the line $x = 1$ to Q. So, whether we take the limit as $x \to 1$ first, or as $n \to \infty$ first, we end up, in the same place, at Q.

We have seen that continuity is preserved under uniform convergence. The following is also true, although we don't yet have the machinery to prove it. We will prove it later in Chapter 7.

Theorem 5.19 *Suppose that $f_n : [a, b] \to R$, for $n = 1, 2, 3, ...$, and that $\int_a^b f_n(x)dx$ exists for each n. Then, if f_n converges uniformly to $f(x)$ on $[a, b]$, it follows that $\int_a^b f(x)dx$ exists and that $\int_a^b f_n(x)dx \to \int_a^b f(x)dx$.*

So integration is preserved under uniform limits of sequences of functions. Unfortunately, differentiation is not preserved under uniform limits, as the following example shows.

Example 5.20 *In Example 5.6 we showed that $f_n(x) \to f(x)$. The sequence of derivatives of the functions $f_n(x) = \frac{\sin nx}{\sqrt{n}}$ did not converge to $f'(x) = 0$. This is true even though the sequence of functions, $f_n(x)$, converges uniformly to $f(x)$ on $[0, 1]$. (The proof of this requires a minor modification of the proof given in Example 5.10.)*

Uniform convergence of functions is a nice thing to have since we can do a lot with it. The following theorem gives at least one condition that will guarantee that a sequence of functions converges uniformly on a set.

Theorem 5.21 *(**Dini's Theorem**) Suppose that K is compact, and that $\{f_n(x)\}$ is a sequence of continuous real valued functions defined on K. If for each fixed $x \in K$, the sequence is decreasing and if $\lim f_n(x) = f(x)$ on K, where $f(x)$ is continuous, then $\{f_n(x)\}$ converges uniformly to f on K.*

Proof. Let $g_n(x) = f_n(x) - f(x)$. Since $\{f_n(x)\}$ decreases to $f(x)$, $g_n(x) \geq 0$. Now, g_n, being the difference of continuous functions on K, must also be continuous on K. Since $f_n \to f$, $g_n \to 0$. Also, since $\{f_n(x)\}$ is decreasing, $f_{n+1} \leq f_n$, from which it follows that $g_{n+1} = f_{n+1} - f \leq f_n - f = g_n$. So, $\{g_n\}$ is a decreasing sequence of continuous functions. Now, choose $\varepsilon > 0$, and let $K_n = \{x \in K |\ g_n(x) \geq \varepsilon\} = g_n^{-1}[\varepsilon, \infty)$. Since $[\varepsilon, \infty)$ is closed in R, and g_n is continuous on K, $g_n^{-1}[\varepsilon, \infty)$ is closed in K by Corollary 4.31. But a closed subset of a compact set is compact, by Theorem 4.40. So each K_n is compact. Since g_n is decreasing, $K_1 \supset K_2 \supset K_3 \supset \ldots$ (Verify!). Now, choose any $z \in K$, the compact space we are working in. Since $g_n(z)$ decreases to 0, we can eventually make $g_n(z)$ less than ε. Thus $z \notin K_n$ after a certain point, which means that $z \notin \cap K_n$. Since z was any element in K and we showed it is not in $\cap K_n$, nothing in K is in $\cap K_n$, which means that $\cap K_n = \varnothing$. Thus, by Corollary 4.62, there is an integer N such that K_N is empty, and since the sequence $\{K_n\}$ is nested, as we pointed out above, K_n is empty for all $n > N$. What this means is that for $n > N$, there are no points in $K_n = \{x \in K |\ g_n(x) \geq \varepsilon\}$, and so $g_n(x) = |g_n(x)| < \varepsilon$ for all $x \in K$, and $n > N$. But $g_n = f_n - f$. This implies that $|f_n(x) - f(x)| < \varepsilon$ for all $x \in K$, and $n > N$, and this is the definition of $\{f_n(x)\}$ converging uniformly to $f(x)$ on K, which is what we wanted to prove. ∎

Remark 5.22 *Dini's theorem is also true when the sequence of functions is increasing to $f(x)$. Thus, Dini's theorem may be restated as, "If a monotone sequence of continuous functions converges to a continuous function on a compact set, then the convergence is uniform on that set." We will have you show this for increasing sequences in the exercises.*

Definition 5.23 *If $f_n : X \to R$, for $n = 1, 2, 3, \ldots$, where X is any set, we say that $\{f_n\}$ is **uniformly Cauchy on X**, if, for each $\varepsilon > 0$, there exists an $N > 0$, such that for any positive integers $m, n > N$, $|f_n(x) - f_m(x)| < \varepsilon$ for all $x \in X$.*

Remark 5.24 *This definition is telling us that for each x, the sequence $\{f_n(x)\}$ is Cauchy. However, it is telling us more. It is telling us that the same N works for each x.*

Example 5.25 *Suppose we have the functions $f_n(x) = \dfrac{\sin 3nx}{n^3}$, where $n = 1, 2, 3, \ldots$. This sequence is uniformly Cauchy on R. To verify this directly from the definition, let m and n be greater than $\sqrt[3]{\dfrac{2}{\varepsilon}}$. We have:* $|f_n(x) - f_m(x)| = \left|\dfrac{\sin 3mx}{m^3} - \dfrac{\sin 3nx}{n^3}\right| \leq \left|\dfrac{\sin 3mx}{m^3}\right| + \left|\dfrac{\sin 3nx}{n^3}\right| \leq \dfrac{1}{m^3} + \dfrac{1}{n^3} < \dfrac{\varepsilon}{2} + \dfrac{\varepsilon}{2} = \varepsilon$, *regardless of what x is.*

The following theorem is not unexpected.

Theorem 5.26 *If $\{f_n\}$ is a sequence of real valued functions defined on a set X, then $\{f_n\}$ converges uniformly to some real value function f defined on X if and only $\{f_n\}$ is uniformly Cauchy on X.*

Proof. ($\Longrightarrow$)If $\{f_n\}$ converges uniformly to f on X, then for each $\varepsilon > 0$, there is an $N > 0$ such that $n > N$ implies that

$$|f_n(x) - f(x)| < \varepsilon/2, \text{ for all } x \in X. \tag{5.6}$$

Now, suppose that both $m, n > N$. Then (5.6) holds for both m and n. So we have

$$\begin{aligned} |f_n(x) - f_m(x)| &= |f_n(x) - f(x) + f(x) - f_m(x)| \\ &\leq |f_n(x) - f(x)| + |f(x) - f_m(x)| \\ &< \varepsilon/2 + \varepsilon/2 \quad \text{(by 5.6)} \\ &= \varepsilon, \text{ for all } x \in X. \end{aligned}$$

We have shown that for $m, n > N$, $|f_n(x) - f_m(x)| < \varepsilon$ for all $x \in X$, which is the definition of the sequence $\{f_n\}$ being uniformly Cauchy on X.

($\Longleftarrow$) Now we are assume that $\{f_n(x)\}$ is uniformly Cauchy on X. For any $\varepsilon > 0$ there exists an $N > 0$, such that for any positive integers $m, n > N$,

$$|f_n(x) - f_m(x)| < \varepsilon/2 \tag{5.7}$$

for all $x \in X$. Now fix x in X. Then (5.7) says that the sequence $\{f_n(x)\}$ is Cauchy for this fixed x, and since R is complete, this sequence converges. The same is true for every $x \in X$. Define for each $x \in X$, $f(x) = \lim f_n(x)$. Now f is a function from X to R. To show that the sequence of functions $f_n(x)$ converges uniformly to $f(x)$ on X, fix n such that $n > N$ and rewrite (5.7) as

$$f_m(x) - \varepsilon/2 < f_n(x) < f_m(x) + \varepsilon/2, \text{ for all } x \in X. \tag{5.8}$$

Now, take the limit of (5.8) as m goes to infinity, realizing that $\lim f_m(x) = f(x)$ (by definition of $f(x)$) to get

$$f(x) - \varepsilon/2 \leq f_n(x) \leq f(x) + \varepsilon/2, \text{ for all } x \in X \text{ (by Theorem 2.50)}$$

This can be written as

$$|f_n(x) - f(x)| \leq \varepsilon/2 < \varepsilon \text{ for each } n > N, \text{ and for all } x \in X.$$

This is the definition of $\{f_n(x)\}$ converging uniformly to $f(x)$ on X ∎

When we introduced complete metric spaces, we talked about $C[0, 1]$, the set of all real valued continuous functions defined on the closed interval $[0, 1]$, and mentioned that it was complete. More generally, if (X, d_1) is any compact metric space, and $C(X)$ is the set of all real valued continuous functions defined on X, we can define a metric, d, on $C(X)$. For $f, g \in C(X)$, $d(f, g) = \max_{x \in X} |f(x) - g(x)|$. We know this maximum exists by Corollary 4.52.

Theorem 5.27 *If X is compact, then $C(X)$ is a complete metric space.*

Proof. Suppose that $\{f_n\}$ is Cauchy in this metric. What this means is that for any $\varepsilon > 0$, there exists an N such that if $m, n > N$,

$$\begin{aligned} d(f_m, f_n) &< \varepsilon, \text{ which is equivalent to} \\ \max_{x \in X} |f_m(x) - f_n(x)| &< \varepsilon, \text{ which is equivalent to} \\ |f_m(x) - f_n(x)| &< \varepsilon \text{ for all } x \in X, \end{aligned}$$

and this tells us that the sequence $\{f_n\}$ is uniformly Cauchy on X. But, by Theorems 5.26 and 5.13, this sequence must converge uniformly to a continuous real valued function $f(x)$ defined on X; that is, for any $\varepsilon > 0$, there is an N such that $n > N$ implies that $|f_n(x) - f(x)| < \varepsilon$, for all $x \in X$. But if $|f_n(x) - f(x)| < \varepsilon$, for all $x \in X$ and $n > N$, then $\max_{x \in X} |f_n(x) - f(x)| < \varepsilon$, for $n > N$, or, said another way, $d(f_n, f) < \varepsilon$ for $n > N$. This says that f_n converges to f in the metric space $C[0, 1]$.

We started with a Cauchy sequence $\{f_n\}$ in this metric space $C[0, 1]$, and showed it converged to $f(x) \in C([0, 1])$. So, $C[0, 1]$ is complete. ■

EXERCISES

1. Show that if a sequence, $\{f_n\}$ of real valued functions defined on a set X converges uniformly to a function f on X, it must also converge pointwise to f on X.

2. Let $f(x)$ be any real valued function. Define $f_n(x) = \dfrac{f(x)}{n}$. Show that the sequence $\{f_n(x)\}$ converges pointwise to $f(x) = 0$ on R and that if $f(x)$ is bounded, the convergence is uniform on R.

3. Let $f_n(x) = \dfrac{1}{2^{(x-n)^2}}$. Show that $f_n(x)$ converges pointwise to (the continuous function) $f(x) = 0$ on R, but the convergence is not uniform on R. Use a graphing program to graph some of these functions, and explain why you would expect this to be the case.

4. Show that the sequence of functions defined by $f_n(x) = \dfrac{\sin(x^2 + x + 1)}{\sqrt{n + 5}}$ where $n = 1, 2, 3, ...$ converges uniformly to $f(x) = 0$ on R

5. Show that if $\{f_n(x)\}$ and $\{g_n(x)\}$ converge uniformly to $f(x)$ and $g(x)$ respectively on a set E, then $\{f_n(x) + g_n(x)\}$ converges uniformly to $f+g$ on E.

6. Show that if $\{f_n(x)\}$ and $\{g_n(x)\}$ are sequences of bounded real valued functions which converge uniformly to $f(x)$ and $g(x)$ on a set E, then the sequence $\{f_n(x)g_n(x)\}$ converges uniformly to $f(x)g(x)$ on E if both $|f_n(x)|$ and $|g_n(x)|$ are less than some fixed positive constant K for all x.

7. Show that if a sequence, $\{f_n\}$, of bounded real valued functions converges uniformly to a function f on a set E, then f is bounded on E. In fact, $\{f_n(x)\}$ is uniformly bounded on E.

8. Prove Theorem 5.13 if all the functions go from a set X to a metric space (Y, d).

9. Let $f(x) = \dfrac{1}{x^3}$. Let $f_n(x) = \begin{cases} f(x) & \text{if } |f(x)| \le n \\ n & \text{if } f(x) > n \\ -n & \text{if } f(x) < -n \end{cases}$. Show that $f_n(x)$ converges to $f(x)$ pointwise on R. Notice also each of the f_n is bounded, but the limit function is not.

10. Suppose that $f_n(x) = g_n(x) = x + \dfrac{1}{n}$. Show that $\{f_n(x)\}$ and $\{g_n(x)\}$ converge uniformly to $f(x) = x$ and $g(x) = x$ on R^+ respectively, but $\{f_n(x)g_n(x)\}$ does not converge uniformly to $f(x)g(x) = x^2$ on R^+. Show the convergence is uniform on any bounded set.

11. Try to make up a sequence of functions on $[0, 1]$ that converges pointwise to a continuous function f defined on $[0, 1]$, but where each function has a point of discontinuity on $[0, 1]$.

12. To what function, $f(x)$, does the sequence $\{f_n\}$ converge, when $f_n(x) = \sin^n(x)$, $x \in [0, \pi/2]$? Is the convergence uniform on this interval? How do you know? Answer the same questions for $f_n(x) = \sin^n(x/2)$.

13. Show that the sequence $\{f_n\}$ defined by $f_n(x) = \dfrac{x^n}{1 + nx^n}$ converges uniformly on $[0, \infty)$. Is the same true for $f_n(x) = \dfrac{x^n}{1 + 2x^n}$ on $[0, \infty)$?

14. Using Example 5.5, show that the sequence $x_n = 1 - \dfrac{1}{n}$ converges to 1 but that $\{f_n(x_n)\}$ does not converge to $f(1)$.

15. Suppose that f_n is a sequence of continuous functions from a metric space (M_1, d_1) to a metric space (M_2, d_2) and that f_n converges uniformly to f on M_1. Show that for any sequence $\{x_n\}$ which converges to x, it follows that $f_n(x_n) \to f(x)$.

16. Verify that the sets $K_1, K_2, K_3,...$ in Dini's theorem are nested.

17. Using the version of Dini's theorem we gave, show that Dini's theorem is true when the sequence of continuous functions is increasing to f where f is continuous.

Chapter 6

Countability (An Informal Approach)

Recall that a sequence in a set X is a function from $Z^+ \to X$. The range of the sequence is the set of values the function takes on, or said more simply, the values that occur in the sequence. So, for instance, the range of the sequence $\{(-1)^n\}$ is just the set $\{-1, 1\}$.

Definition 6.1 *A set, T, is called **countable**, if it is the range of a sequence.*

Example 6.2 *Any finite set $\{x_1, x_2, x_3, ..., x_n\}$ is countable, since it is the range of the sequence $x_1, x_2, x_3, ..., x_n, x_1, x_2, x_3, ..., x_n, ...$, where we keep repeating the same n elements over and over.*

Example 6.3 *The set of natural numbers is countable since it is the range of the sequence $1, 2, 3, ...$ Similarly, the set of even integers is countable since it is the range of the sequence $f(n) = 2n$, defined on the natural numbers. Also, the set of multiples of 3 is countable since it is the range of the sequence $f(n) = 3n$.*

Theorem 6.4 *Any subset S, of a countable set T, is countable.*

Proof. Since T is countable, it is the range of some sequence,

$$t_1, t_2, t_3, ... \tag{6.1}$$

If $S = T$, then of course, S is countable. If $S \neq T$, then T has elements S doesn't have. We simply eliminate those elements in (6.1) that S doesn't have, and we are left with a subsequence of (6.1) (which of course is a sequence) whose range is S. So, S is countable. ∎

Theorem 6.5 *If $\{O_n\}$, where $n = 1, 2, 3, ...$, is a collection of sets each of which is countable, then $\cup O_n$ is countable. (So, a countable union of countable sets is countable.)*

Proof. Since each O_n is countable, we can list, for each n, the sequence that O_n is the range of. Using the notation $O_1 : a_{11}, a_{12}, a_{13}, ...$ to mean that O_1 is the range of the sequence $a_{11}, a_{12}, a_{13},...$, and similarly for the other O_n, we have

$$O_1 : a_{11}, a_{12}, a_{13}, ...$$
$$O_2 : a_{21}, a_{22}, a_{23}, ...$$
$$O_3 : a_{31}, a_{32}, a_{33}, ...$$
$$....$$

We can sequence the union of the O_n as follows: $a_{11}, a_{12}, a_{21}, a_{13}, a_{22}, a_{31}, ...$, where we first list all the a_{ij} such that $i + j = 2$, then where $i + j = 3$, then where $i + j = 4$, and so on. We are essentially running down the diagonals of the above matrix. We account for each element of the union in this way, and we have succeeded in listing them. So this is a sequence. ∎

Theorem 6.6 *The set of rational numbers is countable.*

Proof. Let O_1 denote the set of rational numbers with denominator 1, O_2, the set of rationals with denominator 2, etc. So,

$$O_1 : \frac{1}{1}, \frac{2}{1}, \frac{3}{1}, ...$$
$$O_2 : \frac{1}{2}, \frac{2}{2}, \frac{3}{2}, ...$$
$$O_3 : \frac{1}{3}, \frac{2}{3}, \frac{3}{3}, ...$$
$$...$$

Then the sets O_1, O_2, O_3 etc., are all individually countable since each set can be listed. Therefore, by the previous theorem, $\cup O_n$ is countable. But $\cup O_n$ is clearly the set of rational numbers. So, the set of rational numbers is countable. ∎

A natural question to ask is, "Are there uncountable sets?" That is, are there sets which cannot be sequenced? The surprising answer is yes, and there are many.

Theorem 6.7 *The set of real numbers in $[0, 1]$ is not countable.*

Proof. Every real number x between 0 and 1 can be written as a decimal, $x = .a_1 a_2 a_3...$, where the a_i are single digits between 0 and 9. Furthermore, every such decimal represents a real number between 0 and 1. To prove the theorem, we do a proof by contradiction: Suppose that the set of real numbers between 0 and 1 was the range of a sequence $\{x_1, x_2, x_3, ...\}$. Then we could write each x_i as a decimal and set up the following table:

$$x_1 \;:\; .a_{11} a_{12} a_{13}...$$
$$x_2 \;:\; .a_{21} a_{22} a_{23}...$$
$$x_3 \;:\; .a_{31} a_{32} a_{33}...$$
$$...$$

Now, since this represents *all* the real numbers, one way to get a contradiction would be to make up a real number x that is not in this listing. Doing that is easy. Form the number x as follows: The first digit (after the decimal point) of x is 1 if $a_{11} \neq 1$, and 2 if $a_{11} = 1$. Since the first digit of x is different from the first digit of x_1, $x \neq x_1$. The second digit of x will be 1 if $a_{22} \neq 1$, and 2 if $a_{22} = 1$. Again, since x differs in the second digit from x_2, $x \neq x_2$. We continue in this manner, defining for each $j = 1, 2, 3, ...$, the jth digit in the decimal of x to be 1 if $a_{jj} \neq 1$, and 2 if $a_{jj} = 1$. By our construction, $x \neq x_1, x \neq x_2$, etc. Since x is a decimal, it represents a number in $[0, 1]$. But by our construction of x, it is not equal to any of the x_i, and these represent *all* of $[0, 1]$. We have our contradiction. Therefore, the set of real numbers in $[0, 1]$ cannot be countable. (For the astute reader who thinks, "But some numbers can be represented in more than one way, so this number x you constructed might really be one of the x_i," we mention that this can only happen when a number ends in all zeros or nines,(for example, .2399999..., or .2400000..., which are the same) and the number we created consists of only 1s and 2s. So, this cannot happen here.) ∎

Theorem 6.8 *The set of irrational numbers in $[0, 1]$ is not countable.*

Proof. Suppose that the set A, of irrational numbers in $[0, 1]$, is countable. Since the set B, of all rational numbers in $[0, 1]$, is countable by Theorem 6.6, $[0, 1] = A \cup B$ would be countable by Theorem 6.5. But we just proved $[0, 1]$ is not countable. So, our assumption that the set of irrational numbers in $[0, 1]$ is countable led to a contradiction, which means, the set of irrational numbers in $[0, 1]$ is uncountable. ∎

EXERCISES

1. Show the Z is countable. Then show $Z \times Z$ is countable. Finally, show $Z \times Z \times Z \times ..... \times Z$ (n factors of Z) is countable.

2. Show that if $A \subset B$ and A is not countable, then B is not countable. So in particular, R is not countable since it contains the uncountable set $[0,1]$.

3. Every finite subset of Z^+ is a subset of $S_n = \{1, 2, 3, ...n\}$ for some n. Use this to show that the collection of finite subset of Z^+ is countable.

4. Prove that the set of polynomials in x with integer coefficients is countable.

5. A real number is called algebraic if it is root of a polynomial with integer coefficients. Using the fact that a polynomial of degree n has at most n real roots, and using the result of the previous exercise, show that the set of algebraic numbers is countable.

6. A number which is not algebraic is called transcendental. Prove, using the previous exercise, that transcendental numbers exist. (It is interesting to note that producing the first transcendental number took a long time. Even though there are an uncountable number of transcendental numbers (Why?), the first one produced was produced by the mathematician Joseph Liouville in 1851. It was $\sum_{k=1}^{\infty} 10^{-k!}$.)

7. Any disjoint collection of open intervals in R is countable.

8. Prove that every open interval, I, on the real line is the countable disjoint union of open intervals. [Hint: For each $x \in I$, let I_x the "largest" open subinterval of I containing x. Show these are disjoint.]

9. Show that any monotone function, $y = f(x)$, has a countable number of discontinuities, by observing that the only kinds of discontinuities are jumps in the graph, which provides us with intervals on the y-axis where no function values occur and that these intervals are disjoint.

10. Give an example of an uncountable collection of disjoint closed sets.

11. Show that it is impossible to list the rational numbers in increasing order. That is we cannot find a listing of the rationals such that $r_1 < r_2 < r_3....$

12. Show that the number of sequences consisting of $0s$ and $1s$ is uncountable. [Hint: Consider binary representation of numbers in $[0, 1]$].

13. Using the Nested Interval Property we can prove $I = [0, 1]$ is uncountable. We begin by assuming I is countable, and hence can be sequenced as $I = \{x_1 x_2, x_3, ...\}$. We pick a closed interval interval $I_1 \subset I$ not containing x_1. Then we pick an interval $I_2 \subset I_1$ not containing x_2, and so on. Finish the proof.

6.1. SETS OF MEASURE 0

14. Prove that $[a, b]$ is uncountable for all a and b where $a < b$. [Hint: The function $f(x) = \dfrac{x-a}{b-a}$ takes $[a, b]$ onto $[0, 1]$]

6.1 Sets of Measure 0

Sets of measure zero play a big role in the study of the Riemann integral, and in the theory of the Lebesgue integral.

Definition 6.9 *The **measure** of a bounded interval is defined to be its length.*

So, if $m(E)$ represents the measure of an interval E, the above is telling us that $m([a,b]) = m([a,b)) = m((a,b]) = m((a,b)) = b - a$.

Definition 6.10 *A subset S of the set of real numbers is said to have **measure** 0 if for any $\varepsilon > 0$ we can find a countable collection of open intervals, $\{O_n\}$, which covers S and such that*
$$\sum_n m(O_n) < \varepsilon.$$

Recall that S is covered by $\{O_n\}$ when $S \subset \cup O_n$.

Example 6.11 *Any finite set of real numbers, $S = \{x_1, x_2, x_3, ..., x_n\}$, has measure 0. To see this, suppose we have fifty points, and we want to cover the set of points by a collection of open intervals with measure less than 1. We need only enclose each of the fifty points by an open interval with length less than $1/50$. More generally, if we want the sum of the measures of the intervals to be less than ε, enclose each x_i, for $i = 1, 2, 3, ..., n$, by an open interval O_i, with center x_i and length $< \varepsilon/n$. Then we clearly have covered S by $\{O_i\}$, and the sum of the lengths of the intervals is $< n(\varepsilon/n) = \varepsilon$.*

Example 6.12 *Using the fact that the null set is a subset of any set, it is easy to show that the null set has measure 0. To see this, choose any $\varepsilon > 0$ and let $O_1 = (0, 1/n)$, where n is so large that $1/n < \varepsilon$. Since $\varnothing \subset O_1$ and $m(O_1)$ is less than ε, the null set has measure 0 by the definition of measure 0.*

For the next theorem on measure, which is a critical one, we will need a fact about geometric series. Recall that a geometric series is a series of the form $a + ar + ar^2 + ar^3 + ...$, where each term is the previous term multiplied by some fixed number r. The following is a well known fact about geometric series.

Theorem 6.13 *A geometric series converges (that is, has a finite sum) when $|r| < 1$. And its sum, in this case, is $\dfrac{a}{1-r}$.*

Example 6.14 *The series $1 + \left(-\dfrac{1}{3}\right) + \dfrac{1}{9} + \left(-\dfrac{1}{27}\right) + ...,$ is a geometric series with first term $a = 1$, and $r = -1/3$. Since $|-1/3| < 1$, this series converges, and its sum is $\dfrac{1}{1-(-1/3)} = \dfrac{3}{4}$.*

Example 6.15 *The series $\varepsilon/2 + \varepsilon/4 + \varepsilon/8 + ...,$ is a geometric series with the first term $a = \varepsilon/2$, and $r = 1/2$. Since $|r| < 1$, this series converges, and its sum is $\dfrac{\varepsilon/2}{1-1/2} = \varepsilon$.*

Theorem 6.16 *Every countable set of real numbers has measure 0.*

Proof. Call the countable set $S = \{x_1, x_2, x_3, ...\}$. Pick any $\varepsilon > 0$. Enclose x_1 by an open interval, O_1, with length $< \varepsilon/2$. Enclose x_2 with an open interval, O_2, having length less than $\varepsilon/4$, etc., enclosing each x_i in an open interval with length less that $\varepsilon/2^i$. Now, since each $x_i \in S$ is in O_i, $\{O_i\}$ covers S, and by the previous example, the sum of the lengths of the intervals in $\{O_i\}$ is less than ε. So this set has measure 0. ∎

Corollary 6.17 *The set of rational numbers has measure 0.*

Proof. The set of rational numbers is countable, by Theorem 6.6. ∎

Remark 6.18 *There are uncountable sets with measure 0. The Cantor Set is one of them. We refer the reader to the Internet to learn more about this set.*

Theorem 6.19 *If B is a set of real numbers with measure 0, and if $A \subset B$, then A also has measure 0.*

Proof. Since B has measure 0, for any $\varepsilon > 0$ there is a countable set of open intervals covering B, the sum of whose lengths is less than ε. But since A is a subset of B, A is covered by this same set of open intervals. So, A has measure 0. ∎

Corollary 6.20 *Any subset of the set of rational numbers has measure 0.*

Proof. Since the set of rational numbers has measure 0 by Corollary 6.17, any subset of the set of rational numbers has measure 0 by this theorem. ∎

6.1. SETS OF MEASURE 0

Theorem 6.21 *If $E_1, E_2, E_3, ...,$ is a sequence of sets of real numbers all of which have measure 0, then $\cup E_n$ has measure 0.*

Proof. Pick any $\varepsilon > 0$. Since E_1 has measure 0, there is a countable set of open intervals, $O_1^1, O_1^2, O_1^3...$ covering E_1, such that the sum of the lengths of these intervals is less than $\varepsilon/2$. Similarly, since E_2 has measure 0, there is a countable set of open intervals, $O_2^1, O_2^2, O_2^3...$ covering E_2, such that the sum of the lengths of these intervals is less than $\varepsilon/4$. We continue, covering E_i with a countable set of open intervals, $O_i^1, O_i^2, O_i^3...$, the sum of whose lengths is less than $\varepsilon/2^i$. Now, $\bigcup_1^\infty E_n \subset \bigcup_{i,j \in Z^+} O_i^j$ and the sum of the lengths of the O_i^j is less than $\varepsilon/2 + \varepsilon/4 + \varepsilon/8 + ... = \varepsilon$ (by Example 6.15). So, $\bigcup_1^\infty E_n$ has measure 0. ∎

EXERCISES

1. Show that if A and B have measure 0, then so does $(A - B) \cup (B - A)$.

2. Give an example of a set of measure 0 whose closure doesn't have measure 0.

Chapter 7

The Riemann Integral

7.1 Preliminaries

There are four different approaches to the Riemann integral that one sees in the more popular books on analysis. All are equivalent, in the sense that they all lead to the same value of the integral, and to the same theorems. All of the approaches use one or more of the following definitions.

Definition 7.1 *Suppose that $[a, b]$ is any closed interval. By a **partition** P of $[a, b]$ we mean a set of points $P = \{x_0, x_1, x_2, ..., x_n\}$, where*

$$x_0 = a < x_1 < x_2 < ... < x_n = b.$$

In Figure 7.1 we see a typical partition of $[a, b]$.

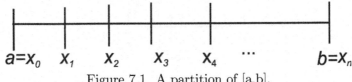

Figure 7.1. A partition of [a,b].

Example 7.2 *If $[a, b] = [1, 4]$, the following set of points $\{1, 1.5, 3, 4\}$ constitutes a partition of $[1, 4]$. So $x_0 = 1$, $x_1 = 1.5$, $x_2 = 3$ and $x_3 = 4$.*

A partition divides an interval into subintervals. We will need to use their lengths in our development.

Definition 7.3 *If $\{x_0, x_1, x_2, ..., x_n\}$ is a partition of $[a, b]$, we define $\Delta x_1 = x_1 - x_0, \Delta x_2 = x_2 - x_1, \Delta x_3 = x_3 - x_2$, and so on. Clearly, $\Delta x_i = x_i - x_{i-1}$ represents the length of the ith subinterval.*

Definition 7.4 If $P = \{x_0, x_1, x_2, ..., x_n\}$ is a partition of $[a, b]$, the length of the longest subinterval in the partition is denoted by $\|P\|$. That is, $\|P\| = \max \Delta x_i$ for $i = 1, 2, 3, ..., n$. $\|P\|$ is called the **norm** of the partition. We observe that if $\|P\| < \varepsilon$, then the length of each subinterval formed by the partition is $< \varepsilon$.

Example 7.5 In Example 7.2, $\Delta x_1 = x_1 - x_0 = 1.5 - 1, or\ .5$, $\Delta x_2 = x_2 - x_1 = 3 - 1.5$, or 1.5, and $\Delta x_3 = x_3 - x_2 = 4 - 3$, or 1. $\|P\| = \max\{.5, 1.5, 1\} = 1.5$.

Definition 7.6 We say that a partition P^* of $[a, b]$ is finer than a partition P of $[a, b]$, or, is a **refinement** of P, if $P \subset P^*$. That is, if P^* contains the points of P and possibly more.

Example 7.7 In Example 7.2, if we let $P^* = \{1, 1.5, 2, 2.5, 3, 3.7, 4\}$, P^* is a refinement of P.

Definition 7.8 If P_1 and P_2 are two partitions of $[a, b]$, we call $P_1 \cup P_2$ the **common refinement** of P_1 and P_2.

Example 7.9 $P_1 = \{1, 3, 4\}$ and $P_2 = \{1, 2, 3.5, 4\}$, are partitions of $[1, 4]$. The common refinement of these is $P_1 \cup P_2 = \{1, 2, 3, 3.5, 4\}$. See Figure 7.2.

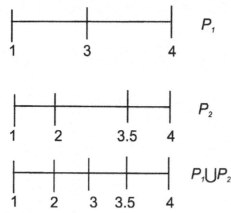

Figure 7.2. Two partitions and their common refinement.

Observe that $P_1 \cup P_2$ refines both P_1 and P_2.

Definition 7.10 A real valued function, $f(x)$, defined on a set S, is called **bounded** if its range is a bounded set.

7.1. PRELIMINARIES

Definition 7.11 *Suppose that f is a bounded real function defined on $[a,b]$ (where R carries the usual metric), and suppose that $P = \{x_0, x_1, x_2, ..., x_n\}$ is any partition of $[a,b]$. For $i = 1, 2, 3, ..., n$, define $m_i = \inf f(x)$ on $[x_{i-1}, x_i]$, and $M_i = \sup f(x)$ on $[x_{i-1}, x_i]$. Define $U(P, f)$, and $L(P, f)$ as follows: $U(P, f) = \sum_{1}^{n} M_i \Delta x_i$ and $L(P, f) = \sum_{1}^{n} m_i \Delta x_i$. $U(P, f)$ and $L(P, f)$ are called **the upper and lower sums**, respectively, for P.*

In Figures 7.3 and 7.4 we illustrate an upper and lower sum for a nonnegative increasing function and a given partition. (It is the shaded area). Since the function is increasing, $\sup f(x)$ on $[x_{i-1}, x_i]$ occurs at the right endpoint, x_i, and $\inf f(x)$ on $[x_{i-1}, x_i]$ occurs at the left endpoint, x_{i-1}. So $M_i = f(x_i)$, while $m_i = f(x_{i-1})$, for each $i = 1, 2, 3, ..., n$. Each term $M_i \Delta x_i = f(x_i) \Delta x_i$ represents the area of a rectangle with height $f(x_i)$ and base Δx_i, while each term $m_i \Delta x_i = f(x_{i-1}) \Delta x_i$ represents the area of a rectangle with height $f(x_{i-1})$ and base Δx_i. Each upper sum provides us with a sum of areas of rectangles which is greater than the area under the curve $y = f(x)$ from $x = a$ to $x = b$, while each lower sum provides us with a sum of areas of rectangles which is less than the area under the curve $y = f(x)$ from $x = a$ to $x = b$, when f is nonnegative. Each partition determines a unique upper sum and a unique lower sum. Of course, there are infinitely many partitions of $[a,b]$, and hence infinitely many upper and lower sums.

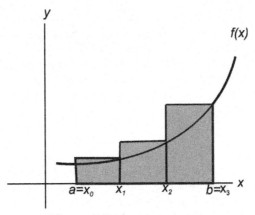

Figure 7.3. An upper sum.

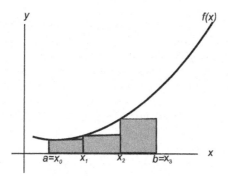

Figure 7.4. A lower sum.

In Figure 7.5, we show a blow up of the interval $[x_0, x_1]$. The large rectangle enclosing all the others is the one that occurs in Figure 7.3. The sum of the areas of the shaded rectangles represents the contribution to the upper sum obtained by adding new division points to $[x_0, x_1]$ (that is, using a finer partition). The picture indicates that the upper sum seems to decrease (or at least, doesn't increase), and is closer to the area under the curve than the original upper sum was. We will soon prove that the upper sums always decrease or remain the same as we refine the partition.

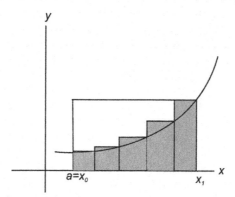

Figure 7.5. Rectangles corresponding to an upper sum.

A similar picture for lower sums seems to indicate that the lower sums (for a continuous nonnegative function) increase (to a value that appears to be the area under the curve), as we refine the partition. In fact, the lower sums always increase or remain the same as we refine the partition, and this is true for any function, continuous or not, increasing or not, as we soon shall see.

7.1. PRELIMINARIES

Definition 7.12 *Let f be a bounded real valued function on $[a,b]$. If P is a partition of $[a,b]$ and t_i is any point in $[x_{i-1}, x_i]$, then the sum $\sum_{i=1}^{n} f(t_i)\Delta x_i$ is called a **Riemann sum** for $f(x)$ with partition P. We denote any such sum by $S(P, f)$.*

Example 7.13 *Suppose that we have the function $f(x) = x^2$ and that we partition $[1,4]$ as follows: $P = \{1, 2.5, 3.7, 4\}$. We use as the t_i the midpoints of each subinterval. So $t_1 = 1.75, t_2 = 3.1, $ and $t_3 = 3.85$. The Riemann sum corresponding to this partition and these values of t_i, is: $S(P, f) = f(t_1)\Delta x_1 + f(t_2)\Delta x_2 + f(t_3)\Delta x_3 = f(1.75)(1.5) + f(3.1)(1.2) + f(3.85)(.3) = (1.75)^2(1.5) + (3.1)^2(1.2) + (3.85)^2(.3) = 20.573$.*

When a function is above the x-axis as in Figure 7.5, each term, $f(t_i)\Delta x_i$ in a Riemann sum represents an area of a rectangle with height $f(t_i)$ and base Δx_i. When the function is partially above and partially below the x-axis on $[a,b]$ as in Figure 7.6, each positive term, $f(t_i)\Delta x_i$ in the Riemann sum represents an area of a rectangle above the x-axis with height $f(t_i)$ and base Δx_i while each each negative term in the Riemann sum represents the negative of an area of a rectangle which is below the x-axis. The negative terms and positive terms combine to give you a number which could be positive, negative or 0.

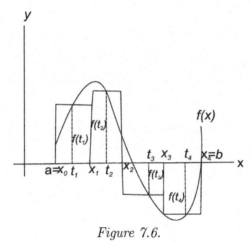

Figure 7.6.

Remark 7.14 *Since the supremum of a set always has elements of the set close to it, M_i can be approximated by $f(t_i)$ for some $t_i \in [x_{i-1}, x_i]$, from which it follows that any upper sum can be well approximated by a Riemann sum. Similarly, any lower sum can be well approximated by a Riemann sum.*

The next two theorems will be needed in the next section. The first theorem tells us that for any partition P of $[a, b]$ and any choice of points $t_i \in [x_{i-1}, x_i]$, the numerical value of the Riemann sum is between the lower sum and upper sum corresponding to that partition. The second theorem tells us that as we refine a partition the upper sums decrease or stay the same, and the lower sums increase or stay the same.

Theorem 7.15 *Using the notation in Definitions 7.11 and 7.12, we have for any bounded function f, and any fixed partition P that*

$$L(P, f) \leq S(P, f) \leq U(P, f).$$

Proof. Since $m_i = \inf f(x)$ on $[x_{i-1}, x_i]$ and $M_i = \sup f(x)$ on $[x_{i-1}, x_i]$, we have for any t_i in $[x_{i-1}, x_i]$, that

$$m_i \leq f(t_i) \leq M_i, \text{ for } i = 1, 2, 3, ..., n. \tag{7.1}$$

Multiplying each inequality in (7.1) by Δx_i, which is nonnegative, we get that

$$m_i \Delta x_i \leq f(t_i) \Delta x_i \leq M_i \Delta x_i, \text{ for } i = 1, 2, 3, ..., n.$$

Summing these inequalities we get that

$$\sum_1^n m_i \Delta x_i \leq \sum_1^n f(t_i) \Delta x_i \leq \sum_1^n M_i \Delta x_i,$$

or just that

$$L(P, f) \leq S(P, f) \leq U(P, f).$$

∎

Theorem 7.16 *Let f be a bounded real valued function on $[a,b]$. If $P = \{x_0, x_1, x_2, ..., x_n\}$ is any partition of $[a, b]$ and we refine P by adding a point $p \in [a, b]$ to this partition to get a new partition, P^*, then $U(P^*, f) \leq U(P, f)$ and $L(P^*, f) \geq L(P, f)$.*

Proof. We will prove the result for the upper sums, and leave the proof of the lower sums to you. Also, we will assume that p comes from the particular interval $[x_1, x_2]$, and leave the more general proof to you since it is virtually identical regardless of which interval p is contained in. Let $M_2^* = \sup f(x)$ on $[x_1, p]$ and let $M_2^{**} = \sup f(x)$ on $[p, x_2]$. We know that

$$M_2^* \leq M_2 \text{ and } M_2^{**} \leq M_2$$

7.1. PRELIMINARIES

(since we are taking the supremum over a subset), and so

$$\begin{aligned} M_2^*(p - x_1) &\leq M_2(p - x_1), \text{ and} \\ M_2^{**}(x_2 - p) &\leq M_2(x_2 - p). \end{aligned}$$

Adding these inequalities, we get that

$$\boxed{M_2^*(p - x_1) + M_2^{**}(x_2 - p)} \leq M_2(p - x_1) + M_2(x_2 - p). \tag{7.2}$$

Now,

$$\begin{aligned} U(P, f) &= \sum_1^n M_i \Delta x_i \\ &= M_1(x_1 - x_0) + \boxed{M_2(x_2 - x_1)} + M_3(x_3 - x_2) + ... + M_n(x_n - x_{n-1}) \\ &= M_1(x_1 - x_0) + \boxed{M_2(p - x_1) + M_2(x_2 - p)} + M_3(x_3 - x_2) + ... + M_n(x_n - x_{n-1}) \\ &\geq M_1(x_1 - x_0) + \boxed{M_2^*(p - x_1) + M_2^{**}(x_2 - p)} + M_3(x_3 - x_2) + ... + M_n(x_n - x_{n-1}) \\ &\qquad\qquad\qquad\qquad\qquad\qquad\qquad\qquad\qquad\qquad \text{(by (7.2))} \\ &= U(P^*, f). \end{aligned}$$

From this it follows that $U(P, f) \geq U(P^*, f)$, or equivalently, that $U(P^*, f) \leq U(P, f)$. ∎

Corollary 7.17 *If P is a partition of $[a, b]$ and P^* is a refinement of P, then $U(P^*, f) \leq U(P, f)$ and $L(P^*, f) \geq L(P, f)$.*

Proof. We simply add one point at a time to go from P to P^*. Each time we add a point, the upper sums decrease or remain the same. So when we are done, $U(P^*, f) \leq U(P, f)$. Similarly, since each time you add a point to P the lower sums increase or remain the same, $L(P^*, f) \geq L(P, f)$. ∎

The next theorem tells us that any lower sum is less or equal to any upper sum.

Theorem 7.18 *Let f be a bounded real valued function defined on $[a, b]$. If P_1 and P_2 are any partitions of $[a, b]$, then $L(P_1, f) \leq U(P_2, f)$.*

Proof. Consider the common refinement $P_1 \cup P_2$, of P_1 and P_2. Since $P_1 \cup P_2$ refines both P_1 and P_2, and since the upper sums increase (or stay the same) while the lower sums decrease (or stay the same) as we refine, we have, $L(P_1, f) \leq L(P_1 \cup P_2, f) \leq U(P_1 \cup P_2, f) \leq U(P_2, f)$. So, $L(P_1, f) \leq U(P_2, f)$. ∎

CHAPTER 7. THE RIEMANN INTEGRAL

Corollary 7.19 *If we let $\{L(P,f)\}$ and $\{U(P,f)\}$ be the sets of lower and upper sums, respectively, for the function f as P runs through all partitions of $[a,b]$, we have that $\sup\{L(P,f)\} \leq \inf\{U(P,f)\}$.*

Proof. It follows from the theorem that any lower sum is bounded above by any upper sum. So we have, for any fixed partition P_0, that $L(P_0,f) \leq U(P,f)$ for any partition P.*of* $[a.b]$ Thus, $L(P_0,f)$ is a lower bound for $\{U(P,f)\}$ and hence, it is less than or equal to the greatest lower bound of $\{U(P,f)\}$. That is,

$$L(P_0,f) \leq \inf\{U(P,f)\}. \tag{7.3}$$

Since P_0 was arbitrary, (7.3) tells us that the fixed number $\inf\{U(P,f)\}$ is an upper bound for the set of all lower sums, so it is greater than or equal to the least upper bound of the set of lower sums. That is,

$$\sup\{L(P,f)\} \leq \inf\{U(P,f)\}.$$

∎

Definition 7.20 *If f is a bounded real valued function defined on $[a,b]$, we define the **upper integral of** $f(x)$ on $[a,b]$ to be $\inf\{U(P,f)\}$, and the **lower integral of** $f(x)$ on $[a,b]$ to be $\sup\{L(P,f)\}$, where P runs through all partitions of $[a,b]$.*

Notation 7.21 *We denote the upper integral of $f(x)$ on $[a,b]$ by $\overline{\int_a^b} f(x)dx$, and the lower integral by $\underline{\int_a^b} f(x)dx$.*

Remark 7.22 *The previous corollary says that $\underline{\int_a^b} f(x)dx \leq \overline{\int_a^b} f(x)dx$. It is particularly important when they are equal.*

The next theorem is a technical lemma which we will need in the next section.

Theorem 7.23 *If $P = \{x_0, x_1, x_2, ..., x_n\}$ is any partition of $[a,b]$, then for each $i = 1, 2, 3, ..., n$, $M_i - m_i = \sup\{f(x) - f(y)\}$ where x, y come from $[x_{i-1}, x_i]$.*

Proof. Recall from Theorems 2.26 and 2.29 that for any sets A and B, $\sup(A+B) = \sup A + \sup B$, and that $-\inf(B) = \sup(-B)$. Here, we will let $A = \{f(x)\}$ and $B = \{-f(y)\}$, where x and y come from $[x_{i-1}, x_i]$. We have

$$\begin{aligned} M_i - m_i &= \sup\{f(x)\} - \inf\{f(y)\} \\ &= \sup A + (-\inf\{f(y)\}) \\ &= \sup A + \sup\{-f(y)\} \\ &= \sup A + \sup B \\ &= \sup(A+B) \\ &= \sup\{f(x) - f(y)\}. \end{aligned}$$

this last step following from the definition of $A+B$. ∎

EXERCISES

1. Suppose that $f(x) = x^2$ and that P is the partition of $[1, 2]$ obtained by dividing it into 4 equal subintervals. What is the upper sum for this partition? What is the lower sum for this partition? What is the Riemann sum obtained by using as the t_i, the midpoints of the intervals?

2. Suppose that $f(x) = x$ defined only on $[0, 1]$. Suppose that we partition $[0, 1]$ into n equal subintervals. Show that $U(P, f) = \dfrac{n+1}{2n}$. Then use this result to find $\overline{\int_0^1 x dx}$. Explain why $\overline{\int_0^1 x dx} = \lim \dfrac{n+1}{2n}$. [Hint: The formula $\sum\limits_{i=1}^{n} i = \dfrac{n(n+1)}{2}$ will help.]

3. Show that if $c \geq 0$, $\overline{\int_a^b cf(x)dx} = c\overline{\int_a^b f(x)dx}$.

4. Show that for any bounded functions $f(x)$ and $g(x)$ defined on $[a, b]$,
$\overline{\int_a^b f(x) + g(x)dx} \leq \overline{\int_a^b f(x)dx} + \overline{\int_a^b g(x)dx}$.

5. Show that for any bounded function $f(x)$ defined on $[a, b]$,
$\overline{\int_a^b f(x)dx} = \overline{\int_a^c f(x)dx} + \overline{\int_c^b f(x)dx}$ where c is strictly between a and b.

7.2 Intuitive Review of the Riemann Integral

If you have taken a calculus course, you have probably seen the development of the Riemann integral to some extent. One starts with a bounded function $f(x)$ defined on $[a, b]$, and then does the following: (1) Partition the interval $[a, b]$ using $P = \{x_0, x_1, x_2, ..., x_n\}$. (2) Pick a point t_i in $[x_{i-1}, x_i]$ for each $i = 1, 2, 3, ..., n$, and let T be this set of points. (3) Form the Riemann sum $S(P, f) = \sum\limits_{i=1}^{n} f(t_i)\Delta x_i$. Take the limit of $S(P, f)$ as the norm of the partition goes to zero. If this limit exists, and is independent of the points t_i chosen in $[x_{i-1}, x_i]$, then we call this limit the definite integral of f over $[a, b]$ and denote this limit by $\int_a^b f(x)dx$.

Remark 7.24 *We have never defined a limit of this type. So what does this mean? We need to be more formal here, and we soon will be.*

Intuitively, when $f(x)$ is nonnegative and continuous on $[a, b]$, then as the norm of the partition gets smaller and smaller, the Riemann sums $S(P, f)$, approach the area above the x-axis and under the curve $f(x)$ on the interval $[a, b]$. You can refer to Figure 7.5 and refine the partition and draw more rectangles to see this. This area interpretation of $\int_a^b f(x)dx$, is what most people remember.

When f is partially above the x-axis and partially below the x-axis, $\int_a^b f(x)dx$ gives a difference of areas. More specifically, it takes all of the area below the curve which is above the x-axis and subtracts all the area which is above the curve but below the x-axis. See Figure 7.7. In that figure $\int_a^b f(x)dx = A_1 + A_2 - A_3$. Each of A_1, A_2 and A_3, are nonnegative numbers representing the areas of the shaded regions. While it is nice to have an intuitive view of the integral, the analyst is concerned with a different question: When does the integral exist (irrespective of any geometric interpretation)? We will soon answer this. First, we give the formal definition of $\int_a^b f(x)dx$. (We will work with bounded real valued functions where R carries the usual metric).

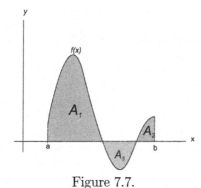

Figure 7.7.

Definition 7.25 *Suppose that f is a bounded real valued function defined on $[a, b]$. Then we say that f is **Riemann integrable on** $[a, b]$ if there is a number A which has the following property: For any $\varepsilon > 0$ there is a $\delta > 0$, such that for every partition P of $[a, b]$ with norm less than δ, and every choice of points t_i in $[x_{i-1}, x_i]$,*

$$|S(P, f) - A| < \varepsilon.$$

*The number A is called the **definite integral** of $f(x)$ over $[a, b]$ and is denoted by $\int_a^b f(x)dx$.*

We are asking a lot here. We have to approach the same number regardless of which points we use for t_i. The first question is, does $\int_a^b f(x)dx$ ever exist? Our

7.2. INTUITIVE REVIEW OF THE RIEMANN INTEGRAL

prior geometric discussion of this concept seems to indicate that when the function is continuous on $[a, b]$, this limit will exist and have an interpretation involving areas. We state our first *plausible* theorem, saving the proof until later.

Theorem 7.26 *If $f(x)$ is continuous on $[a, b]$, then $\int_a^b f(x)dx$ exists.*

Of course, knowing that the integral exists might not satisfy everyone. Most scientists would want to know how to evaluate it, since the integral is so useful in the sciences. The following Fundamental Theorem of Calculus tells us how we do it, and will be proved later.

Theorem 7.27 *If $f(x)$ is continuous on $[a, b]$, then $\int_a^b f(x)dx = F(b) - F(a)$, where $F(x)$ is any antiderivative of $f(x)$.*

This theorem is quite remarkable since it links antiderivatives to finding differences of areas (and actual areas when $f(x) \geq 0$ on $[a, b]$). A typical calculus problem follows.

Example 7.28 *Find the area between the curve $f(x) = x^2$ and the x-axis from $x = 1$ to $x = 3$.*

Solution: Since $f(x) \geq 0$ on $[1, 3]$, the area will be $\int_a^b f(x)dx = \int_1^3 x^2 dx = \left.\frac{x^3}{3}\right|_1^3 = \frac{3^3}{3} - \frac{1^3}{3} = \frac{26}{3}$.

Definition 7.29 *For convenience, we define $\int_a^a f(x)dx$ to be 0.*

EXERCISES

1. Show that the value A in the definition of Riemann integral, is unique.

2. Show that if $f(x) = c$ on $[a, b]$ where c is a constant, then $\int_a^b f(x)dx = c(b - a)$.

3. Show that if $m \leq f(x) \leq M$ on $[a, b]$ and f is Riemann integrable on $[a, b]$, then $m(b - a) \leq \int_a^b f(x)dx \leq M(b - a)$.

4. Let $f(x) = \frac{1}{x^3}$. Let $f_n(x) = \begin{cases} f(x) & \text{if } |f(x)| < n \\ n & \text{if } f(x) > n \\ -n & \text{if } f(x) < -n \end{cases}$. Show that though $\{f_n(x)\}$ converges to $f(x)$ pointwise on $[-1, 1]$, $\int_{-1}^1 f_n(x)dx$ does not converge to $\int_{-1}^1 f(x)dx$.

5. Appealing to the geometric interpretation of integral as the difference of areas, what should the value of $\int_0^{2\pi} \sin x \, dx$ be? Verify your answer by using Theorem 7.27.

7.3 Lebesgue's Theorem

Some natural questions to ask are: (1) Are there discontinuous functions whose Riemann integrals exist? The answer is "Yes." (2) How discontinuous can a function be while still being integrable? The answer is "Very." In fact, the mathematician Lebesgue proved the following deep result.

Theorem 7.30 *(Lebesgue) Suppose that f is bounded on $[a,b]$, then f is Riemann integrable on $[a,b]$ if and only if it is continuous everywhere on $[a,b]$ except on a set of measure 0.*

We will prove this in the Appendix.

Example 7.31 *The function whose graph is shown in Figure 7.8 is Riemann integrable on $[a,b]$.*

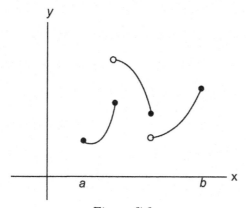

Figure 7.8.

Also, the function $f(x)$ defined by

$$f(x) = \begin{cases} 5 & \text{if } x = 0 \\ \frac{1}{2} & \text{if } x \in [1/2, 1) \\ \frac{1}{4} & \text{if } x \in [1/4, 1/2) \\ \frac{1}{8} & \text{if } x \in [1/8, 1/4) \\ & \text{etc.} \end{cases}$$

(part of whose graph is shown in Figure 7.9), has infinitely many discontinuities in $[0,1]$, and all the discontinuities occur at rational numbers. Since the set of rational numbers is countable, the set of discontinuities of this function, being a subset of the rational numbers, is countable, and since countable sets have measure 0 (Theorem 6.16), this function is Riemann integrable on $[0,1]$.

7.3. LEBESGUE'S THEOREM

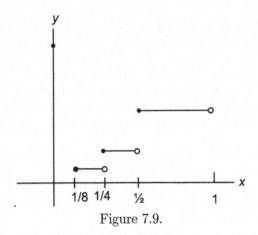

Figure 7.9.

Example 7.32 We have already seen in Example 4.25 that the function

$$f(x) = \begin{cases} 0 & \text{if } x \text{ is rational} \\ 1 & \text{if } x \text{ is irrational} \end{cases}$$

restricted to $[0,1]$, is discontinuous at every point of the interval $[0,1]$. Thus, the set of discontinuities has measure $m[0,1] = 1$. Therefore, this function is not Riemann integrable on $[0,1]$.

Example 7.33 By contrast, the function

$$f(x) = \begin{cases} 0 & \text{if } x \text{ is irrational} \\ 1/q & \text{if } x = p/q \text{ in lowest terms} \end{cases}$$

restricted to $[0,1]$, is continuous at every irrational number in $[0,1]$ (Example 4.26). Hence, the discontinuities occur at the rational numbers in the interval, which we know has measure 0 by Corollary 6.20. So this function does have a Riemann integral on $[0,1]$. A more interesting question to ask is, "What is the value of the integral?" Does this function have an antiderivative that is easy to compute? Can we use the Fundamental Theorem of Calculus? (The answer to the last two questions is "No", but proving it is not so easy.) There is a theorem that says that if a function f is the derivative of another function F, then f has the intermediate value property. That is, if c is between $f(a)$ and $f(b)$, then there is a z, where $a < z < b$ such that $f(z) = c$. Since this function doesn't take on all values between 0 and $1/2$, for example, it can't be the derivative of another function. Said another way, f has no antiderivative on $[0,1]$. A proof of this Intermediate Value Theorem for derivatives may be found in Apostol [1], page 112.

Theorem 7.30 has many useful corollaries. We give some now.

Corollary 7.34 *If f is Riemann integrable on $[a,b]$, then f is Riemann integrable on any subinterval $[c,d]$ of $[a,b]$.*

Proof. Since f is Riemann integrable on $[a,b]$, the set of discontinuities of f on $[a,b]$ has measure 0. The set of discontinuities of f on $[c,d]$ is a subset of the set of discontinuities of f on $[a,b]$. So the set of discontinuities of f on $[c,d]$ also has measure 0 by Theorem 6.19, and therefore f is Riemann integrable on $[c,d]$ by Theorem 7.30. ∎

Corollary 7.35 *If φ is a bounded continuous (real valued) function defined on the range of $f(x)$, where $f : [a,b] \to R$ and f is Riemann integrable on $[a,b]$, then $(\varphi \circ f)$ is Riemann integrable on $[a,b]$.*

Proof. If f is continuous at a point $x_0 \in [a,b]$, then because φ is continuous on the range of $f(x)$, we have that $(\varphi \circ f)$ is continuous at x_0 by Theorem 4.20. Therefore, using the contrapositive, if $(\varphi \circ f)$ is not continuous at x_0 then f is not continuous at x_0. Said another way, the set of discontinuities of $(\varphi \circ f)$ is a subset of the set of discontinuities of f. But f is Riemann integrable on $[a,b]$, so the set of discontinuities of f has measure 0. Therefore the set of discontinuities of $(\varphi \circ f)$ also has measure 0, by Theorem 6.19, and $\varphi \circ f$ is Riemann integrable on $[a,b]$ by Theorem 7.30. ∎

Corollary 7.36 *If f is Riemann integrable on $[a,b]$, so are the functions $f^2(x)$ and $|f(x)|$.*

Proof. We need only take $\varphi(x) = x^2$ in the previous corollary to prove the first result, and take $\varphi(x) = |x|$ to prove the second result. ∎

Corollary 7.37 *If f and g are Riemann integrable on $[a,b]$, then so are $f+g$, $f-g$, fg, and $c_1 f + c_2 g$, where c_1 and c_2 are constants.*

Proof. We will just give the proof for $f+g$, as the proof of the others is very similar. We know that if f and g are continuous at x_0, then so is $f+g$ by Theorem 4.20. The contrapositive of this (which we know must be true) is that if $f+g$ is not continuous at x_0, then either f or g is not continuous at x_0. Said another way, the set of discontinuities of $f+g$ is a subset of the union of the set of discontinuities of f and the set of discontinuities of g, both of which have measure 0. So the set of discontinuities of $f+g$ has measure 0 by Theorem 6.21, and $f+g$ is Riemann integrable on $[a,b]$. ∎

Corollary 7.38 *If f_n is a sequence of real valued functions converging uniformly to $f(x)$ on $[a,b]$, and if each f_n is Riemann integrable on $[a,b]$, then f is Riemann integrable on $[a,b]$.*

7.3. LEBESGUE'S THEOREM

Proof. We know by Theorem 5.13 that if each f_n is continuous at x_0, then f is continuous at x_0. By using the contrapositive, we have that if f is discontinuous at x_0, then one of the f_n must be discontinuous at x_0. Said another way, the set of discontinuities of f is a subset of the union of the set of discontinuities of the various f_n, each of which has measure zero. Their union, by Theorem 6.21, must have measure 0, and the set of discontinuities of f, being a subset of this set, must also have measure 0. So f is Riemann integrable on $[a,b]$. We will give a full independent proof of this theorem later on. ∎

Corollary 7.39 *If f is continuous on $[a,b]$, then f is Riemann integrable on $[a,b]$.*

Proof. The set of discontinuities is $\varnothing$, which has measure 0. So f is Riemann integrable on $[a,b]$. We will give a full, independent proof of this result later on in the book. ∎

EXERCISES

1. Show that the function $f(x) = \sin(1/x)$ if $x \neq 0$ and 5 if $x = 0$ is Riemann integrable on $[0,1]$.

2. Make up an example of a function which is not Riemann integrable on $[0,1]$ but whose absolute value is.

3. Give two reasons why if $f(x)$ and $g(x)$ are Riemann integrable on $[a,b]$ it does not follow that f/g is Riemann integrable on $[a,b]$.

4. Let $f(x)$ be the function defined in Example 7.33 and let $g(x) = 1$ if $0 < x \leq 1$ and $g(0) = 0$. Show that both $f(x)$ and $g(x)$ are both Riemann integrable on $[0,1]$, but $g(f(x))$ is not. Contrast this to Corollary 7.35.

5. Accepting the fact that the natural logarithm function is continuous on its domain, (since it is differentiable) and the given fact that $f(x) \geq m > 0$ on $[a,b]$, show that if $f(x)$ is Riemann integrable on $[a,b]$, then $h(x) = \ln(f(x))$ is Riemann integrable on $[a,b]$. Where are we using the fact that $f(x) \geq m > 0$?

6. Show that if $f(x)$ and $g(x)$ are Riemann integrable on $[a,b]$ and $f(x) \geq m > 0$ on $[a,b]$, then $M(x) = f(x)^{g(x)}$ is Riemann integrable on $[a,b]$.

7.4 Basic Results on Riemann Integrals

The previous few corollaries tell us that certain functions are integrable, but how do we compute the integrals? The Fundamental Theorem of Calculus, proved next (in a more general setting than is seen in calculus), answers this question for many functions. Once we have proved this, we will have a proof of Theorem 7.27. We need only two facts from calculus— the Mean Value Theorem (Theorem 4.76), and the fact that if a function has a derivative at a point, then it is continuous at the point.

Theorem 7.40 *If $F(x)$ is an antiderivative of $f(x)$ on $[a,b]$, and f is Riemann integrable on $[a,b]$, then*

$$\int_a^b f(x)dx = F(b) - F(a).$$

Proof. Since f is Riemann integrable on $[a,b]$, by definition of being Riemann integrable, for each $\varepsilon > 0$, there is a $\delta > 0$ such that if $||P|| < \delta$, then

$$\left| S(P,f) - \int_a^b f(x)dx \right| < \varepsilon, \tag{7.4}$$

regardless of the points t_1, t_2, etc., chosen from $[x_{i-1}, x_i]$. Now, take any partition $P = \{x_0, x_1, x_2, ..., x_n\}$ with $||P|| < \delta$, and let S be the number $F(b) - F(a)$. We will show that S can be written as a Riemann sum $S(P,f)$ for this partition, and since this partition has norm less than δ, $S(P,f)$ will satisfy (7.4). Now,

$$\begin{aligned}
S &= F(b) - F(a) = F(x_n) - F(x_o) \\
&= (F(x_1) - F(x_0)) + (F(x_2) - F(x_1)) + (F(x_3) - F(x_2)) + ... + (F(x_n) - F(x_{n-1})) \\
&\quad \text{(as we see by canceling like terms)} \\
&= f(t_1)(x_1 - x_0) + f(t_2)(x_2 - x_1) + f(t_3)(x_3 - x_2) + ... + f(t_n)(x_n - x_{n-1}) \\
&\quad \text{(Corollary 4.78 where each } t_i \in [x_{i-1}, x_i]) \\
&= \sum_{i=1}^n f(t_i) \Delta x_i \\
&= S(P,f).
\end{aligned}$$

Thus, $S = S(P,f)$. Substituting this into (7.4), we get $\left| S - \int_a^b f(x)dx \right| < \varepsilon$. Since this is true for each $\varepsilon > 0$, the fixed number $S - \int_a^b f(x)dx$ must be zero, or equivalently, $S = \int_a^b f(x)dx$. This finishes the proof since $S = F(b) - F(a)$. ∎

7.4. BASIC RESULTS ON RIEMANN INTEGRALS

Corollary 7.41 *If c is a constant, then $\int_a^b c\,dx = c(b-a)$.*

Proof. Since cx is an antiderivative of c, we have $\int_a^b c\,dx = cx|_a^b = c(b-a)$. ∎

The above theorem tells us how to integrate certain functions. But there are functions that do not fall under this category, for example, the function we presented in Example 7.33. To integrate some of those functions we will need to take a different approach, and we do that soon.

Here are some preliminary results concerning integrals, none of which are really surprising.

Theorem 7.42 *If f and g are Riemann integrable on $[a,b]$, then so is $c_1 f + c_2 g$, where c_1 and c_2 are constants. Furthermore, $\int_a^b (c_1 f(x) + c_2 g(x))\,dx = c_1 \int_a^b f(x)\,dx + c_2 \int_a^b g(x)\,dx$.*

Proof. (Note that if both c_1 and c_2 equal 0 the theorem holds trivially by Corollary 7.41, so we will assume otherwise.) Let $h = c_1 f + c_2 g$. Now, let P be any partition of $[a,b]$ and consider

$$\begin{aligned}
S(P,h) &= \sum_{1}^{n} h(t_i)\Delta x_i \\
&= \sum_{1}^{n} (c_1 f(t_i) + c_2 g(t_i))\Delta x_i \\
&= \sum_{1}^{n} c_1 f(t_i)\Delta x_i + \sum_{1}^{n} c_2 g(t_i)\Delta x_i \\
&= c_1 \sum_{1}^{n} f(t_i)\Delta x_i + c_2 \sum_{1}^{n} g(t_i)\Delta x_i \\
&= c_1 S(P,f) + c_2 S(P,g).
\end{aligned}$$

So, from this we have that (for any fixed set of t_i's)

$$S(P,h) = c_1 S(P,f) + c_2 S(P,g). \tag{7.5}$$

Pick any $\varepsilon > 0$. Since f is Riemann integrable on $[a,b]$, there is a δ_1 such that if $||P|| < \delta_1$, then

$$\left| S(P,f) - \int_a^b f(x)\,dx \right| < \frac{\varepsilon}{(|c_1| + |c_2|)}. \tag{7.6}$$

Similarly, since g is Riemann integrable on $[a,b]$, there is a δ_2 such that if $||P|| < \delta_2$, then

$$\left| S(P,g) - \int_a^b g(x)\,dx \right| < \frac{\varepsilon}{(|c_1| + |c_2|)}. \tag{7.7}$$

Now let δ be any number less than or equal to both δ_1 and δ_2. Then both (7.6) and (7.7) hold. Now, consider a partition P with norm less than δ, and let $A = c_1 \int_a^b f(x)dx + c_2 \int_a^b g(x)dx$. We will show that $|S(P,h) - A| < \varepsilon$, which by the definition of the Riemann integral will show that $\int_a^b h(x)dx = A$, since ε is arbitrary. Using the definitions of $h(x)$ and A, this last statement reduces to $\int_a^b (c_1 f(x) + c_2 g(x))dx = c_1 \int_a^b f(x)dx + c_2 \int_a^b g(x)dx$, which is what we want to prove. Here are the steps: Since (by 7.5) $S(P,h) = c_1 S(P,f) + c_2 S(P,g)$, we have

$$
\begin{aligned}
|S(P,h) - A| &= \\
&= \left| c_1 S(P,f) + c_2 S(P,g) - \left(c_1 \int_a^b f(x)dx + c_2 \int_a^b g(x)dx \right) \right| \\
&= \left| c_1 S(P,f) + c_2 S(P,g) - c_1 \int_a^b f(x)dx - c_2 \int_a^b g(x)dx \right| \\
&= \left| c_1 S(P,f) - c_1 \int_a^b f(x)dx + c_2(S(P,g) - c_2 \int_a^b g(x)dx \right| \\
&\leq \left| c_1 S(P,f) - c_1 \int_a^b f(x)dx \right| + \left| c_2 S(P,g) - c_2 \int_a^b g(x)dx \right| \\
&= \left| c_1 \left(S(P,f) - \int_a^b f(x)dx \right) \right| + \left| c_2 \left(S(P,g) - c_2 \int_a^b g(x)dx \right) \right| \\
&= |c_1| \left| S(P,f) - \int_a^b f(x)dx \right| + |c_2| \left| S(P,g) - \int_a^b g(x)dx \right| \\
&< |c_1| \cdot \frac{\varepsilon}{(|c_1| + |c_2|)} + |c_2| \cdot \frac{\varepsilon}{(|c_1| + |c_2|)} \quad \text{(by (7.6) and (7.7))} \\
&= \frac{\varepsilon(|c_1| + |c_2|)}{(|c_1| + |c_2|)} \\
&= \varepsilon.
\end{aligned}
$$

∎

Corollary 7.43 *If f and g are Riemann integrable on $[a,b]$, then*
(1) $\int_a^b (f(x) + g(x))dx = \int_a^b f(x)dx + \int_a^b g(x)dx$,
(2) $\int_a^b cf(x)dx = c \int_a^b f(x)dx$, *and*
(3) $\int_a^b (f(x) - g(x))dx = \int_a^b f(x)dx - \int_a^b g(x)dx$.

Proof. For (1), take $c_1 = c_2 = 1$. For (2), take $c_1 = c$ and $c_2 = 0$. For (3), take $c_1 = 1$ and $c_2 = -1$. ∎

If $c \in [a,b]$, then it seems reasonable that $\int_a^b f(x)dx$ should equal $\int_a^c f(x)dx + \int_c^b f(x))dx$. This seems true since any partition of $[a,b]$ containing the point c is the

7.4. BASIC RESULTS ON RIEMANN INTEGRALS

union of a partition of $[a, c]$ and of $[c, b]$ and if the norm of the partition is small enough, the Riemann sums of f on $[a, c]$ and $[c, b]$ are close to the integrals $\int_a^c f(x)dx$ and $\int_c^b f(x))dx$, respectively, assuming these exist. So the next theorem is just a matter of writing out everything our intuition tells us.

Theorem 7.44 *Let f be a bounded function. If $f(x)$ is continuous on $[a, b]$ except on a set of measure 0, and if $c \in [a, b]$, then $\int_a^b f(x)dx = \int_a^c f(x)dx + \int_c^b f(x)dx$.*

Proof. We know each of the three integrals in the above statement exists by Lebesgue's theorem, Theorem 7.30, and the first corollary of that theorem. Since $\int_a^c f(x)dx$ exists, for any $\varepsilon > 0$ there is a $\delta_1 > 0$ such that if P_1 is a partition of $[a, c]$ with $||P_1|| < \delta_1$, then

$$\left| S(P_1, f) - \int_a^c f(x)dx \right| < \varepsilon/2. \tag{7.8}$$

Similarly, since $\int_c^b f(x)dx$ exists, for this same ε there is a $\delta_2 > 0$ such that if P_2 is a partition of $[c, b]$ with $||P_2|| < \delta_2$, then

$$\left| S(P_2, f) - \int_c^b f(x)dx \right| < \varepsilon/2. \tag{7.9}$$

Now, let δ be any number smaller than both δ_1 and δ_2, and pick a partition P of $[a, b]$ with norm $< \delta$. If c is not already in P, adjoin it to P. This cannot increase the norm of the partition since we have simply refined the partition. Now, let $P_1 = P \cap [a, c]$ and $P_2 = P \cap [c, b]$. Then P_1 and P_2 are partitions of $[a, c]$ and $[c, b]$, respectively, and the norms of both P_1 and P_2 are less that δ. So, both (7.8) and (7.9) are true. Observe that with this P, any Riemann sum, $S(P, f)$ for $[a, b]$, can be split into Riemann sums $S(P_1, f)$ for $[a, c]$ and $S(P_2, f)$ for $[c, b]$, such that

$$S(P, f) = S(P_1, f) + S(P_2, f). \tag{7.10}$$

Now, for simplicity, let $A = \int_a^c f(x)dx + \int_c^b f(x)dx$. By (7.8) and (7.9), we have

$$\begin{aligned} &|S(P, f) - A| \\ &= \left| S(P, f) - \left(\int_a^c f(x)dx + \int_c^b f(x)dx \right) \right| \\ &= \left| S(P_1, f) + S(P_2, f) - \int_a^c f(x)dx - \int_c^b f(x)dx \right| \quad \text{(by 7.10)} \\ &\leq \left| S(P_1, f) - \int_a^c f(x)dx \right| + \left| S(P_2, f) - \int_c^b f(x)dx \right| \\ &< \varepsilon/2 + \varepsilon/2 \\ &= \varepsilon. \end{aligned}$$

To summarize, we found that for any $\varepsilon > 0$ there is a $\delta > 0$ such that for any partition P of $[a,b]$ with norm $< \delta$, it follows that $|S(P,f) - A| < \varepsilon$, and this shows that $A = \int_a^b f(x)dx$. But $A = \int_a^c f(x)dx + \int_c^b f(x)dx$. So, $\int_a^b f(x)dx = \int_a^c f(x)dx + \int_c^b f(x)dx$. ∎

Theorem 7.45 *If $f(x) \geq 0$ on $[a,b]$ and f is Riemann integrable on $[a,b]$, then $\int_a^b f(x)dx \geq 0$.*

Proof. Suppose not. That is, suppose that $\int_a^b f(x)dx = A < 0$. Now, using the definition of "f is Riemann integrable on $[a,b]$," we have that for any $\varepsilon > 0$, there is a $\delta > 0$ such that if $\|P\| < \delta$, then $|S(P,f) - A| < \varepsilon$. We take the specific value $-A$ for ε. So there is a $\delta > 0$, such that if $\|P\| < \delta$, then $|S(P,f) - A| < -A$, which is equivalent to $A < S(P,f) - A < -A$. This implies, by adding A to each inequality, that $S(P,f) < 0$. But this can't be since $S(P,f) = \sum_1^n f(t_i)\Delta x_i$ and each term, $f(t_i)\Delta x_i$, is ≥ 0 since f is ≥ 0 and each Δx_i is positive. This contradiction, that $S(P,f) < 0$, tells us that our assumption that $\int_a^b f(x)dx = A < 0$ was false, and so $\int_a^b f(x)dx \geq 0$. ∎

Theorem 7.46 *If $f(x) \leq g(x)$ on $[a,b]$, and both f and g are Riemann integrable on $[a,b]$, then $\int_a^b f(x)dx \leq \int_a^b g(x)dx$.*

Proof. Let $h(x) = g(x) - f(x)$. Since $f(x) \leq g(x)$ on $[a,b]$, $h(x) \geq 0$ on $[a,b]$. So, by the previous theorem, $\int_a^b h(x)dx \geq 0$. But $\int_a^b h(x)dx = \int_a^b g(x)dx - \int_a^b f(x)dx$ by Corollary 7.43, so $\int_a^b g(x)dx - \int_a^b f(x)dx \geq 0$, which implies that $\int_a^b f(x)dx \leq \int_a^b g(x)dx$. ∎

Theorem 7.47 *If f is Riemann integrable on $[a,b]$, then so is $|f|$ and $\left|\int_a^b f(x)dx\right| \leq \int_a^b |f(x)|\,dx$.*

Proof. Using the last theorem, the fact that $-|f(x)| \leq f(x) \leq |f(x)|$, and that these functions are Riemann integrable on $[a,b]$ (Corollary 7.36), we get that $\int_a^b -|f(x)|\,dx \leq \int_a^b f(x)dx \leq \int_a^b |f(x)|\,dx$, or equivalently, that $-\int_a^b |f(x)|\,dx \leq \int_a^b f(x)dx \leq \int_a^b |f(x)|\,dx$. This tells us that $\left|\int_a^b f(x)dx\right| \leq \int_a^b |f(x)|\,dx$. (This follows since $-b \leq a \leq b$ is equivalent to saying that $|a| \leq b$. Here $a = \int_a^b f(x)dx$, and $b = \int_a^b |f(x)|\,dx$. See Proposition 1.50 part(5)). ∎

We now give some examples to illustrate how one could integrate a function which is not continuous on $[a,b]$.

7.4. BASIC RESULTS ON RIEMANN INTEGRALS

Example 7.48 *Suppose that*

$$f(x) = \begin{cases} 5 & \text{if } x = 1 \\ x & \text{if } 1 < x \leq 2 \end{cases}.$$

Because in $[1,2]$, f is discontinuous only at 1, by Lebesgue's theorem (Theorem 7.30), it is integrable on $[1,2]$. What we propose to show, is that the discontinuity really doesn't matter as far as the evaluation of the integral goes. That is, we can ignore it, and assume that $f(x) = x$ on the entire interval $[1,2]$, so that $\int_1^2 f(x)dx = \int_1^2 xdx = \frac{x^2}{2}\Big|_1^2 = \frac{3}{2}$. (This is intuitively clear since the integral represents the area under the curve, and changing the function at one, or even a finite number of points, should not affect the area.)

So let us split the interval $[1,2]$ into $[1,x_1] \cup [x_1,2]$ where $1 < x_1 \leq 2$. We know, by Theorem 7.44, that $\int_1^2 f(x)dx = \int_1^{x_1} f(x)dx + \int_{x_1}^2 f(x)dx$. Now, since $0 \leq f(x) \leq 5$, we have by Theorems 7.45 and 7.46, that

$$0 \leq \int_1^{x_1} f(x)dx \leq \int_1^{x_1} 5dx = 5x\Big|_1^{x_1} = 5x_1 - 5. \qquad (7.11)$$

We know that $\int_{x_1}^2 f(x)dx = \int_{x_1}^2 xdx = 2 - \frac{1}{2}x_1^2$, which, for reasons we will soon show, we write as,

$$2 - \frac{1}{2}x_1^2 \leq \int_{x_1}^2 f(x)dx \leq 2 - \frac{1}{2}x_1^2.$$

Adding this to (7.11) we get that

$$2 - \frac{1}{2}x_1^2 \leq \int_1^{x_1} f(x)dx + \int_{x_1}^2 f(x)dx \leq 2 - \frac{1}{2}x_1^2 + 5x_1 - 5,$$

or just that

$$2 - \frac{1}{2}x_1^2 \leq \int_1^2 f(x)dx \leq 2 - \frac{1}{2}x_1^2 + 5x_1 - 5, \qquad (7.12)$$

for any $x_1 \in (1,2]$. Now, as the norm of any partition of $[1,2]$ goes to 0, x_1 goes to 1. By taking the limit of (7.12) as $x_1 \to 1^+$, we get

$$\frac{3}{2} \leq \int_1^2 f(x)dx \leq \frac{3}{2},$$

so that $\int_1^2 f(x)dx = \frac{3}{2}$, which is the same as $\int_1^2 xdx$.

This type of argument generalizes to any finite number of discontinuities. So if, for example, we have that

$$g(x) = \begin{cases} 5 & \text{if } x = 1 \\ x^3 & \text{if } 1 < x < 2 \\ 7 & \text{if } x = 2 \end{cases},$$

then we can ignore the discontinuities at 1 and 2, and simply state that $\int_1^2 g(x)dx = \int_1^2 x^3 dx = \frac{15}{4}$. And if

$$h(x) = \begin{cases} 5 & \text{if } x = 1 \\ x^3 & \text{if } 1 < x < 1.5 \\ 2x & \text{if } 1.5 \leq x < 2 \\ 9 & \text{if } x = 2 \end{cases},$$

which has discontinuities at 1, 1.5 and 2, then $\int_1^2 h(x)dx = \int_1^{1.5} x^3 dx + \int_{1.5}^2 2x dx = 2.7656$.

We have illustrated the following theorem, whose proof we won't give in full generality (see Exercise 5).

Theorem 7.49 *Let f and g be bounded functions defined on $[a, b]$. If f is Riemann integrable on $[a, b]$ and $f = g$ except at a finite number of points in $[a, b]$, then g is Riemann integrable on $[a, b]$ and $\int_a^b f(x)dx = \int_a^b g(x)dx$.*

Remark 7.50 *This theorem can be generalized to: "If f is Riemann integrable on $[a, b]$ and $f = g$ except at a set of measure 0, then if $\int_a^b g(x)dx$ exists, $\int_a^b f(x)dx = \int_a^b g(x)dx$." In the generalization, we need to require that $\int_a^b g(x)dx$ exists. For if $g(x) = 0$ when x is rational and 1 when x is irrational, then $g(x)$, being discontinuous everywhere, is not Riemann integrable on $[0, 1]$. This is true even though $g(x)$ equals the function $f(x) = 1$ on $[0, 1]$ except on a set of measure 0, and $f(x)$ is integrable on $[0, 1]$.*

We are ready to prove a theorem we promised to prove earlier in this primer.

Theorem 7.51 *If $\{f_n\}$ is a sequence of functions converging uniformly to f on $[a, b]$ and each f_n is Riemann integrable on $[a, b]$, then so is $f(x)$, and $\int_a^b f_n(x)dx \to \int_a^b f(x)dx$.*

7.4. BASIC RESULTS ON RIEMANN INTEGRALS

Proof. We first note that Corollary 7.37 covers the existence of this integral, so we turn to the second half of the proof. Pick an $\varepsilon > 0$. Since f_n converges uniformly to f on $[a, b]$, there is an $N > 0$ such that $n > N$ implies that

$$|f_n(x) - f(x)| < \frac{\varepsilon}{2(b-a)}. \tag{7.13}$$

To show that the sequence, $\left\{\int_a^b f_n(x)dx\right\}$, of real numbers converges to the real number $\int_a^b f(x)dx$, we look at $\left|\int_a^b f_n(x)dx - \int_a^b f(x)dx\right|$. We have,

$$\left|\int_a^b f_n(x)dx - \int_a^b f(x)dx\right|$$
$$= \left|\int_a^b (f_n(x) - f(x))dx\right| \text{ (by Corollary 7.43)}$$
$$\leq \int_a^b |f_n(x) - f(x)|\, dx, \text{ for } n > N \text{ (by Theorem 7.47)}$$
$$\leq \int_a^b \frac{\varepsilon}{2(b-a)} dx, \text{ for } n > N \text{ (by (7.13))}$$
$$= \frac{\varepsilon}{2(b-a)} \cdot (b-a), \text{ for } n > N \text{ (by Corollary 7.41)}$$
$$= \varepsilon/2$$
$$< \varepsilon.$$

We have shown that for any $\varepsilon > 0$, there is an N such that $n > N$ implies that $\left|\int_a^b f_n(x)dx - \int_a^b f(x)dx\right| < \varepsilon$, and this shows that the sequence of integrals $\left\{\int_a^b f_n(x)dx\right\}$ converges to $\int_a^b f(x)dx$. ∎

EXERCISES

1. Suppose that $f(x) = x^2$ if $x \in (1, 2]$ but $f(1) = 10$. Compute $\int_1^2 f(x)dx$.

2. Suppose that $\{x_0, x_1, x_2, ..., x_n\}$ is any partition of $[a, b]$. Then for any Riemann integrable function, f, on $[a, b]$, $\int_a^b f(x)dx = \int_{x_0}^{x_1} f(x)dx + \int_{x_1}^{x_2} f(x)dx + ... + \int_{x_{n-1}}^b f(x)dx$.

3. If $f(x) = 0$ everywhere on $[a, b]$ except at b, then $\int_a^b f(x)dx = 0$. Similarly, if $f(x) = 0$ everywhere on $[a, b]$ except at a then $\int_a^b f(x)dx = 0$.

4. If $f(x) = 0$ on $[a,b]$ except at a finite number of points, $x_1, x_2, x_3, ..., x_n$, show that $\int_a^b f(x)dx = 0$.

5. Show that if $f(x)$ and $g(x)$ are both Riemann integrable on $[a,b]$, and if $f(x) = g(x)$ except at a finite number of points, then $\int_a^b f(x)dx = \int_a^b g(x)dx$. [Hint: Apply the previous theorem to $f(x) - g(x)$.]

6. Suppose that f is Riemann integrable on $[a,b]$ and that $f \geq 0$ $[a,b]$, then $\int_a^b f(x)dx \geq \int_c^d f(x)dx$ where $[c,d] \subset [a,b]$. Give an example to show that this is not true when $f(x)$ takes on negative values.

7. It is a fact that if $f : [a,b] \to R$ is continuous at c (where R and $[a,b]$ carry the usual metric), and $f(c) > 0$, then there is a interval $(c-p, c+p)$ containing c on which $f(x)$ is positive. (See Exercise 5 of section 4.3.) Using this fact, prove that If the Riemann integral of a nonnegative continuous real valued function defined on $[a,b]$ is 0, then f must be 0.

8. If $\int_a^b h(x)dx = 0$ where $h(x)$ is continuous on $[a,b]$ (but not necessarily nonnegative), then $h(x)$ must be zero somewhere in $[a,b]$.

9. Show that the function $d \colon C[0,1] \times C[0,1] \to R$ defined by
$$d(f,g) = \sqrt{\int_a^b |f(x) - g(x)|dx}$$ is a metric. [Hint: Using exercise 7 together with the fact that $\sqrt{a+b} \leq \sqrt{a} + \sqrt{b}$ when a and b are nonnegative will help.]

10. True or False: Using the function from Example 7.32, there is a sequence $\{P_n\}$ of partitions of $[0,1]$ such that $S(P_n, f)$ converges to 0.

11. Suppose that for each $c \in [a,b]$ it is true that $\int_c^b f(x)dx$ exists. Is is true that $\int_a^b f(x)dx$ exists?

7.5 Other Approaches to the Riemann Integral

An examination of different books shows that there are several different ways of approaching the Riemann integral. The definition we gave earlier we will call Definition 1, and that is:

Definition 1 of Riemann Integrable: Suppose that f is a bounded real valued function defined on $[a,b]$. We say that f is Riemann integrable on $[a,b]$, if there is a number A which has the following property: For any $\varepsilon > 0$ there is a $\delta > 0$, such that

7.5. OTHER APPROACHES TO THE RIEMANN INTEGRAL

for every partition P of $[a,b]$ with norm less than δ and every choice of points t_i in $[x_{i-1}, x_i]$,

$$|S(P,f) - A| < \varepsilon.$$

(The number A is called the Riemann integral of $f(x)$ over $[a,b]$ and is denoted by $\int_a^b f(x)dx$.)

Definition 2 of Riemann Integrable: Suppose that f is a bounded real valued function defined on $[a,b]$. We say that f is Riemann integrable on $[a,b]$ if there is a number A which has the following property: For any $\varepsilon > 0$ there is a partition P_ε of $[a,b]$, with the property that if P is a refinement of P_ε, then $|S(P,f) - A| < \varepsilon$ for every choice of points t_i in $[x_{i-1}, x_i]$. (Again, the number A is called the Riemann integral of $f(x)$ over $[a,b]$ and is denoted by $\int_a^b f(x)dx$.)

Definition 3 of Riemann Integrable: A bounded real valued function f, defined on $[a,b]$, is Riemann integrable if $\overline{\int_a^b} f(x)dx = \underline{\int_a^b} f(x)dx$. That is, if $\inf U(P,f) = \sup L(P,f)$. In this case, the common value of $\overline{\int_a^b} f(x)dx$ and $\underline{\int_a^b} f(x)dx$ is called the integral of $f(x)$ on $[a,b]$ and is denoted by $\int_a^b f(x)dx$.

Finally, we have,

Definition 4 of Riemann Integrable: A bounded real valued function defined on $[a,b]$ is Riemann integrable if for any $\varepsilon > 0$ there is a partition P_ε of $[a,b]$ such that $U(P_\varepsilon, f) - L(P_\varepsilon, f) < \varepsilon$. (And once we show this definition implies Definition 3, the value of the integral is the common value that we get in Definition 3.)

We will soon show that Definitions (1) – (4) are equivalent, so that the value one gets for the integral using any approach is the same, and we get the same theorems. Definition 4 seems particularly easy to apply. Let us show what we can get immediately from Definition (4). (We promised to prove this earlier.)

Theorem 7.52 *If $f : [a,b] \to R$ is continuous, then f is Riemann integrable on $[a,b]$.*

Proof. Since f is continuous on $[a,b]$, it is uniformly continuous (Corollary 4.74). So, given any $\varepsilon > 0$ there is a $\delta > 0$, such that if p and q are any points in $[a,b]$ with $|p - q| < \delta$, then

$$|f(p) - f(q)| < \varepsilon/(b-a). \tag{7.14}$$

Now, take a partition of $[a,b]$, $P = \{x_0, x_1, x_2, ..., x_n\}$, with norm less than δ. Consider $U(P,f) - L(P,f) = \sum_{i=1}^n M_i \Delta x_i - \sum_{i=1}^n m_i \Delta x_i = \sum_{i=1}^n (M_i - m_i)\Delta x_i$. Now, since a continuous function has an absolute maximum and minimum on any closed interval, (Theorem

4.51) $M_i = f(p_i)$ and $m_i = f(q_i)$ for some p_i, and $q_i \in [a, b]$. So,

$$\begin{aligned}
& U(P, f) - L(P, f) \\
&= \sum_{i=1}^{n}(M_i - m_i)\Delta x_i \\
&= \sum_{i=1}^{n}(f(p_i) - f(q_i))\Delta x_i \\
&\leq \sum_{i=1}^{n}\frac{\varepsilon}{(b-a)}\Delta x_i \quad \text{(by (7.14), since the norm of the partition is less than } \delta\text{)} \\
&= \frac{\varepsilon}{(b-a)}\sum_{i=1}^{n}\Delta x_i \\
&= \frac{\varepsilon}{(b-a)}(b-a) \quad \left(\text{since }\sum_{i=1}^{n}\Delta x_i = b - a\right) \\
&= \varepsilon.
\end{aligned}$$

Since we have found, for any $\varepsilon > 0$, a partition, P, which makes $U(P, f) - L(P, f) < \varepsilon$, f is Riemann integrable by Definition (4) of Riemann Integral. ∎

EXERCISES

In answering these questions, you should use the equivalence of the four different approaches to Riemann integration given in this section (even though we haven't yet proved they are equivalent).

1. Using upper and lower sums, show that if $f(x) = c$ on $[a, b]$ where c is a constant, then $\int_a^b f(x)dx = c(b - a)$.

2. Show, using upper and lower sums, that if $m \leq f(x) \leq M$ on $[a, b]$ and f is Riemann integrable on $[a, b]$, then $m(b - a) \leq \int_a^b f(x)dx \leq M(b - a)$.

3. Another way to prove the Fundamental Theorem of Calculus is to show that $F(b) - F(a)$, being a Riemann sum (as we saw in the text), is between any upper and lower sum. Finish this train of thought.

4. A function real valued function is called odd if $f(-x) = -f(x)$. Show that if f is Riemann integrable on $[-a, a]$ where a is positive, then $\int_{-a}^{a} f(x)dx = 0$.

5. Show that the $\int_0^1 f(x)dx$ of the function given in Example 7.33 is 0.

7.6. THE EQUIVALENCE OF...

6. Suppose that $f(x) = 0$ everywhere on $[a,b]$ except at a. Show, using lower sums, that $\int_a^b f(x)dx = 0$.

7. A real valued function, $s(x)$, defined on $[a,b]$ is called a **step function** if there is a partition $P = \{x_0, x_1, x_2, ..., x_n\}$ such that $s(x)$ is constant on each open subinterval $I_i = (x_{i-1}, x_i)$.

    (a) If a step function $s(x)$ takes on the values $c_1, c_2, c_3, ..., c_n$ on intervals $I_1, I_2, I_3, ..., I_n$ in $[a,b]$, then it is Riemann integrable on $[a,b]$ and $\int_a^b s(x)dx$
    $$= \sum_{i=1}^{n} c_i \cdot (\text{length of } I_i).$$

    (b) Suppose that $f(x)$ is Riemann integrable on $[a,b]$. Then $\int_a^b f(x)dx = \inf \int_a^b s(x)dx$ where $s(x)$ is a step function greater than or equal to $f(x)$ on $[a.b]$. It is also $\sup \int_a^b \widetilde{s}(x)dx$ where $\widetilde{s}(x)$ is a step function less than or equal to $f(x)$ on $[a,b]$.

7.6 The Equivalence of the Different Definitions of Riemann Integrable

We have examined four different definitions of the Riemann integral that one finds in books. In this section we show that they are equivalent. It would probably be worthwhile to write these definitions down and refer to them as you read.

Definition 1 of Riemann Integrable$\Longrightarrow$Definition 2 of Riemann Integrable: If f is Riemann integrable according to Definition 1, then for any $\varepsilon > 0$ there is a $\delta > 0$ such that, if P is a partition of $[a,b]$ where $||P|| < \delta$, then

$$|S(P,f) - A| < \varepsilon \qquad (7.15)$$

(for all choices of $t_i \in [x_{i-1}, x_i]$, where $i = 1, 2, ..., n$). Fix such a partition P_ε. Then if P is finer than P_ε, $||P|| < \delta$, so (7.15) holds, and we have that f is Riemann integrable according to Definition 2.

We now show the equivalence of Definitions (2), (3), and (4), by showing that Definition (3) implies Definition (2), which implies Definition (4), which in turn implies Definition (3). After doing this, we will separately show the equivalence of Definitions (1) and (2).

Definition 3 of Riemann Integrable$\Longrightarrow$Definition 2 of Riemann Integrable: First, we do it intuitively. We are assuming that $\inf\{U(P,f)\} = \sup\{(L(P,f)\} = A$, and we know by Theorem 7.15 that for any partition, P, of $[a,b]$

$$L(P,f) \leq S(P,f) \leq U(P,f). \qquad (7.16)$$

Since each $S(P, f)$ is between $U(P, f)$ and $L(P, f)$, and both of these can be made close to A, $S(P, f)$ can be made close to A. Now, we formalize this.

Since $A = \sup\{L(P, f)\}$, for any $\varepsilon > 0$ there is a partition P_1 of $[a, b]$ such that
$$A - \varepsilon < L(P_1, f) \text{ (by Theorem 2.20)}.$$

And since the lower sums increase or stay the same as we refine P_1, we have that
$$A - \varepsilon < L(P, f), \text{ if } P \text{ refines } P_1. \tag{7.17}$$

Similarly, there is a partition P_2 of $[a, b]$ such that
$$U(P_2, f) < A + \varepsilon,$$
and since the upper sums decrease or remain the same as we refine P_2, we have
$$U(P, f) < A + \varepsilon, \text{ if } P \text{ refines } P_2. \tag{7.18}$$

Now, let $P_3 = P_1 \cup P_2$. Then P_3 refines both P_1 and P_2, and any refinement P of P_3 does the same. So if P refines P_3, we have, by (7.17) and (7.18) (and 7.16), that
$$A - \varepsilon < L(P, f) \leq S(P, f) \leq U(P, f) < A + \varepsilon,$$
from which it follows that
$$A - \varepsilon < S(P, f) < A + \varepsilon,$$
or equivalently that
$$|S(P, f) - A| < \varepsilon, \text{ if } P \text{ refines } P_3.$$

We have shown that there is a partition P_3 of $[a, b]$, such that P finer that P_3 implies that $|S(P, f) - A| < \varepsilon$, and this shows that f is Riemann integrable according to Definition 2.

Definition 2 of Riemann Integrable$\Longrightarrow$Definition 4 of Riemann Integrable: The intuitive argument goes something like this: Since the Riemann sums $S(P, f)$ get close to A as you refine P, they get close to each other. But by Remark 7.14, the upper and lower sums for P are close to Riemann sums of $f(x)$ (which are close to the $\int_a^b f(x) dx$ as we refine) so the upper and lower sums for P should be close to each other. Writing this out precisely is more difficult than you might think. We do that now.

Suppose that $\varepsilon > 0$ is given. According to Definition 2 of Riemann integrable, there is a partition $P_1 = \{x_0, x_1, x_2, ..., x_n\}$ such that if P is finer than P_1, then
$$|S(P, f) - A| < \varepsilon/3, \tag{7.19}$$

7.6. THE EQUIVALENCE OF...

for all choices of points t_i in $[x_{i-1}, x_i]$, where $i = 1, 2, 3, ..., n$. Let $T = \{t_1, t_2, ..., t_n\}$ be one set of points chosen for the t values, let $T' = \{t'_1, t'_2, ..., t'_n\}$ be another such set of t values, and let $S_1(P, f)$ and $S_2(P, f)$ be the Riemann sums for points in T and T' respectively. Then, by (7.19) we have,

$$|S_1(P_1, f) - A| < \varepsilon/3 \tag{7.20}$$

and

$$|S_2(P_1, f) - A| < \varepsilon/3 \tag{7.21}$$

(since P_1 is a refinement of itself). Using (7.20) and (7.21), we have that

$$\begin{aligned} &|S_1(P_1, f) - S_2(P_1, f)| \\ = \ &|S_1(P_1, f) - A + A - S_2(P_1, f)| \\ \leq \ &|S_1(P_1, f) - A| + |A - S_2(P_1, f)| \\ < \ &\varepsilon/3 + \varepsilon/3 \\ = \ &2\varepsilon/3, \end{aligned}$$

or that,

$$|S_1(P_1, f) - S_2(P_1, f)| < 2\varepsilon/3. \tag{7.22}$$

This shows that the Riemann sums are close to each other for this partition. Now, for the partition P_1, $M_i - m_i = \sup f(x) - \inf f(y)$, where x and y come from $[x_{i-1}, x_i]$, and this, by Theorem 7.23, is equal to $\sup\{f(x) - f(y)\}$. Since $M_i - m_i = \sup\{f(x) - f(y)\}$, where x and y come from $[x_{i-1}, x_i]$, we have from Theorem 2.20 that there are points t_i and t'_i in $[x_{i-1}, x_i]$ such that

$$M_i - m_i - \frac{\varepsilon}{3(b-a)} < f(t_i) - f(t'_i), \text{ for each } i = 1, 2, 3, ..., n,$$

or equivalently,

$$M_i - m_i < f(t_i) - f(t'_i) + \frac{\varepsilon}{3(b-a)}, \text{ for each } i = 1, 2, 3, ..., n. \tag{7.23}$$

We are almost there. Let us consider $U(P_1, f) - L(P_1, f)$. Observing that $\sum_{i=1}^{n} \Delta x_i =$

$b - a$, we have

$$\begin{aligned}
& U(P_1, f) - L(P_1, f) \\
=\ & |U(P_1, f) - L(P_1, f)| \\
=\ & \left| \sum_{i=1}^{n} (M_i - m_i) \Delta x_i \right| \\
<\ & \left| \sum_{i=1}^{n} \left(f(t_i) - f(t_i') + \frac{\varepsilon}{3(b-a)} \right) \Delta x_i \right| \quad \text{(by (7.23))} \\
=\ & \left| \sum_{i=1}^{n} f(t_i) \Delta x_i - \sum_{i=1}^{n} f(t_i') \Delta x_i + \sum_{i=1}^{n} \frac{\varepsilon}{3(b-a)} \Delta x_i \right| \\
=\ & \left| S_1(P_1, f) - S_2(P_1, f) + \frac{\varepsilon}{3(b-a)} \sum_{i=1}^{n} \Delta x_i \right| \\
\leq\ & |S_1(P_1, f) - S_2(P_1, f)| + \left| \frac{\varepsilon}{3(b-a)} \sum_{i=1}^{n} \Delta x_i \right| \quad \text{(triangle inequality)} \\
<\ & \frac{2\varepsilon}{3} + \frac{\varepsilon}{3(b-a)} (b-a) \quad \text{(by (7.22))} \\
=\ & \varepsilon.
\end{aligned}$$

We have found a partition P_1 of $[a, b]$ that makes $U(P_1, f) - L(P_1, f) < \varepsilon$, so f is Riemann integrable on $[a, b]$ according to Definition 4.

Definition 4 of Riemann Integrable$\Longrightarrow$Definition 3 of Riemann Integrable: Figure 7.10 shows an upper and lower sum that are within ε of each other. The shaded area represents their difference.

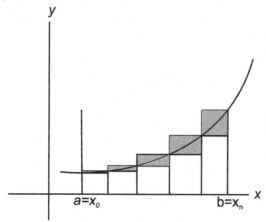

Figure 7.10.

7.6. THE EQUIVALENCE OF...

As we refine the partitions, the upper sums decrease or stay the same, and the lower sums increase or remain the same, which means the shaded area gets smaller and smaller. This seems to indicate that the infimum of the upper sums and the supremum of lower sums are the same. Now, we prove it.

According to Definition 4 of Riemann Integrable, there is a partition P_ε of $[a,b]$ such that $U(P_\varepsilon, f) - L(P_\varepsilon, f) < \varepsilon$, or equivalently

$$U(P_\varepsilon, f) < L(P_\varepsilon, f) + \varepsilon. \tag{7.24}$$

Now,

$$\begin{aligned}
\overline{\int_a^b} f(x)dx &= \inf U(P,f) \\
&\leq U(P_\varepsilon, f) \\
&< L(P_\varepsilon, f) + \varepsilon \text{ (by (7.24))} \\
&\leq (\sup L(P,f)) + \varepsilon \\
&= \underline{\int_a^b} f(x)dx + \varepsilon
\end{aligned}$$

Thus, for any $\varepsilon > 0$ we have $\overline{\int_a^b} f(x)dx \leq \underline{\int_a^b} f(x)dx + \varepsilon$, which implies by Lemma 2.23 that $\overline{\int_a^b} f(x)dx \leq \underline{\int_a^b} f(x)dx$. But we know from Corollary 7.19 that $\underline{\int_a^b} f(x)dx \leq \overline{\int_a^b} f(x)dx$. So $\underline{\int_a^b} f(x)dx = \overline{\int_a^b} f(x)dx$, and this tells us that f is Riemann integrable on $[a,b]$ according to Definition 3.

In summary, we have shown that Definition 3 implies Definition 2 and that Definition 2 implies Definition 4 and Definition 4 implies Definition 3. So this means Definitions 2, 3 and 4 are equivalent. We have already shown that Definition (1) implies Definition (2). We now show the converse.

Definition (2) implies Definition (1): Before we start, we make some observations. In what follows, $P_\varepsilon = \{y_0, y_1, y_2, ..., y_n\}$ is a partition of $[a.b]$, and $P = \{x_0, x_1, x_2, ..., x_m\}$ is another partition of $[a,b]$. f is a bounded real valued function defined on $[a,b]$, and $M = \sup |f(x)|$ on $[a,b]$. $N_i = \sup f(x)$ on $[y_{i-1}, y_i]$ and $M_i = \sup f(x)$ on $[x_{i-1}, x_i]$. We will need some lemmas.

Lemma 7.53 *Suppose that f is nonnegative on $[a,b]$. Then the contribution to $U(P,f)$ from the intervals of P totally contained within intervals of P_ε, is less than or equal to $U(P_\varepsilon, f)$.*

Proof. We will focus on one interval in the partition P_ε and show that the result holds. The theorem is then proved by doing this for each subinterval of P_ε containing intervals of P, and summing the results.

Below, in Figure 7.11, we see a subinterval $[y_{i-1}, y_i]$ of P_ε, and some of the intervals of P that are contained in this interval, which we call I_1, I_2 and I_3

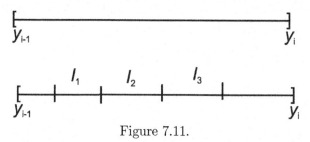

Figure 7.11.

The term in $U(P_\varepsilon, f)$ that arises from the interval $[y_{i-1}, y_i]$ is $N_i(y_i - y_{i-1})$, where N_i is the supremum of f on $[y_{i-1}, y_i]$. The terms in $U(P, f)$ arising from the intervals I_1, I_2 and I_3, are N_{i_1}(length of I_1), N_{i_2}(length of I_2), and N_{i_3}(length of I_3), where N_{i_1}, N_{i_2}, and N_{i_3} are the respective suprema of f on I_1, I_2 and I_3. Since I_1, I_2 and I_3 are subsets of $[y_{i-1}, y_i]$, each of N_{i_1}, N_{i_2}, and N_{i_3} is $\leq N_i$. So,

$$\begin{aligned}
& N_{i_1}(\text{length of } I_1) + N_{i_2}(\text{length of } I_2) + N_{i_3}(\text{length of } I_3) \\
\leq\ & N_i(\text{length of } I_1) + N_i(\text{length of } I_2) + N_i(\text{length of } I_3) \\
=\ & N_i(\text{length of } I_1 + \text{length of } I_2 + \text{length of } I_3) \\
\leq\ & N_i(y_i - y_{i-1}) \quad (\text{since } N_i \geq 0).
\end{aligned}$$

This says that the contribution to $U(P, f)$ from the intervals I_1, I_2 and I_3, is less than or equal to $N_i(y_i - y_{i-1})$, which is the corresponding contribution in $U(P_\varepsilon, f)$ made by the interval $[y_{i-1}, y_i]$. Now, we do this for each subinterval in the partition P_ε containing intervals of P, and sum the results. So the sum of these terms, which constitute the contribution of these terms to $U(P, f)$, is less than or equal to the corresponding terms of $U(P_\varepsilon, f)$. There are other terms of $U(P_\varepsilon, f)$, but they are all nonnegative. So this sum is less than or equal to $U(P_\varepsilon, f)$. ∎

Lemma 7.54 *Suppose that P_ε has n intervals. Suppose that the norm of P is $\leq \frac{\varepsilon}{2nM}$. Then the contribution to $U(P, f)$ from the intervals of P not totally contained in subintervals of P_ε, is less than or equal to $\varepsilon/2$.*

Proof. Since we are only looking at intervals not totally contained in intervals of P_ε, any such interval must contain a partition point of P_ε. We see below, in Figure 7.12 some of these intervals.

7.6. THE EQUIVALENCE OF...

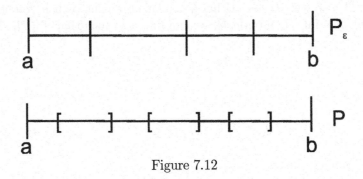

Figure 7.12

There are at most n such intervals and each term in $U(P, f)$ arising from these intervals, is $\leq M$(the length of the interval)$\leq M\frac{\varepsilon}{2nM}$. So the contribution from the at most n intervals to $U(P, f)$ is $\leq nM\frac{\varepsilon}{2nM} = \frac{\varepsilon}{2}$. ∎

We are now ready to prove that:

Definition (2) implies Definition (1): We assume first that f is nonnegative on $[a, b]$.

Theorem 7.55 *If f is ≥ 0, then Definition (2) implies Definition (1).*

Proof. Since Definition (2) is equivalent to Definition (3), the Riemann integral is the infimum of the upper sums. Therefore, there is a partition P_ε of $[a, b]$ such that

$$U(P_\varepsilon, f) < \int_a^b f(x)dx + \frac{\varepsilon}{2}. \tag{7.25}$$

Now, let δ_1 be small, say, less than $\frac{\varepsilon}{2nM}$ where n is the number of partition points of P_ε, and consider any partition P with norm less than δ_1. Let S_1 be the contribution to $U(P, f)$ arising from these intervals totally contained in an interval of P_ε, and S_2 the contribution from the remaining intervals. According to Lemma 7.53, $S_1 < U(P_\varepsilon, f)$. By Lemma 7.54, $S_2 \leq \frac{\varepsilon}{2}$. Adding these inequalities, and realizing that $S_1 + S_2 = U(P, f)$, we get that

$$\begin{aligned} U(P, f) &< U(P_\varepsilon, f) + \frac{\varepsilon}{2} \\ &< \left[\int_a^b f(x)dx + \frac{\varepsilon}{2}\right] + \frac{\varepsilon}{2} \quad \text{(by 7.25)} \\ &= \int_a^b f(x)dx + \varepsilon. \end{aligned}$$

So, $U(P, f) < \int_a^b f(x)dx + \varepsilon$. Using a similar argument (which you should write out), if we let $m = \inf |f(x)|$ on $[a, b]$, we can find a δ_2 such that if $||P|| < \delta_2$, then

$$L(P, f) > \int_a^b f(x)dx - \varepsilon.$$

Now, let δ be smaller than both δ_1 and δ_2. Then, if $||P|| < \delta$ we have,

$$\int_a^b f(x)dx - \varepsilon < L(P, f) \leq S(P, f) \leq U(P, f) < \int_a^b f(x)dx + \varepsilon,$$

from which it follows that

$$\int_a^b f(x)dx - \varepsilon < S(P, f) < \int_a^b f(x)dx + \varepsilon,$$

or equivalently, that

$$\left| S(P, f) - \int_a^b f(x)dx \right| < \varepsilon.$$

We have shown that for any $\varepsilon > 0$ there is a δ such that if $||P|| < \delta$, then $\left| S(P, f) - \int_a^b f(x)dx \right| < \varepsilon$, and this is the definition of Riemann integrable according to Definition 1. ∎

We have shown that Definition (2) implies Definition (1) for nonnegative functions. How do we go from there to the same result for functions, f, that take on negative values? The next lemma answers that.

Lemma 7.56 *If f is Riemann integrable according to Definition (2) then so is $f + c$ for any constant c and $\int_a^b (f(x) + c)dx = \int_a^b f(x)dx + c(b - a)$.*

Proof. Observe first that if we chose a point t_i in the interval $[x_{i-1}, x_i]$ and compare the function values for f and $f + c$, we would find that $(f + c)(t_i) = f(t_i) + c$. So for any partition P of $[a, b]$,

$$\begin{aligned}
S(P, f + c) &= \sum_{i=1}^m (f(t_i) + c)(x_i - x_{i-1}) \\
&= \sum_{i=1}^m (f(t_i))(x_i - x_{i-1}) + \sum_{i=1}^m (c)(x_i - x_{i-1}) \\
&= \sum_{i=1}^m (f(t_i))(x_i - x_{i-1}) + c \sum_{i=1}^m (x_i - x_{i-1}) \\
&= \sum_{i=1}^m (f(t_i))(x_i - x_{i-1}) + c(b - a). \quad (7.26)
\end{aligned}$$

7.6. THE EQUIVALENCE OF...

Since $S(P, f) = \sum_{i=1}^{m}(f(t_i)(x_i - x_{i-1})$, from the previous chain of equalities we have that

$$S(P, f+c) = S(P, f) + c(b-a), \tag{7.27}$$

which in words, says that any Riemann sum for f and the corresponding Riemann sum for $f+c$ differ by the constant $c(b-a)$. Now, if $\int_a^b f(x)dx$ exists according to Definition (2), for any $\varepsilon > 0$ there is a partition P_ε such that for any refinement P of P_ε,

$$\left| S(P, f) - \int_a^b f(x)dx \right| < \varepsilon. \tag{7.28}$$

It follows from (7.28) and (7.27) that

$$\left| S(P, f+c) - \left(\left(\int_a^b f(x)dx\right) + c(b-a)\right) \right|$$
$$= \left| S(P, f) + c(b-a) - \left(\left(\int_a^b f(x)dx\right) + c(b-a)\right) \right|$$
$$= \left| S(P, f) - \int_a^b f(x)dx \right| < \varepsilon,$$

or said another way, $\int_a^b (f(x)+c)dx = \int_a^b f(x)dx + c(b-a)$ (since the Riemann sums for $f+c$ approach $\left(\int_a^b f(x)dx\right) + c(b-a)$ as we refine P). ∎

Finally, we have,

Theorem 7.57 *For any Riemann integrable function f defined on $[a, b]$, Definition (2) implies Definition (1).*

Proof. Add a constant c to f so that $f + c$ is nonnegative. Then by the previous lemma, $f + c$ is integrable according to Definition (2), and so by Theorem 7.55, $f + c$ is Riemann integrable according to Definition (1). Thus, there is a $\delta > 0$, such that if P is a partition with $||P|| < \delta$, then

$$\left| S(P, f+c) - \left(\left(\int_a^b f(x)dx\right) + c(b-a)\right) \right| < \varepsilon,$$

or equivalently, such that

$$\left| S(P, f) - \int_a^b f(x)dx \right| < \varepsilon.$$

So, f is Riemann integrable according to Definition (1). ∎

We can summarize our work from this section in one big theorem.

Theorem 7.58 *A bounded real valued function is Riemann integrable on $[a,b]$ if and only if any of the following conditions hold:*

(1) There is a number A which has the following property: For any $\varepsilon > 0$, there is a $\delta > 0$, such that for any partition, P, of $[a,b]$ with $||P|| < \delta$, $|S(P,f) - A| < \varepsilon$.

(2) There is a number A which has the following property: For any $\varepsilon > 0$, there is a partition, P_ε, of $[a,b]$ with the property that if P is a refinement of P_ε, $|S(P,f) - A| < \varepsilon$.

(3) $\overline{\int_a^b} f(x)dx = \underline{\int_a^b} f(x)dx$.

(4) For any $\varepsilon > 0$, there is a partition P of $[a,b]$ such that $U(P,f) - L(P,f) < \varepsilon$.

The number A is what we call the Riemann integral of $f(x)$ over $[a,b]$, which is denoted by $\int_a^b f(x)dx$, and $\overline{\int_a^b} f(x)dx = \underline{\int_a^b} f(x)dx = A$.

7.7 Riemann-Stieltjes Integrals

There is another integral, known as the Riemann -Stieltjes integral, whose development parallels the development of the Riemann integral, yet it is more general and has uses in the sciences and in probability. There are also four different definitions of the Riemann-Stieltjes integral, of which three out of the four are equivalent. However, we will concentrate on only one approach here.

Definition 7.59 *We say that a real valued function α is **increasing**, if whenever $x < y$, it follows that $\alpha(x) \leq \alpha(y)$.*

Example 7.60 *The function $f(x) = x^3$ is increasing, as is the function $f(x) = 2$, since the definition of increasing allows that $\alpha(x) = \alpha(y)$ when $x < y$.*

Definition 7.61 *Suppose that f and α are bounded real valued functions defined on $[a,b]$ and that α is increasing. Suppose that $P = \{x_0, x_1, x_2, ..., x_n\}$, is a partition of $[a,b]$, and that $\Delta \alpha_i = \alpha(x_i) - \alpha(x_{i-1})$ (which is necessarily ≥ 0 since α is increasing). Suppose that we pick points $t_i \in [x_{i-1}, x_i]$, and form the sum*

$$\sum_{i=1}^n f(t_i) \Delta \alpha_i.$$

Such a sum is called a **Riemann-Stieltjes sum**, and it is abbreviated $S(P, f, \alpha)$. Of course, there are infinitely many such sums.

Example 7.62 *Let $[a,b] = [1,4]$, $f(x) = 3x + 1$ and let $\alpha(x) = x^2$. Then α is increasing on $[a,b]$. Let $P = \{1,2,3,4\}$ and suppose that we use the right endpoint of*

7.7. RIEMANN-STIELTJES INTEGRALS

each subinterval made by the partition for our t_i. What is the Riemann-Stieltjes sum that corresponds to this partition and this choice of points for t_i?

According to our choice for the t_i, we have $t_1 = 2, t_2 = 3$, and $t_3 = 4$. Now, $\sum_{i=1}^{n} f(t_i)\Delta\alpha_i = f(2)(\alpha(2) - \alpha(1)) + f(3)(\alpha(3) - \alpha(2)) + f(4)(\alpha(4) - \alpha(3)) = 7(3) + 10(5) + 13(7) = 162$.

Definition 7.63 *Suppose that f and α are as in the previous definition. Suppose that there is a fixed number A having the following property: For each $\varepsilon > 0$ there is a $\delta > 0$, such that, for every partition P of $[a,b]$ with norm less than δ and every choice of points t_i in $[x_{i-1}, x_i]$,*

$$|S(P, f, \alpha) - A| < \varepsilon.$$

Then A is called the Riemann-Stieltjes integral of $f(x)$ with respect to α over $[a,b]$ and is denoted by $\int_a^b f(x) d\alpha(x)$, or more simply, as $\int_a^b f d\alpha$.

Remark 7.64 *Notice the wording, "A is called the Riemann-Stieltjes" integral, which seems to indicate that there can only be one value for A. That is true, and we ask you to prove that in the exercises. Notice also that when $\alpha(x) = x$, the Riemann-Stieltjes integral reduces to the ordinary Riemann integral, since $\sum_{i=1}^{n} f(t_i)\Delta\alpha_i = \sum_{i=1}^{n} f(t_i)(\alpha(x_i) - \alpha(x_{i-1})) = \sum_{i=1}^{n} f(t_i)(x_i - x_{i-1}) = \sum_{i=1}^{n} f(t_i)\Delta x_i$, which is a Riemann sum.*

Definition 7.65 *For any partition P of $[a,b]$, define $U(P, f, \alpha) = \sum_{i=1}^{n} M_i \Delta \alpha_i$, and $L(P, f, \alpha) = \sum_{i=1}^{n} m_i \Delta \alpha_i$, where $M_i = \sup f(x)$ on $[x_{i-1}, x_i]$ and $m_i = \inf f(x)$ on $[x_{i-1}, x_i]$.*

Remark 7.66 *The definition we have decided to work with is the analog of Definition 1 of Riemann integral. The other definitions of Riemann integrable are exactly the same as the definitions for Riemann integrable except that wherever one sees $S(P, f)$, one replaces it with $S(P, f, \alpha)$. For example, the definition of Riemann-Stieltjes integral according to Definition 3 would be "A bounded real valued function defined on $[a,b]$ is Riemann-Stieltjes integrable on $[a,b]$ with respect to α if $\overline{\int_a^b} f(x) d\alpha = \underline{\int_a^b} f(x) d\alpha$; that is, if $\inf U(P, f, \alpha) = \sup L(P, f, \alpha)$. In this case, the common value of $\overline{\int_a^b} f(x) d\alpha$ and $\underline{\int_a^b} f(x) d\alpha$ is called the Riemann-Stieltjes integral of $f(x)$ with respect to α on* [a, b] and is denoted by $\int_a^b f(x)d\alpha$. As we have pointed out, Definitions 2,3, and 4 for Riemann-Stieltjes integrals are equivalent.

Although we haven't talked about how to compute Riemann-Stieltjes integrals, we can prove some basic theorems similar to the ones we proved for Riemann integrals. Because most of the proofs are virtually identical to the corresponding proofs for Riemann integrals (only, wherever we see Δx_i we replace it with $\Delta \alpha_i$), we omit the proofs and simply state what is true, leaving out the details.

Theorem 7.67 *If f and g are Riemann-Stieltjes integrable on $[a, b]$, then so is $c_1 f + c_2 g$, where c_1 and c_2 are constants. Furthermore, $\int_a^b (c_1 f(x) + c_2 g(x))d\alpha = c_1 \int_a^b f(x)d\alpha + c_2 \int_a^b g(x)d\alpha$. In particular, for any constant, c, $\int_a^b cf\, d\alpha = c \int_a^b f\, d\alpha$.*

Theorem 7.68 *Suppose that $c \in (a, b)$. Then if $\int_a^c f d\alpha$ and $\int_c^b f d\alpha$ exist, so does $\int_a^b f d\alpha$, and $\int_a^b f d\alpha = \int_a^c f d\alpha + \int_c^b f d\alpha$.*

Theorem 7.69 *If $f \geq 0$ on $[a, b]$, and if $\int_a^b f d\alpha$ exists, then $\int_a^b f d\alpha \geq 0$.*

Theorem 7.70 *If $f(x) \leq g(x)$ on $[a, b]$, and if $\int_a^b f d\alpha$ and $\int_a^b g d\alpha$ both exist, then $\int_a^b f d\alpha \leq \int_a^b g d\alpha$.*

Theorem 7.71 *If P^* is a refinement of P, then $U(P^*, f, \alpha) \leq U(P, f, \alpha)$, and $L(P, f, \alpha) \leq L(P^*, f, \alpha)$.*

Theorem 7.72 *If $\int_a^b f d\alpha$ exists, then so does $\int_a^b |f|\, d\alpha$ and $\left|\int_a^b f d\alpha\right| \leq \int_a^b |f|\, d\alpha$.*

Theorem 7.73 *$L(P, f, \alpha) \leq S(P, f, \alpha) \leq U(P, f, \alpha)$ for any partition P of $[a, b]$.*

EXERCISES

1. Prove Theorem 7.67.

2. Prove Theorem 7.69.

3. Prove Theorem 7.70.

4. Show that the value of A is the definition of Riemann-Stieltjes integral is unique.

5. Let $f(x) = \begin{cases} 0 & \text{if } -1 \leq x < 0 \\ 1 & \text{if } 0 \leq x \leq 1 \end{cases}$ and let $\alpha(x) = \begin{cases} 0 & \text{if } -1 \leq x \leq 0 \\ 1 & \text{if } 0 < x \leq 1 \end{cases}$.

(a) Show that if we take $P = \{-1, 0, 1\}$, $U(P, f, \alpha) = L(P, f, \alpha) = 1$, and so f is Riemann-Stieltjes integrable with respect to α according to Defintion 2 (since Definitions 2, 3, and 4 of Riemann-Stieltjes integrable are equivalent). Also, show that for any refinement, P^*, of P, $S(P^*, f, \alpha) = 1$.

(b) Suppose that we use the partiion, P^{**} obtained by dividing $[-1, 1]$ it into equal lengths of size $1/n$, and then excluding the point 0. Form $S(P^{**}, f, \alpha)$ and use the point t in the interval $[-1/n, 1/n]$ to be $-1/n$. Then $S(P^{**}, f, \alpha) = 0$. Thus, there are partitions, P^{**} of arbitrarily small norm which make $S(P^{**}, f, \alpha) = 0$. If we now include the point 0 in the partition, we can get get another partition of norm less than $1/n$, where $S(P^{**}, f) = 1$. So the Riemann-Stieltjes integral does not exist according to Definition 1 and Defintions 1 and 2 for Riemann-Stieltjes integrals are not equivalent.

7.8 Evaluation of Riemann-Stieltjes Integrals

One of the most important things to know is when a Riemann-Stieltjes integral exists. Our first theorem, whose proof we omit, has a proof that is virtually identical to the proof of the analogous result for Riemann integrals. Wherever we have $b - a$, we substitute instead, $\alpha(b) - \alpha(a)$ and we use the equivalence of Definitions (2),(3), and (4) for Riemann-Stieltjes integrals, which we have pointed out is true.

Theorem 7.74 *If f is continuous on $[a, b]$, and α is increasing on $[a, b]$, then $\int_a^b f d\alpha$ exists.*

The next theorem relates evaluation of the Riemann-Stieltjes integral to the Riemann integral.

Theorem 7.75 *Suppose that $\int_a^b f d\alpha$ exists, and that α has a continuous derivative on $[a, b]$. Then $\int_a^b f d\alpha = \int_a^b f \alpha' dx$.*

Before giving the proof, we illustrate.

Example 7.76 *Let $f(x) = 3x + 1$, and let $\alpha(x) = x^2$. Then $\int_1^4 f d\alpha = \int_1^4 (3x + 1)(2x)dx = \int_1^4 (6x^2 + 2x)dx = 2x^3 + x^2\big|_1^4 = 141$.*

Proof. (of Theorem 7.75) By definition, the Riemann-Stieltjes sums for f, $S(P, f, \alpha)$, get close to $\int_a^b f d\alpha$ as the norm of the partition goes to zero, and the *Riemann* sums, $S(P, f\alpha')$, for $f\alpha'$, get close to $\int_a^b f\alpha' dx$. The idea of the proof is to realize that the

Riemann sums, $S(P, f\alpha')$, and the Riemann-Stieltjes sums, $S(P, f, \alpha)$, can be made close to each other as we take partitions with smaller and smaller norms, and so the integrals they get close to must be close to each other (and in fact can be made as close to each other as we wish, which implies they are equal).

Now, since the Riemann-Stieltjes integral $\int_a^b f d\alpha$ exists, we can find for any $\varepsilon > 0$, a δ_1 such that

$$|S(P, f, \alpha) - A| < \varepsilon/2, \tag{7.29}$$

where $A = \int_a^b f d\alpha$ and $||P|| < \delta_1$. Let us examine $S(P, f, \alpha)$ more closely. $S(P, f, \alpha) = \sum_{i=1}^n f(t_i)(\alpha(x_i) - \alpha(x_{i-1}))$. Applying the Mean Value Theorem to $\alpha(x)$ on each interval $[x_{i-1}, x_i]$, we get that $\alpha(x_i) - \alpha(x_{i-1}) = \alpha'(v_i)(x_i - x_{i-1})$, for some $v_i \in (x_{i-1}, x_i)$. So,

$$\begin{aligned} S(P, f, \alpha) &= \sum_{i=1}^n f(t_i)(\alpha(x_i) - \alpha(x_{i-1})) \\ &= \sum_{i=1}^n f(t_i)(\alpha'(v_i)) \Delta x_i. \end{aligned}$$

This would be a Riemann sum if the t_i were equal to v_i, but they are not necessarily the same. Let us look at the Riemann sum $S(P, f\alpha') = \sum_{i=1}^n f(t_i)(\alpha'(t_i))\Delta x_i$, for $f\alpha'$ where the v_i and t_i are the same, and consider the difference between the Riemann-Stieltjes sum, $S(P, f, \alpha)$, and this Riemann sum, $S(P, f\alpha')$.

$$\begin{aligned} |S(P, f, \alpha) - S(P, f\alpha')| &= \left| \sum_{i=1}^n f(t_i)(\alpha'(v_i))\Delta x_i - \sum_{i=1}^n f(t_i)(\alpha'(t_i))\Delta x_i \right| \\ &= \left| \sum_{i=1}^n f(t_i)((\alpha'(v_i) - \alpha'(t_i))\Delta x_i \right|. \end{aligned} \tag{7.30}$$

Since α' is continuous on the compact set $[a, b]$, it is uniformly continuous so if the distance between any two points, say v_i and t_i is small, say less than δ_2, we can make $((\alpha'(v_i) - \alpha'(t_i))) < \varepsilon/2M(b-a)$ where M is the sup of $|f(x)|$ on $[a, b]$. Let P be a partition of $[a, b]$ with norm less than $\delta = \min\{\delta_1, \delta_2\}$. Then, using (7.30) we

7.8. EVALUATION OF RIEMANN-STIELTJES INTEGRALS

have (taking note of the fact that $\sum_{i=1}^{n} \Delta x_i = b - a$),

$$\begin{aligned}
|S(P, f, \alpha) - S(P, f\alpha')| &= \left| \sum_{i=1}^{n} f(t_i)((\alpha'(v_i) - \alpha'(t_i))\Delta x_i \right| \\
&\leq \sum_{i=1}^{n} |f(t_i)(\alpha'(v_i) - \alpha'(t_i))\Delta x_i| \quad \text{(triangle inequality)} \\
&= \sum_{i=1}^{n} |f(t_i)||(\alpha'(v_i) - \alpha'(t_i))||\Delta x_i| \quad \text{(since } |ab| = |a||b|\text{)} \\
&< \sum_{i=1}^{n} M \frac{\varepsilon}{2M(b-a)} \Delta x_i, \text{ since } \Delta x_i \text{ is positive} \\
&= \sum_{i=1}^{n} \frac{\varepsilon}{2(b-a)} \Delta x_i \\
&= \frac{\varepsilon}{2(b-a)} \sum_{i=1}^{n} \Delta x_i \\
&= \frac{\varepsilon}{2(b-a)} (b-a) \\
&= \frac{\varepsilon}{2}.
\end{aligned}$$

So we have
$$|S(P, f, \alpha) - S(P, f\alpha')| < \frac{\varepsilon}{2}, \tag{7.31}$$

which says that every Riemann-Stieltjes sum, $S(P, f, \alpha)$, can be made close to the Riemann sum $S(P, f\alpha')$, where $t_i = v_i$. Now for the finishing touches:

$$\begin{aligned}
|S(P, f\alpha') - A| &= |S(P, f\alpha') - S(P, f, \alpha) + S(P, f, \alpha) - A| \\
&\leq |S(P, f\alpha') - S(P, f, \alpha)| + |S(P, f, \alpha) - A| \\
&\leq \frac{\varepsilon}{2} + \frac{\varepsilon}{2} \quad \text{(by (7.31) and (7.29))} \\
&= \varepsilon.
\end{aligned}$$

We have shown that, for any $\varepsilon > 0$, there is a $\delta > 0$, such that for any partition with norm less than δ, we have that the absolute value of the difference between a Riemann sum $S(P, f\alpha')$ for $f\alpha'$, and A, can be made less that ε. Thus, the Riemann integral $\int_a^b f\alpha' dx$ is equal to A. But recall that $A = \int_a^b f d\alpha$. So, $\int_a^b f\alpha' dx = \int_a^b f d\alpha$. ∎

The Riemann-Stieltjes integral is quite general, and our integrators can be discontinuous functions. Integrators which are step functions (defined below) play a strong role in probability. Riemann-Stieltjes integrals can handle both continuous and discrete probability distributions as well as mixed distributions.

Definition 7.77 *A real valued function, α, defined on $[a, b]$, is called a **step function** if there is a partition $P = \{x_0, x_1, x_2, ..., x_n\}$ such that α is constant on each open subinterval (x_{i-1}, x_i). The value of the function at the partition points, is irrelevant. Below, in Figure 7.13 we see the graph of a step function.*

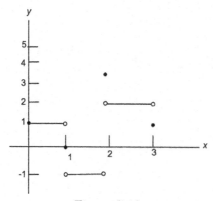

Figure 7.13.

Definition 7.78 *Using the notation of the previous definition, if we let $\alpha(x_k^+) = \lim_{x \to x^+} \alpha(x)$ and $\alpha(x_k^-) = \lim_{x \to x^-} \alpha(x)$, then the **jump** of α at x is defined to be $\alpha(x_k^+) - \alpha(x_k^-)$. The jump at $x_0 = a$ is defined to be $\alpha(x_0^+) - \alpha(x_0)$, while the jump at $x_n = b$ is defined to be $\alpha(x_n) - \alpha(x_n^-)$.*

Example 7.79 *Let*
$$\alpha(x) = \begin{cases} 1 & \text{if } 0 \leq x < 1 \\ 3 & \text{if } 1 < x < 2 \\ 4 & \text{if } 2 < x < 3 \\ 2 & \text{if } x = 1 \\ 3.5 & \text{if } x = 2 \\ 5 & \text{if } x = 3 \end{cases}$$

The graph of this step function, $\alpha(x)$, is given below in Figure 7.14.

7.8. EVALUATION OF RIEMANN-STIELTJES INTEGRALS

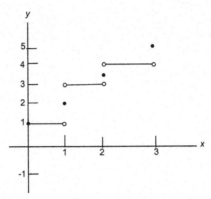

Figure 7.14.

The jump at $x = 0$ is $\alpha(0^+) - \alpha(0) = 0$. The jump at $x = 1$ is $\alpha(1^+) - \alpha(1^-) = 2$. The jump at $x = 2$ is $\alpha(2^+) - \alpha(2^-) = 1$. The jump at $x = 3$ is $\alpha(3) - \alpha(3^-) = 1$.

The following theorem, which we won't prove, tells us how to compute Riemann-Stieltjes integrals when the integrator is a step function.

Theorem 7.80 *Using the notation of the previous definition, suppose that α is a step function on $[a, b]$, and f is a real valued function defined on $[a, b]$. If at least one of f and α is continuous at each point x_k of the partition, then $\int_a^b f \, d\alpha = \sum_{k=0}^{n} f(x_k) \cdot (jump$ of α at $x_k)$.*

We refer the reader to Apostol's book [1] for the proof.

Corollary 7.81 *Finite sums of numbers can be written as Riemann-Stieltjes integrals.*

Proof. If $\sum_{k=1}^{n} a_k$ is any finite sum, then take $f(x) = a_k$ if $k-1 < x \leq k$, $k = 1, 2, ..., n$, and $f(0) = 0$. Let $\alpha(x) = [x]$ (defined to be the greatest integer less than or equal to x). Then it follows from the theorem that $\int_0^n f \, d\alpha = \sum_{k=0}^{n} (f(x_k)) \cdot$ (jump of α at x_k) $= \sum_{k=0}^{n} a_k \cdot$ (jump of α at x_k) $= \sum_{k=0}^{n} a_k$, since the jump at each x_k, other than x_0, is 1. ∎

Example 7.82 Let $f(x) = x^2$ and let $\alpha(x)$ be the function defined in Example 7.79. Then $\int_0^3 f d\alpha = f(0) \cdot$ (*the jump at* 0) $+ f(1) \cdot$ (*the jump at* 1) $+ f(2) \cdot$ (*the jump at* 2) $+ f(3) \cdot$ (*the jump at* 3) $= 0 + 1 \cdot 2 + 4 \cdot 1 + 9 \cdot 1 = 15$.

EXERCISES

1. Show that if $\alpha(x) = k$ (constant), then any bounded function $f(x)$ defined on $[a, b]$ is Riemann-Stieltjes integrable with respect to α.

2. Evaluate $\int_2^3 x^3 d(x^4)$.

3. Evaluate $\int_0^{\pi/2} \cos x \, d(\sin x)$.

4. Let $f(x) = \begin{cases} 0 & \text{if } x \in [0, \frac{1}{2}) \\ 1 & \text{if } x \in [\frac{1}{2}, 1] \end{cases}$. Let $\alpha(x)$ be the same except that $\alpha(\frac{1}{2}) = 0$. Evaluate $\int_0^1 f d\alpha$.

5. Let $f(x) = \begin{cases} 3 & \text{if } x \leq 0 \\ 3 - 2x & \text{if } 0 < x < 1 \\ 1 & \text{if } x \geq 1 \end{cases}$. Let $\alpha(x) = \begin{cases} 0 & \text{if } x \leq 0 \\ 3 & \text{if } 0 < x < 1 \\ 5 & \text{if } x \geq 1 \end{cases}$. Compute $\int_{-3}^3 f d\alpha$.

6. If α and β are increasing functions with continuous derivatives, show that $\int_a^b f d(\alpha + \beta) = \int_a^b f d\alpha + \int_a^b f d\beta$ assuming the integrals on the right exist.

7. Write out the proof of Theorem 7.74.

Chapter 8

Introduction to the Lebesgue Integral and Measure

8.1 Overview

In chapter 6 we defined the measure, or size, of a bounded interval, to be its length. The following extension of this definition is natural and will be used shortly.

Definition 8.1 *The measure of a set consisting of a disjoint union of intervals, is the sum of the lengths of the intervals. Thus, the measure of the set $E = [0,1) \cup (2,3) \cup [5,8]$ would be $1 + 1 + 3$ or 5.*

Notation 8.2 *If E is a disjoint union of intervals, we denote the measure of E by $m(E)$, consistent with our previous notation for measure of intervals.*

When studying the Riemann integral and which functions were integrable, Lebesgue knew that if a bounded function had a finite number of discontinuities, this would not affect the existence of the integral. But if it had too many discontinuities, it did. For example, the function $f(x) = 0$ if x is rational and 1 if x is irrational, defined on $[0,1]$, is discontinuous everywhere, and thus has no Riemann integral on this interval. We see what goes wrong with this example when we examine the Riemann sums. Any Riemann sum $\sum_{i=1}^{n} f(t_i)\Delta x_i$, will take on the value $b - a$ if we choose all the t_i to be irrational numbers in $[x_{i-1}, x_i]$, and will be 0 if we choose all the t_i to be rational. Thus, the sums cannot approach a fixed number regardless of the points we choose for the t_i as the norm of the partition goes to 0, which is a requirement for Riemann integrability. It is this "very discontinuous" behavior of $f(x)$ that is at fault. The function can vary by large amounts within a small interval, something

204   CHAPTER 8. INTRODUCTION TO THE LEBESGUE INTEGRAL

that cannot happen with a continuous function on $[a,b]$. Lebesgue thought about this problem deeply, and decided to try and approach integration differently. Rather than partition the domain where the function values can vary wildly, he thought it might be better to partition the range of the function instead, and look at the sets on the x-axis which map into these subintervals of the range, formed by the partition. We need to observe first, that given any interval I on the y-axis, the function values that lie in I is the set $f^{-1}(I)$. If the norm of the partition of the range is small, then the function values in these intervals on the y-axis cannot vary that much, and so in some sense, we have tamed the wild swings a function can have by considering these sets. Just as we did with Riemann integrals, we will try to approximate the area under a curve for a nonnegative continuous function using these sets, and see where it leads.

Example 8.3 *Look at the graph of the continuous function shown in Figure 8.1.*

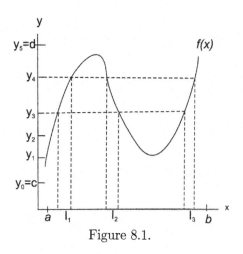

Figure 8.1.

The function with this graph is defined on $[a,b]$ and is bounded, so we can find an interval (c,d) such that $c < f(x) < d$. We partition $[c,d]$ as shown in the figure, and examine the pre-image of the half open intervals $[y_{i-1}, y_i)$, which we call E_i ($i = 1,2,3,4,5$). Thus, $E_1 = f^{-1}[y_0, y_1)$, $E_2 = f^{-1}[y_1, y_2)$, $E_3 = f^{-1}[y_2, y_3)$, $E_4 = f^{-1}[y_3, y_4)$ and $E_5 = f^{-1}[y_4, y_5)$. Figure 8.1 shows that $E_4 = f^{-1}[y_3, y_4)$ is a union of intervals, $I_1 \cup I_2 \cup I_3$. So, $m(E_4) = m(I_1) + m(I_2) + m(I_3)$ by Definition 8.1, and for any t_4 in $[y_3, y_4)$, the quantity $t_4 m(E_4) = t_4 m(I_1) + t_4 m(I_2) + t_4 m(I_3)$. Remembering that the measure of an interval is its length, the sum of terms on the right side of this equality can be interpreted as the sum of the areas of three rectangles (shaded), each whose height is t_4 and whose bases have measures equal to the lengths of the intervals I_1, I_2 and I_3, respectively, as shown in Figure 8.2.

8.1. OVERVIEW

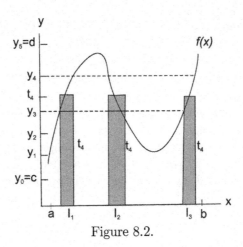

Figure 8.2.

So, for each t_i in $[y_{i-1}, y_i)$ we choose, we can interpret $t_i m(E_i)$ as a sum of areas of rectangles in this picture. If we form the sum, $\sum_{i=1}^{n} t_i m(E_i)$, this sum seems to be a sum of areas of rectangles filling out an area close to the area under the curve. So we would expect that as the norm of the partition (of the range) goes to zero, we should get the area under the curve, which is the same thing the Riemann integral gives us for nonnegative continuous functions that are bounded, like the one above.

Definition 8.4 *We call a sum of the form* $\sum_{i=1}^{n} t_i m(E_i)$, *a **Lebesgue sum**.*

We are now ready to be more formal. Lebesgue defined his integral as follows:

Definition 8.5 *Begin with a bounded function $f(x)$ defined on $[a, b]$, and find numbers c and d such that $c < f(x) < d$.*

(1) Partition $[c, d]$ using partition points $y_0, y_1, y_2, ..., y_n$.

(2) Pick a point t_i in each interval $[y_{i-1}, y_i)$.

(3) Form the (Lebesgue) sum $\sum_{i=1}^{n} t_i m(E_i)$ where $E_i = f^{-1}([y_{i-1}, y_i))$.

*(4) Now, let the norm of the partition approach zero. If the sums from (3) approach a fixed number, regardless of the points t_i chosen in $f^{-1}([y_{i-1}, y_i))$, then this fixed number is called the **Lebesgue integral of the function** $f(x)$ **on** $[a, b]$, and is denoted by $(L) \int_a^b f(x) dx$.*

The discussion before this definition seems to indicate that in the case of continuous functions, the value we get for our Lebesgue integral is the same as the value we get for the Riemann integral. The only way the two approaches differ is in how we accumulate areas of rectangles. In the Riemann approach, we partition $[a, b]$, pick a point t_i in each subinterval, $[x_{i-1}, x_i]$, and draw a rectangle whose height is $f(t_i)$ and whose base has length $\Delta x_i = x_i - x_{i-1}$. Thus, each rectangle can have a different height. In Lebesgue's approach, each term in the sum consists of a sum of areas of rectangles with the same height. The difference between Riemann's approach and Lebesgue's approach is similar to the following example, where you want to count the money in your pocket. Say you start pulling out the coins one at a time, to get a dime, a nickel, another nickel, a dime and then two quarters in a row, and then another nickel. The Riemann approach adds your coins as they are taken out (adds the area of the rectangles one at a time) to get the value of your money (in cents): $10 + 5 + 5 + 10 + 25 + 25 + 5$, while Lebesgue counts this way: He counts how many dimes there are, how many nickels there are and how many quarters there are. So his count looks like this: $10(2) + 5(3) + 25(2)$. (The analogy with Lebesgue sums is that we add the areas of rectangles with a given height, and do this for each specific height chosen.)

Lebesgue's idea is a good one. But we quickly run into a problem with his definition. What if the E_i are not intervals or finite unions of intervals? Then, how do we measure them? And what is the geometric interpretation of $t_i m(E_i)$ in this case? This led Lebesgue to develop a theory of measure for all kinds of sets on the real line. His measure theory was a successful attempt to measure the size of sets whose structures were not as simple as the union of intervals. We saw a bit of this approach in section 6.2, where we defined sets of measure 0. In his more general theory, he called a set E **measurable** if he could measure its "size" in some predetermined way (whose details we save for the end of the chapter). We denote the measure of any measurable set E by $m(E)$, just as we have done so far. What Lebesgue found, was that he could measure the "size" of the following sets on the real line: all open sets, all closed sets, sets formed from performing any or all of the following operations on measurable sets: countable unions, countable intersections, differences and complements. So, for example, if F_1, F_2, F_3, F_4, and F_5 are measurable sets, then so is $F = ((F_1 \cup F_2) - F_3)^c \cap F_4 \cap F_5$. Sets of measure 0, are of course, measurable, having measure 0.

Remark 8.6 *A special case of his theory says that if $[a, b]$ is considered as a metric subspace of R, then any open set in $[a, b]$ is measurable.*

Remark 8.7 *Lebesgue proved that if $\{E_i\}$ is a countable collection of sets which are pairwise disjoint (that is, $E_i \cap E_j = \varnothing$ if $i \neq j$), then $m(\cup E_i) = \sum m(E_i)$. We will*

8.1. OVERVIEW

need this soon.

Lebesgue found with his definition of "size", or measure, that if A and B are measurable sets with the measure of B finite, and if $A \subset B$, then the measure of $B-A$ is the measure of B minus the measure of A. In symbols, $m(B-A) = m(B) - m(A)$.

Example 8.8 *The set $[0,1]$ is measurable and has measure equal to its length, which is 1. The set E_1 of rationals in $[0,1]$, has measure 0, as we saw in Corollary 6.17. If we let E_2 be the set of irrational numbers in $[0,1]$, then $E_2 = [0,1] - E_1$ and it follows from the discussion in the previous paragraph, that $m(E_2) = m([0,1]) - m(E_1) = 1 - 0 = 1$. So the set of irrationals in $[0,1]$ has measure 1! Who would have thought we could measure the set of rationals and the set of irrationals included in an interval in this way?*

We can now do a more sophisticated example which illustrates how to integrate a function the Lebesgue way.

Example 8.9 *Let us look at the example we discussed earlier, namely, $f(x) = 0$ if x is rational and 1 if x is irrational, defined only on $[0,1]$. We already pointed out that this function has no Riemann integral (and is discontinuous everywhere on $[0,1]$), but we will now show that it does have a Lebesgue integral. We begin by enclosing the range of the function in an interval, say $[-.1, 1.1]$, and partition this in any way we want, using a partition P where $P = \{y_0, y_1, y_2, y_3, ..., y_n\}$, where $y_0 = -.1$ and $y_n = 1.1$. We make P fine enough so that 0 and 1 are in different intervals of the partition. We form the sets $E_i = f^{-1}([y_{i-1}, y_i))$. Now, the E_i are disjoint and if $1 \in [y_{i-1}, y_i)$, then $E_i = f^{-1}([y_{i-1}, y_i)) = \{\text{irrational numbers in } [0,1]\}$, and hence by Example 8.8, has measure 1. Thus, for any point $t_i \in [y_{i-1}, y_i)$, $t_i m(E_i) = t_i$. If $0 \in [y_{i-1}, y_i)$, then $E_i = f^{-1}([y_{i-1}, y_i)) = \{\text{rational numbers in } [0,1]\}$, and we know that $m(E_i) = 0$. So for any $t_i \in [y_{i-1}, y_i)$, $t_i m(E_i) = 0$. All other E_i are empty since the range of this function is only $\{0,1\}$. So any Lebesgue sum equals t_i where t_i is the sole interval on the y-axis containing 1. Now, as $||P|| \to 0$, the interval enclosing 1 gets smaller and smaller, and $t_i \to 1$. It follows that the sums $\sum_{i=1}^{n} t_i m(E_i)$ approach 1. Thus the Lebesgue integral of this function is 1, while the Riemann integral doesn't exist!*

Lebesgue succeeded in integrating this highly discontinuous (and non Riemann integrable) function, which is quite exciting! His idea is bearing fruit! The results of this approach are far reaching. There are many new functions we can integrate using Lebesgue's approach that we cannot integrate using Riemann's approach.

Getting back to the definition of the Lebesgue integral, in order to define it we must be able to measure $E_i = f^{-1}([y_{i-1}, y_i))$, and we were able to do this in the last example. The need to measure preimages of half open intervals led to the following definition.

Definition 8.10 *We say that a function $f : [a, b] \to R$ is **measurable**, if for every half open interval $[p, q)$ in R, $f^{-1}([p, q))$ is a measurable set.*

Thus, in order to define the Lebesgue integral, we must be working with *measurable functions*! Since our initial discussion of this integral seemed to indicate that for a continuous function the Lebesgue integral could be defined, and would give us the area under the curve, our first theorem is expected.

Theorem 8.11 *Every continuous real valued function defined on $[a, b]$ is measurable.*

Proof. $f^{-1}[p, q) = f^{-1}((p, q) \cup \{p\}) = f^{-1}((p, q)) \cup f^{-1}(\{p\})$. Now, by Theorem 4.27, $f^{-1}((p, q))$ is open in $[a, b]$ and hence measurable by Remark 8.6. Also, $f^{-1}(\{p\})$ is closed (Corollary 4.31) and hence measurable, and so $f^{-1}([p, q)) = f^{-1}((p, q)) \cup f^{-1}(\{p\})$, being the union of two measurable sets, is measurable. ∎

In contrast to Riemann's approach (which uses closed intervals), Lebesgue's approach uses half-open intervals. This is a technical device needed to guarantee that the sets E_i are disjoint, a necessity in his theory of measure to insure accuracy, and to guarantee that the measure of the union of the sets is the sum of the measures of the sets. To illustrate, if $A = (0, 2) \cup (1, 2)$, it would not be true that the measure of A is the sum of the measures of the intervals $(0, 2)$ and $(1, 2)$, since there is overlap. We need the intervals representing A to be disjoint in order for the sum of their lengths to give the measure of A. (See Remark 8.7.)

Remark 8.12 *One of the observations we will soon need is the following: If $f : [a, b] \to R$ is measurable such that $c < f(x) < d$, and if $E_i = f^{-1}[y_{i-1}, y_i)$, for $i = 1, 2, 3, ..., n$, then $\bigcup_{i=1}^{n} E_i = \bigcup_{i=1}^{n} f^{-1}[y_{i-1}, y_i) = f^{-1}\left(\bigcup_{i=1}^{n} [y_{i-1}, y_i)\right) = f^{-1}(c, d) = [a, b]$. And since the E_i are disjoint, $\sum_{i=1}^{n} m(E_i) = m[a, b]$ by Remark 8.7.*

The previous theorem tells us that the set of measurable functions contains all the continuous functions. Thus, Lebesgue's theory could handle integrals of continuous functions defined on a closed interval, as well as highly discontinuous functions as Example 8.9 shows. One can, if one wishes, prove all kinds of theorems similar to

8.1. OVERVIEW

those we proved for Riemann sums with this definition of the Lebesgue integral. For example, we can define **lower Lebesgue sums** by

$$\sum_{i=1}^{n} y_{i-1} m(E_i),$$

and **upper Lebesgue sums** by

$$\sum_{i=1}^{n} y_i m(E_i),$$

where $E_i = f^{-1}([y_{i-1}, y_i))$. One can also talk about **Lebesgue sums**, which are sums of the form

$$\sum_{i=1}^{n} t_i m(E_i)$$

where $t_i \in [y_{i-1}, y_i)$ and prove that the upper Lebesgue sums decrease or remain the same while the lower Lebesgue sums increase or remain the same, as we refine the partition. We can also easily prove that

$$\sum_{i=1}^{n} y_{i-1} m(E_i) \leq \sum_{i=1}^{n} t_i m(E_i) \leq \sum_{i=1}^{n} y_i m(E_i). \tag{8.1}$$

From this it follows (just as it does for Riemann integrals) that any lower sum is less than or equal to any upper sum, and therefore the supremum of the lower sums is less than or equal to the infimum of the upper sums.

We have an advantage with Lebesgue sums that we didn't have with Riemann sums. With Riemann sums, there was no guarantee that the infimum of the upper sums and the supremum of the lower sums, for a particular function, were the same (which if they were, would imply that the function is Riemann integrable). In fact, to prove the Riemann integrability for continuous functions, we had to use the sophisticated notion of uniform continuity. But for Lebesgue integrals, this is not an issue. The infimum of the upper sums and the supremum of the lower sums is *always* the same for bounded measurable functions defined on $[a, b]$, which is surprising. Let us see why.

We take the difference between the upper and lower sums associated with a partition, P. That difference (which is nonnegative) is

$$\sum_{i=1}^{n} y_i m(E_i) - \sum_{i=1}^{n} y_{i-1} m(E_i)$$
$$= \sum_{i=1}^{n} (y_i - y_{i-1}) m(E_i)$$
$$\leq ||P|| \sum_{i=1}^{n} m(E_i)$$
$$= ||P|| m([a,b]) \text{ (by Remark 8.12)}$$
$$= ||P||(b-a).$$

From this chain of relationships it follows that $0 \leq \sum_{i=1}^{n} y_i m(E_i) - \sum_{i=1}^{n} y_{i-1} m(E_i) \leq ||P||(b-a) \to 0$ as the norm of the partition goes to zero, and therefore the upper and lower sums *always* approach each other. Since $\inf \sum_{i=1}^{n} y_i m(E_i)$, $\sup \sum_{i=1}^{n} y_{i-1} m(E_i)$, and $\sum_{i=1}^{n} t_i m(E_i)$, are all squeezed between any lower sum and any upper sum (just as in the case of the Riemann integral), and the difference between any upper and corresponding lower sum goes to zero as the norm of the partition goes to zero, the infimum of the upper sums and the supremum of the lower sums are equal to each other as well as to $\lim_{||P|| \to 0} \sum_{i=1}^{n} t_i m(E_i)$ since it is squeezed between them. Since $\lim_{||P|| \to 0} \sum_{i=1}^{n} t_i m(E_i)$ exists, the Lebesgue integral of f on $[a,b]$ exists. **From this informal discussion, it follows that for a bounded measurable function defined on [a,b], the Lebesgue integral will always exist and can be described as the infimum of the Lebesgue upper sums, the supremum of the Lebesgue lower sums, or the limit as the norm of the partition goes to zero of the Lebesgue sums!** Contrast this to Riemann integrals which will only exist if the function is continuous except on a set of measure zero. Also, our discussion shows that we could alternatively define the Lebesgue integral for a bounded measurable function defined on $[a,b]$, as the common value of the infimum of the Lebesgue upper sums and the supremum of the Lebesgue lower sums.

Remark 8.13 *We can form Riemann sums for any bounded function defined on $[a,b]$. For Lebesgue sums we must use bounded **measurable** functions.*

8.1. OVERVIEW

Remark 8.14 *We should mention that the definition of the Lebesgue integral seems to indicate that the integral depends on the interval $[c,d]$, containing the range. That is not true. One can show that no matter which interval we take for $[c,d]$, the answer for the integral is the same. The proof of this relies on the fact that if $[c,d]$ and $[c,d']$ are two such intervals containing the range, where $d' < d$, all the sets E_i after d' are empty and so contribute nothing to the sum. So any Lebesgue sum on $[c,d']$ is one on $[c,d]$ (provided we take d' to be a partition point of the partition of $[c,d]$). A similar statement holds if we shrink the interval containing the range of f on the left side to give us an interval $[c',d'] \subset [c,d]$.*

The next theorem is important in the general development, but we won't prove it.

Theorem 8.15 *If $f(x)$ and $g(x)$ are defined on $[a,b]$ and $g(x)$ is a bounded measurable function, then if $f(x) = g(x)$ except on a set of measure zero, f is also measurable and $(L)\int_a^b f(x)dx = (L)\int_a^b g(x)dx$.*

Example 8.16 *Suppose that we enumerate the rational numbers to get a sequence, $r_1, r_2, r_3, \ldots$. Define $f_n(x) = 1$ everywhere except at $r_1, r_2, r_3, \ldots, r_n$, where $f_n(x) = 0$. Thus, $f_n(x) = 1$ on $[0,1]$ except for the first n rational numbers (which is a set of measure 0). Since each $f_n(x) = 1$ (except on a set of measure zero) and since the constant function $g(x) = 1$ is measurable, being continuous, we have, by the previous theorem, that each $f_n(x)$ is measurable and $(L)\int_0^1 f_n(x)dx = (L)\int_0^1 g(x)dx = (L)\int_0^1 1 dx = 1$ (which we get by direct computation using the definition of the Lebesgue integral for the function 1 on $[0,1]$). As $n \to \infty$, f_n approaches $f(x)$, where $f(x)$ is the function $f(x) = 0$ if x is rational and $f(x) = 1$ if x is irrational. Now, it is not true that Riemann integrals of $f_n(x)$ on $[0,1]$ converge to the Riemann integral of $f(x)$ (since $\int_0^1 f(x)dx$ doesn't even exist!). Yet, $(L)\int_0^1 f_n(x)dx \to (L)\int_0^1 f(x)dx$, since all the integrals have value 1. This kind of theorem, where the integrals of a sequence of functions converge to the integral of the limit function, is very desirable and happens under very simple conditions for the Lebesgue integral, which do not involve uniform convergence at all. See Theorem 8.24.*

Now, while we discussed this whole concept of Lebesgue integration for a bounded measurable function defined on a closed interval $[a,b]$, there is nothing special about using a closed interval for the domain. In fact, one can do *exactly* what we did with any bounded measurable function defined on a measurable set, E of finite measure. (Here measurable means that $f^{-1}[y_{i-1}, y_i)$ is a measurable set for each half-open interval, $[y_{i-1}, y_i)$). Namely, partition any interval containing the range, form Lebesgue sums as we did before, and take the limit as the norm of the partition goes to zero.

This limit will always exist (and the proof of this mimics the informal one given on page 210 when $E = [a, b]$). This gives us what we call the Lebesgue integral of $f(x)$ over E, which we denote by $(L) \int_E f(x)dx$. So we can integrate functions over the set E of rational numbers, say, in $[0, 1]$, or the set E of irrational numbers in $[0, 1]$, or any bizarre measurable set with finite measure. Lebesgue's idea brings us so far. We state for reference what we just observed.

Theorem 8.17 *If $f(x)$ is a bounded measurable function defined on a measurable set E with finite measure, then $(L) \int_E f(x)dx$ will always exist. Furthermore, we can write $(L) \int_E f(x)dx$ as $\inf \sum_{i=1}^{n} y_i m(E_i)$ or as $\sup \sum_{i=1}^{n} y_{i-1} m(E_i)$.*

We first give the flavor of a few theorems one proves when one studies the Lebesgue integral, and then move into the modern approach to the Lebesgue integral that one often sees in books.

Theorem 8.18 *Suppose that E is a measurable set of real numbers with finite measure, and that $f : E \to R$ is a bounded measurable function defined on E. If $c \leq f(x) \leq d$, then*

$$c \cdot m(E) \leq (L) \int_E f(x)dx \leq d \cdot m(E).$$

Proof. The proof is fairly straightforward. For any $p > 0$, we can enclose the range of $f(x)$ by the interval $(c - \frac{1}{p}, d + \frac{1}{p})$. Let us assume at first that p is fixed. Thus, for each t_i occurring in a Lebesgue sum, we have that

$$c - \frac{1}{p} < t_i < d + \frac{1}{p}.$$

Multiplying this chain of inequalities by $m(E_i)$, we have that for each $i = 1, 2, 3, ..., n$ (and this fixed p),

$$(c + \frac{1}{p}) \cdot m(E_i) < t_i m(E_i) < (d + \frac{1}{p}) \cdot m(E_i),$$

and summing from $i = 1$ to n we get

$$(c + \frac{1}{p}) \cdot \sum_{i=1}^{n} m(E_i) < \sum_{i=1}^{n} t_i m(E_i) < (d + \frac{1}{p}) \sum_{i=1}^{n} m(E_i).$$

But from Remark 8.7, since $\cup E_i = E$ ($i = 1, 2, 3, ..., n$), and the E_i are pairwise disjoint, we have that

$$(c + \frac{1}{p}) \cdot m(E) < \sum_{i=1}^{n} t_i m(E_i) < (d + \frac{1}{p}) \cdot m(E).$$

8.1. OVERVIEW

Now, take the limit of everything as the norm of the partition goes to zero, to get

$$(c+\frac{1}{p}) \cdot m(E) \leq (L)\int_E f(x)dx \leq (d+\frac{1}{p}) \cdot m(E),$$

which is true for any fixed p. Now, take the limit of this expression as $p \to \infty$, and we get the result we wanted to prove. ∎

Corollary 8.19 *If we (Lebesgue) integrate a bounded measurable function over a set of measure 0, then we get 0.*

Proof. This follows since both $c \cdot m(E) = 0$ and $d \cdot m(E) = 0$. ∎

Of course, it follows from this that if we integrate a bounded measurable function over a set of rational numbers, we will get 0.

Corollary 8.20 *If E has finite measure, then $\int_E c\,dx$, where c is a constant equals $cm(E)$.*

Proof. Just take $c = d$ in the theorem. ∎

It follows from this that $(L)\int_E 5dx$, where E is the set of irrational numbers in $[0, 1]$, is 5. This follows since $m(E) = 1$ by Example 8.8.

Corollary 8.21 *If f is a bounded measurable function defined on a measurable set E of finite measure, and $f(x) \geq 0$, then $(L)\int_E f(x)dx \geq 0$.*

Proof. Take c in the theorem equal to zero. ∎

There are other standard things that we expect to happen with integrals, like the integral of the sum is the sum of the integrals, etc., but we save those theorems for the next section where we prove them using the modern approach to the Lebesgue integral. They can be proved the way we prove the corresponding theorems for Riemann integrals, but we wanted to give the flavor of the modern approach.

Definition 8.22 *Suppose that $\{f_n(x)\}$ is a sequence of real valued functions defined on a set E. We say that $\{f_n(x)\}$ is uniformly bounded if there is a positive constant, M, such that $|f_n(x)| \leq M$ for all n and for all $x \in E$.*

Example 8.23 *The sequence of functions $\{f_n(x)\}$ where $f_n(x) = \sin nx$, is uniformly bounded. In fact, $|f_n(x)| \leq 1$ for each n.*

To finish this section, we state one of the most important theorems on Lebesgue integrals (but cannot prove it since we haven't developed the machinery). It is called Lebesgue's Dominated Convergence theorem.

Theorem 8.24 *(Lebesgue's Dominated Convergence theorem) Suppose that $f_n(x)$ is a uniformly bounded sequence of measurable functions defined on a measurable set E, with finite measure, and suppose that $f_n(x) \to f(x)$. Then*

$$(L)\int_E f_n(x)dx \to (L)\int_E f(x)dx.$$

One should refer back to Example 8.16, as the explanation of why $\int_{[0,1]} f_n(x)dx$ converges to $\int_{[0,1]} f(x)$ can now be given by this theorem– all the functions in that example were bounded by 1.

Remark 8.25 *The conclusion of this theorem can be stated as $\lim \int_E f_n(x)dx = \int_E \lim f(x)dx$, where the integrals are all Lebesgue integrals. That is, we can place the word "limit" after the integral sign, under these conditions.*

Example 8.26 *An example of a sequence of functions which is not uniformly bounded and whose integrals don't converge to the integral of the limit function follows: Consider the functions defined on $[0,1]$: $f_1 = 1$ if x is in $(0,1)$ and 0 otherwise, $f_2 = 2$ if x is in $(0, 1/2)$ and 0 otherwise, $f_3 = 3$ if x is in $(0, 1/3)$ and 0 otherwise, etc., where $f_n(x) = n$ if $x \in (0, \frac{1}{n})$ and 0 otherwise. This sequence is not uniformly bounded, and the Lebesgue integral of each is 1. (It is the same as the Riemann integral.) The limit, f, of the f_n, is 0 for each x in $[0,1]$, and the integral of f over $[0,1]$ is 0. So, the integrals of the f_n don't converge to the integral of f.*

EXERCISES

1. Give a formal definition of a bounded function defined on $[a,b]$ being Lebesgue integrable. Begin with the words,
   "We say that a bounded measurable function, f, defined on $[a,b]$ is Lebesgue integrable on $[a,b]$ if for all $\varepsilon > 0$...."

2. Prove (8.1).

3. We gave an informal proof that the difference between the upper and lower Lebesgue sums goes to zero as the norm of the partition goes to 0 and the infimum of the upper Lebesgue sums is the supremum of the lower Lebesgue sums. Let's try to formalize this.

    (a) Show that the difference $\sum_{i=1}^{n} y_i m(E_i) - \sum_{i=1}^{n} y_{i-1} m(E_i)$ can be made $< \varepsilon$ for any $\varepsilon > 0$.

8.1. OVERVIEW

(b) Since any lower sum is less than or equal to any upper sum, the upper sums are bounded below by any lower sum and hence the infimum of the upper sums exists. Similarly, every lower sum is bounded above by any upper sum, and so the supremum of the lower sums exist. Call the infimum of the upper sums b, and the supremum of the lower sums a. Show that we can make $|a - b| < \varepsilon$ for any $\varepsilon > 0$, and so $a = b$ by Lemma 2.41.

(c) Show that the limit of the Lebesgue sums as the norm of the partition goes to 0 is the common value of a and b. [Hint: Show that if $||P|| < \dfrac{\varepsilon}{2(b-a)}$ then each of $\left|\sum_{i=1}^{n} t_i m(E_i) - \sum_{i=1}^{n} y_{i-1} m(E_i)\right|$ and $\left|\sum_{i=1}^{n} y_{i-1} m(E_i) - b\right|$ can be made $< \dfrac{\varepsilon}{2}$.]

4. Suppose $f(x) = \begin{cases} 1 & \text{if } x = \dfrac{1}{n}, n = 1, 2, 3, \dots \\ 0 & \text{otherwise} \end{cases}$. What is the value of $L\int_0^1 f(x)dx$? What is the value of the Riemann integral of this function?

5. Show that a constant function, $f(x) = c$ defined on a measurable set, E is measurable.

6. Show that every function defined on a set, E, of measure 0 is measurable.

7. Show that the characteristic function, χ_E defined on a set measurable set X, is measurable, if and only if E is a measurable set.

8. Suppose that E_i are measurable sets for each $i = 1, 2, 3, \dots n$, and $E = \bigcup_1^n E_i$. Then if a real valued function, f, is defined on E, and measurable on each E_i, f is measurable on E. As a corollary to this and the previous exercise, get that step functions are measurable.

9. Suppose that f is defined on a measurable set X. Using the fact that a countable union of measurable sets is a measurable set, show that $f^{-1}([a, \infty))$ is a measurable set when f is measurable, and hence by going to complements, that $f^{-1}(-\infty, a)$ is measurable. Then show that $f^{-1}(-\infty, a]$ is a measurable set as well as $f^{-1}(\{a\})$.

10. Let L be the collection of lower sums, and S be the set of integrals of step functions greater than or equal to $f(x)$ on $[a, b]$. Let E be the set of simple measurable functions greater than or equal to $f(x)$ on $[a, b]$ and $\widetilde{E}$ the set of simple measurable functions less than or equal to $f(x)$ on $[a, b]$.

(a) Why is it true that every step function is a simple function?

(b) Why is it true that $\inf E \leq \inf S$, and $\sup L \leq \sup \widetilde{E}$?

(c) We know that $\alpha = (L) \int_a^b f(x)dx = \inf E = \sup \widetilde{E}$ and that $\beta = (R) \int_a^b f(x)dx = \sup L = \inf S$ as we saw in Exercise 7 of Section 7.6. Show that if a function, $f(x)$ is Riemann integrable on $[a, b]$, then $(L) \int_a^b f(x)dx$ is the same as the Riemann integral $\int_a^b f(x)dx$, by showing that $\alpha \leq \beta \leq \alpha$.

8.2 The Modern Approach to the Lebesgue Integral

In this section we will drop the (L) before the integral, as all integrals in this section will be Lebesgue integrals.

Definition 8.27 *Suppose that X is a set and that E is a subset of X. We define the characteristic function of a set, E, to be the real valued function defined on X as follows:* $\chi_E(x) = \begin{cases} 1 & \text{if } x \in E \\ 0 & \text{if } x \notin E \end{cases}$. *We will just abbreviate this as χ_E.*

Suppose that $X = [1, 4]$ and that $E = \{1, 4\}$. Then χ_E takes on the value 1 on the set E and 0 off of E. Its picture is shown in Figure 8.3.

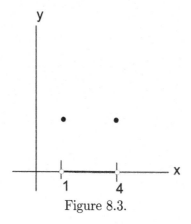

Figure 8.3.

The function χ_E, where $X = [1, 4]$, and $E = [1, 3]$ is shown in Figure 8.4. It takes on the value 1 on $[1, 3]$ and 0 otherwise.

8.2. THE MODERN APPROACH TO THE LEBESGUE INTEGRAL

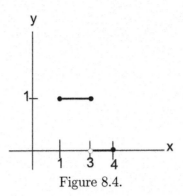

Figure 8.4.

Characteristic functions are very helpful in describing functions which only take on a finite number of values. Suppose that a function f is defined on a set X and only takes on the values 5 and 7. It takes on the value 5 on a set E_1 and the value 7 on the set E_2, where $E_1 \cup E_2 = X$. Then the function can be written as $f(x) = 5\chi_{E_1} + 7\chi_{E_2}$. To see this, pick an $x \in E_1$. Then $5\chi_{E_1}(x) = 5$ and $7\chi_{E_2}(x) = 0$, so $f(x) = 5$. Similarly, if we take an $x \in E_2$, then $5\chi_{E_1}(x) = 0$ and $7\chi_{E_2}(x) = 7$, so $f(x) = 7$. Obviously, $E_1 \cap E_2 = \emptyset$, since if there were an x in the intersection, $f(x)$ would take on two values, 5 and 7, which is impossible.

Definition 8.28 *A **simple measurable function** defined on a measurable subset E of R, is a function of the form $\Psi(x) = \sum_{i=1}^{n} \alpha_i \chi_{E_i}(x)$, where each E_i is a measurable subset of E and the sets E_i are pairwise disjoint.*

In Figure 8.5 we see the graph of the simple measurable function $f(x) = 3\chi_{[1,3]} + 4\chi_{(3,4)}$, defined on $[1,4]$. It takes on the value 3 on $[1,3]$, 4 on $(3,4)$, and 0 elsewhere.

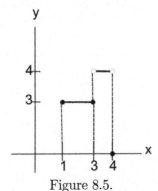

Figure 8.5.

Definition 8.29 If $\Psi(x) = \sum_{i=1}^{n} \alpha_i \chi_{E_i}(x)$ is a simple measurable function defined on a set E, we define $\int_E \Psi(x)dx$ to be $\sum_{i=1}^{n} \alpha_i m(E_i)$.

So, if we integrate the function $f(x) = 3\chi_{[1,3]} + 4\chi_{(3,4)}$ defined on $[1,4]$, we get that $\int_{[1,4]} f(x)dx = \int_{[1,4]} (3\chi_{[1,3]} + 4\chi_{(3,4)})dx$
$= 3(m([1,3]) + 4m((3,4)) = 3(2) + 4(1) = 10$. The integral of $f(x)$ over $[1,4]$ is shown in Figure 8.6. It is the shaded area.

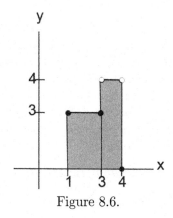

Figure 8.6.

Let us return to Figure 8.1 which we have redrawn below (as Figure 8.7), and examine it in the context of simple functions.

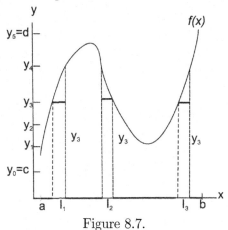

Figure 8.7.

8.2. THE MODERN APPROACH TO THE LEBESGUE INTEGRAL   219

You may remember that when discussing the Lebesgue integral of this function, we enclosed the range of the function by an interval (c, d) and partitioned the interval $[c, d]$. We then let E_i represent $f^{-1}([y_{i-1}, y_i))$. And then we talked about Lebesgue sums, which were sums of the form $\sum_{i=1}^{n} t_i m(E_i)$ where t_i was a point in $[y_{i-1}, y_i)$. Suppose that we take t_i to be a very specific point, namely, y_{i-1}. Then our Lebesgue sum becomes $\sum_{i=1}^{n} y_{i-1} m(E_i)$, which is a lower sum. Let us focus on one term in this sum, the term $y_3 m(E_4)$. If we define $\Phi_3(x) = y_3 \chi_{E_4}$ on the interval $[a, b]$, then the picture of $\Phi_3(x)$ is given in Figure 8.7 above. (It consists of the dark lines.) One sees from the picture that $\Phi_3(x)$ is below the curve $f(x)$ on E_4. That is to be expected, since if $x \in E_4 = f^{-1}([y_3, y_4))$, then $y_3 \leq f(x) < y_4$, which, in particular implies that $y_3 \chi_{E_4}(x) \leq f(x) \chi_{E_4}(x)$. But on E_4, $f(x) \chi_{E_4}(x)$ takes on the value $f(x)$, so that on E_4, $\Phi_3(x) = y_3 \chi_{E_4} \leq f(x)$.

The picture of $\int_{[a,b]} \Phi_3(x) dx$, is shown in Figure 8.8. (It is the shaded area.)

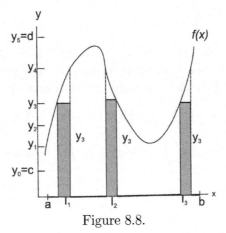

Figure 8.8.

The picture seems to indicate that $\int_{[a,b]} \Phi_3(x) dx \leq \int_{[a,b]} f(x) dx$, for this nonnegative continuous function. What we did above with E_4 we can do for each E_i, namely, define $\Phi_i(x) = y_{i-1} \chi_{E_i}$, for $i = 1, 2, ...n$, and we will have that on E_i, $\Phi_i(x) \leq f(x)$. Now, define a function $\Phi(x) = \sum_{i=1}^{n} y_{i-1} \chi_{E_i}$. Then, as above, $\Phi(x) \leq f(x)$ on $[a, b]$. (For if $x \in [a, b]$, then $x \in E_i$ for some unique i, so $\Phi(x) = y_{i-1} \chi_{E_i}(x)$, which we have seen is less than or equal to $f(x)$ on E_i.) A picture of $\Phi(x)$ is given in Figure 8.9. (It consists of the darkened lines below the curve $f(x)$.)

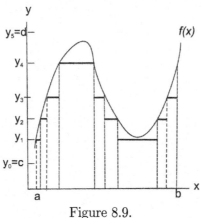

Figure 8.9.

Now, $\int_{[a,b]} \Phi(x)dx = \sum_{i=1}^{n} y_{i-1}m(E_i)$, by definition of the integral of a simple function, and we recognize this sum on the right side of the equality as a lower Lebesgue sum. **Thus, lower Lebesgue sums can be interpreted as integrals of simple measurable functions which are less than or equal to $f(x)$ on $[a,b]$.** The picture of $\int_{[a,b]} \Phi(x)dx = \sum_{i=1}^{n} y_{i-1}m(E_i)$ is given in Figure 8.10, and again, it is apparent from this picture that $\int_{[a,b]} \Phi(x)dx \le \int_{[a,b]} f(x)dx$.

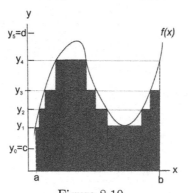

Figure 8.10.

8.2. THE MODERN APPROACH TO THE LEBESGUE INTEGRAL 221

In a similar manner, we can define (on $[a,b]$) the simple measurable function $\Psi(x) = \sum_{i=1}^{n} y_i \chi(E_i)$, and as above, $\Psi(x) \geq f(x)$ on $[a,b]$, and $\int_{[a,b]} \Psi(x)dx \geq \int_{[a,b]} f(x)dx$. We observe that $\int_{[a,b]} \Psi(x)dx = \sum_{i=1}^{n} y_i m(E_i)$ is an upper sum, and so **every upper sum can be interpreted as the Lebesgue integral of a simple measurable function which is greater than or equal to $f(x)$ on $[a,b]$**. Below, in Figure 8.11, we see a picture of $\sum_{i=1}^{n} y_i \chi_{E_i}$ (it consists only of the darkened lines),

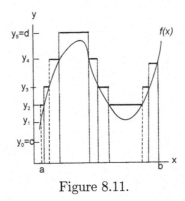

Figure 8.11.

and in Figure 8.12 we see its integral represented as the shaded area.

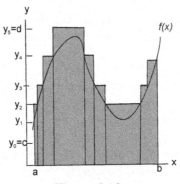

Figure 8.12.

The modern approach to the Lebesgue integral starts here, and defines for any bounded measurable function, $f(x)$, the Lebesgue integral of $f(x)$ (where $f(x)$ is defined and bounded on a measurable set E of finite measure) to be $\int_E f(x)dx = \inf \int_E \Psi(x)dx$, where $\Psi(x)$ is a simple measurable function defined on E which is $\geq f(x)$. So immediately we get that $\int_E f(x)dx \leq \int_E \Psi(x)dx$, where $\Psi(x)$ is any simple measurable function defined on E which is $\geq f(x)$. An alternate approach is to define the Lebesgue integral to be $\sup \int_E \Phi(x)dx$, where $\Phi(x)$ is a simple measurable function defined on E which is less or equal than f. With this approach, we immediately get that $\int_E \Phi(x)dx \leq \int_E f(x)dx$. On sets of finite measure, one can prove $\inf \int_E \Psi(x)dx = \sup \int_E \Phi(x)dx$ if and only if f is bounded and measurable. So the two approaches are equivalent on sets of finite measure. Let us prove some basic theorems about Lebesgue integrals using the modern approach.

We note at once, that if a is a constant, and if $\Psi(x) = \sum_{i=1}^n \alpha_i \chi_{E_i}$ is a simple function, $a\Psi(x)$ is also a simple function and $a \int \Psi(x)dx = a \sum_{i=1}^n \alpha_i m(E_i) = \sum_{i=1}^n a\alpha_i m(E_i) = \int a\Psi(x)dx$. Thus, for integrals of simple functions we can pull constants outside the integral sign. We now generalize this.

Theorem 8.30 *If f and g are bounded measurable functions defined on a set E of finite measure, and a is a constant, then*
(1) $\int_E af(x)dx = a\int_E fdx$, and
(2) $\int_E (f(x) + g(x))dx = \int_E f(x)dx + \int_E g(x)dx$.
(3) If $f \leq g$ on E, then $\int_E fdx \leq \int_E gdx$

Proof. (of (1)): Suppose first that $a > 0$, and that $\Psi(x)$ is any simple function defined on E which is $\geq f(x)$. Then $a\Psi(x)$ is a simple function which is $\geq af(x)$. Conversely, if we start with a simple function which is greater than or equal to $af(x)$ on E, and divide by a, we get a simple function $\geq f(x)$ on E. So the simple functions greater than or equal to af and those greater than or equal to f, are related by multiplication by the constant a (when $a > 0$). So, if we let $A = \{\int_E \Psi(x)dx | \Psi(x) \geq f(x) \text{ on } E\}$ and $B = \{\int_E a\Psi(x)dx | \Psi(x) \geq f(x) \text{ on } E\}$ (where $\Psi(x)$ is simple measurable function on E), then $B = aA$, since for simple functions we can pull constants out of integral signs. So, $\inf B = a \inf A$ (using the proof to part (1) of Theorem 2.29, with obvious modifications, to see this). But $\inf B = \int_E af(x)dx$ and $\inf A = \int_E f(x)dx$. And since $\inf B = a\inf A$, we have that $\int_E af(x)dx = a\int_E fdx$.

8.2. THE MODERN APPROACH TO THE LEBESGUE INTEGRAL

Now, suppose that $a < 0$. Then when we multiply a simple function, $\Phi(x)$ which is less than or equal to f on E, by the negative number a, we get a simple function greater than or equal to af on E, and if we divide a simple function greater than or equal to f by a, we get a function $\Psi(x) \leq f(x)$ on E. Thus, the simple functions less than or equal to f and those which are greater than or equal to af, are related by multiplication by the constant a, when $a < 0$. So, if we let $A = \{\int \Phi(x)dx | \Phi(x) \leq f(x) \text{ on } E\}$ and $B = \{\int a\Phi(x)dx | \Phi(x) \leq f(x) \text{ on } E\}$, then $B = aA$ since we can pull constants out of integral signs for simple functions. Also, since a is negative, $\inf B = a \sup A$ (using part (2) of Theorem 2.29 to help in understanding this). But $\inf B = \int_E af(x)dx$ and $\sup A = \int_E f(x)dx$. And since $\inf B = a \sup A$, we have that $\int_E af(x)dx = a \int_E f dx$.

(Proof of 2) Suppose that
$A = \{\int_E \Psi_1(x)dx \, | \, \Psi_1(x) \text{ is a simple measurable function} \geq f \text{ on } E\}$,
$B = \{\int_E \Psi_2(x)dx | \, \Psi_2(x) \text{ is a simple measurable function} \geq g \text{ on } E\}$, and
$C = \{\int_E \Psi(x)dx | \, \Psi(x) \text{ is a simple measurable function} \geq f + g \text{ on } E\}$.

Then $\inf A = \int_E f(x)dx$, $\inf B = \int_E g(x)dx$, and $\inf C = \int_E (f(x) + g(x))dx$. Since the sum of two simple measurable functions on E is a simple measurable function on E, we have that $A + B \subset C$. So, $\inf(A + B) \geq \inf C = \int_E (f(x) + g(x))dx$. But $\inf(A + B) = \inf A + \inf B$. So substituting into this last equation, we get

$$\int_E f(x)dx + \int_E g(x)dx \geq \int_E (f(x) + g(x))dx. \tag{8.2}$$

To prove the inequality the other way, we let
$A = \{\int_E \Phi_1(x)dx | \, \Phi_1(x) \text{ is a simple measurable function} \leq f \text{ on } E\}$,
$B = \{\int_E \Phi_2(x)dx | \, \Phi_2(x) \text{ is a simple measurable function} \leq g \text{ on } E\}$,
and $C = \{\int_E \Phi(x)dx | \, \Phi(x) \text{ is a simple measurable function} \leq f + g \text{ on } E\}$. Then $A + B \subset C$, as before. Now, $\sup A = \int_E f(x)dx$, $\sup B = \int_E g(x)dx$ and $\sup C = \int (f(x) + g(x))dx$. Since $\sup C \geq \sup(A + B) = \sup A + \sup B$, by substituting in we get that

$$\int_E (f(x) + g(x))dx \geq \int_E f(x)dx + \int_E g(x)dx. \tag{8.3}$$

From (8.2) and (8.3) we get that $\int_E (f(x) + g(x))dx = \int_E f(x)dx + \int_E g(x)dx$.

(Proof of 3) If a simple function defined on E is greater than or equal to g it is greater than or equal to f.

So if we let $A = \{\int_E \Psi_2(x)dx | \, \Psi_2(x) \text{ is a simple measurable function} \geq g \text{ on } E\}$ and $B = \{\int_E \Psi_1(x)dx | \, \Psi_1(x) \text{ is a simple measurable function} \geq f \text{ on } E\}$, then $A \subset B$, and so $\inf A \geq \inf B$.

But $\inf A = \int_E g(x)dx$ and $\inf B = \int_E f(x)dx$, from which the result follows. ∎

Although we have been using measurable sets throughout this chapter, we have never said exactly what Lebesgue's definition of a measurable set is. We do that now.

Lebesgue started with bounded sets. Now, it is a fact, that any bounded open set on the real line can be written uniquely as a disjoint union of open intervals, and we define the measure of such an open set to be the sum of the lengths of the (disjoint) intervals making up that set.

Definition 8.31 *For a bounded set of real numbers, E, we define the **outer measure** of E to be the $\inf\{m(O_\alpha)\}$, where the O_α are open sets containing E, and $m(O_\alpha)$ stands for the measure of O_α. The **inner measure** of E is defined to be the $\sup\{m(F_\alpha)\}$, where the F_α are closed sets contained in E. A set, E, is called **measurable**, if the outer and inner measure of the set is the same, and in this case the common value is called the **(Lebesgue) measure** of the set.*

When it came to unbounded sets, Lebesgue defined the measure as follows: We say that an unbounded set of real numbers, E, is measurable if each of the sets $E \cap [-n, n]$ is measurable. In this case, we define $m(E)$ to be $\lim(E \cap [-n, n])$. For example, R has infinite measure. For $R \cap [-n, n] = [-n, n]$ so $m(R \cap [-n, n]) = 2n$. As $n \to \infty$, $2n \to \infty$, so the measure of R is infinite. Similarly, $m([0, \infty)) = m((-\infty, a]) = \infty$.

As we have pointed out, Lebesgue showed that every open set is measurable as is every closed set, every countable union and intersection of these are measurable, and complements of measurable sets are measurable also. The proof of all these facts is time consuming and somewhat tedious.

Appendix

In what follows, $f(x)$ will be a bounded real valued function defined on $[a,b]$ and $P = \{x_0, x_1, x_2, ..., x_n\}$ will represent any partition of $[a,b]$. For each $i = 1, 2, 3, ..., n$, $M_i = \sup f(x)$ on $[x_{i-1}, x_i]$ while $m_i = \inf f(x)$ on $[x_{i-1}, x_i]$. M will represent $\sup f(x)$ on $[a,b]$, while m will represent $\inf f(x)$ on $[a,b]$. Our goal in this section is to prove Lebesgue's Theorem on the Existence of the Riemann integral, which is Theorem 7.30. We repeat that theorem here for convenience.

Theorem 8.32 *f is Riemann integrable on $[a,b]$ if and only if the set of discontinuities of $f(x)$ on $[a,b]$ has measure 0.*

The key ingredient in the proof is the behavior of $U(P,f) - L(P,f)$, the difference between the upper and lower sums. If we can find partitions that make this arbitrarily small, then f is Riemann integrable by Theorem 7.58. If we can't, then f is not. What we propose to show is that when the set of discontinuities of f on $[a,b]$ has positive measure, $U(P,f) - L(P,f)$ cannot be made arbitrarily small, but when the discontinuities has measure 0, this difference can be made as small as we wish.

Let us examine, for any partition P, $U(P,f) - L(P,f) = \sum_{i=1}^{n}(M_i - m_i)\Delta x_i$. What can prevent this from getting small? For one, the function can have big swings in a subinterval, no matter how small the subinterval is, and so $(M_i - m_i)$ will not be small. (This was the case with the function that was 1 at the rational numbers in $[0,1]$ and 0 at the irrational numbers in $[0,1]$. For that function, $(M_i - m_i)$ was always 1.) One might think that since Δx_i is getting small as the norm of the partition goes to 0, the bigness of $(M_i - m_i)$ would be counteracted by the smallness of the Δx_i and the term $(M_i - m_i)\Delta x_i$ will be small. But we are adding up a bunch of terms of the form $(M_i - m_i)\Delta x_i$, and for partitions with many points this sum can get large. So there is this battle going on between the bigness of $M_i - m_i$ and the smallness of the Δx_i, on intervals where the function has big swings. But since the function is bounded, each term $(M_i - m_i)\Delta x_i \leq (M - m)\Delta x_i$ and the sum of such terms would be $\leq (M - m)$ times the sum of the lengths of these intervals. So if the sum of the lengths of these intervals is small, the sum of these terms will be small. Surely, this must be where Lebesgue got the notion of measure 0. He wanted the sum of the lengths of the intervals to be small where the function had its big swings. Having a set of measure 0, would do the trick. This is the intuition behind Lebesgue's theorem.

The idea in both parts of the proof is to write $U(P,f) - L(P,f) = \sum_{i=1}^{n}(M_i - m_i)\Delta x_i$ as a sum of two terms, one where the function values have "big" swings, and the other where they don't. First, however, we need an observation. If $f(x)$ is not continuous at $c \in [a,b]$, then $\lim_{x \to c} f(x)$ doesn't exist. (f is defined at c) or, $\lim_{x \to c} f(x)$ does exist, but does not equal $f(c)$. In either case, that there is an $\varepsilon > 0$, such that for any open interval $(c-\delta, c+\delta)$ containing c, there are values x and y in the interval $[a,b] \cap (c-\delta, c+\delta)$, where the difference in the function values $f(x) - f(y)$ can be made $\geq \varepsilon$. Thus, if we let $M_\delta(c)$ be the supremum of $f(x)$ on the set $[a,b] \cap (c-\delta, c+\delta)$, and $m_\delta(c)$ be the infimum of $f(x)$ in that set, then $M_\delta(c) - m_\delta(c) \geq f(x) - f(y) \geq \varepsilon$. (Here we are using the fact that $M_\delta(c) - m_\delta(c) = \sup\{f(x) - f(y)\}$ for $x, y \in [a,b] \cap (c-\delta, c+\delta)\}$, which follows from Theorem 7.23.) So intuitively, if c is a point of discontinuity of f, the swings in the function values are "big" for any interval containing c where by "big" we mean $\geq \varepsilon$. Also, since for any $\varepsilon > 0$ we can find an n such that $1/n < \varepsilon$, we have that if $M_\delta(c) - m_\delta(c) \geq \varepsilon$, then $M_\delta(c) - m_\delta(c) \geq \frac{1}{n}$ for some n. The n varies one point of discontinuity to another. All of this shows the following:

Lemma 8.33 *If c is a point of discontinuity of $f(x)$ on $[a,b]$, there is an n such that $M_\delta(c) - m_\delta(c) \geq \frac{1}{n}$ for all $\delta > 0$.*

It follows that

Lemma 8.34 *If $D_n = \{c \in [a,b] | \ M_\delta(c) - m_\delta(c) \geq \frac{1}{n}$ for all $\delta > 0\}$ and D represents the set of discontinuities of $f(x)$ in $[a,b]$, then $D = \cup D_n$.*

Proof. If $c \in D$ is a point of discontinuity of f, then the previous lemma shows that $c \in D_n$ for some n. Thus

$$D \subset \bigcup_{i=1}^{\infty} D_n. \tag{8.4}$$

Conversely, if $d \in \bigcup_{i=1}^{\infty} D_n$, then $d \in D_n$ for some n. This means that for all $\delta > 0$, we can find x, y in the interval $[a,b] \cap (c-\delta, c+\delta)$, such that $f(x) - f(y)$ can be made $\geq \frac{1}{n}$, so f can't be continuous at d. That is, $d \in D$, and we have that:

$$\bigcup_{i=1}^{\infty} D_n \subset D. \tag{8.5}$$

(8.4) and (8.5) show that $D = \bigcup_{i=1}^{\infty} D_n$. ∎

8.2. THE MODERN APPROACH TO THE LEBESGUE INTEGRAL 227

Lemma 8.35 **Lemma 8.36** *D has measure 0 if and only if each D_n has measure 0 ($n = 1, 2, 3, ...$).*

Proof. If D has measure 0, then so does D_n since D_n is a subset of D.

Conversely, if D_n has measure 0 for each n, then so does D since the countable union of sets of measure 0 has measure 0. ∎

Lemma 8.37 *If $c \in D_n$ is a point of $[x_{i-1}, x_i]$, which is not an endpoint, then $M_i - m_i \geq \frac{1}{n}$ for some n.*

Proof. Since c is not an endpoint, there is an interval $I = (c-\delta, c+\delta) \subset [x_{i-1}, x_i] = S$. Since $I \subset S$, $M_\delta(c) \leq M_i$ and $m_\delta(c) \geq m_i$, so that $[m_\delta(c), M_\delta(c)] \subset [m_i, M_i]$. Since $M_\delta(c) - m_\delta(c) \geq \frac{1}{n}$, for some n, we have that $M_i - m_i \geq \frac{1}{n}$. ∎

Lemma 8.38 *If D_n doesn't have measure 0 for some fixed n, then $f(x)$ cannot be Riemann integrable on $[a, b]$.*

Proof. Suppose $P = \{x_0, x_1, x_2, ..., x_n\}$ is a partition of $[a, b]$. Consider the set $D'_n = D_n - P$. No point of D'_n can be an endpoint of any subinterval of the partition, and so if a point in D'_n is in $[x_{i-1}, x_i]$, then $M_i - m_i \geq \frac{1}{n}$ by the previous lemma for some fixed n. Since D_n doesn't have measure 0, and P is finite (and hence of measure 0), D'_n doesn't have measure 0. So, for some fixed $\varepsilon > 0$, and for *any* countable (in particular finite) collection of open intervals covering D'_n, the sum of the lengths of those intervals is at least ε. Consider $U(P, f) - L(P, f)$. Each point of D'_n is in one of the *open* subintervals formed by the partition. So these *open* intervals cover D'_n and the sum of their lengths has to be at least ε. As pointed out, $M_i - m_i \geq \frac{1}{n}$, so each term $(M_i - m_i)\Delta x$ in $U(P, f) - L(P, f)$, containing points of D'_n, is greater than or equal to $\frac{1}{n}\Delta x_i$, and the sum of these terms is greater than or equal to $\frac{1}{n}$(the sum of the lengths of the intervals)$\geq \frac{1}{n} \cdot \varepsilon = \frac{\varepsilon}{n}$. The remaining terms in $U(P, f) - L(P, f)$ are nonnegative since $M_i - m_i$ is. So, $U(P, f) - L(P, f) \geq \frac{\varepsilon}{n}$. Since n and ε are fixed, it is impossible to make $U(P, f) - L(P, f) < \frac{\varepsilon}{n} = \varepsilon'$ and this means, by Theorem 7.58, that f is not Riemann integrable, since for f to be Riemann integrable we have to be able to make $U(P, f) - L(P, f)$ smaller than any $\varepsilon' > 0$. ∎

Corollary 8.39 *(The first half of Lebesgue's theorem) If D, the set of discontinuities of $f(x)$ on $[a, b]$, does not have measure 0, then f is not Riemann integrable on $[a, b]$. Equivalently, if f is Riemann integrable, then the set of discontinuities of f on $[a, b]$ has measure 0.*

Proof. If D does not have measure 0, then by Lemma 8.36, some D_n does not have measure 0. So the corollary follows right away from the lemma. ∎

What we will show now, is that if D has measure 0, then we can always make $U(P,f) - L(P,f)$ as small as we want by taking an appropriate partition, from which it will follow that f is Riemann integrable on $[a,b]$. It is here where we use the notion of compactness.

Lemma 8.40 D_n *is closed, and hence compact.*

Proof. Let d be a limit point of D_n. We will show $d \in D_n$, and that will guarantee D_n is closed by Theorem 3.63. Consider $(d-\delta, d+\delta)$. We know, from Corollary 3.60, there is a sequence of points $c_n \in D_n$ converging to d. Thus, there will be a point c_{n_o} of this sequence in $(d-\delta, d+\delta)$. But since $(d-\delta, d+\delta)$ is open, we can find a $\delta_1 > 0$, such that $(c_{n_o} - \delta_1, c_{n_o} + \delta_1) \subset (d-\delta, d+\delta)$. By virtue of c_{n_o} being in D_n, $M_{\delta_1}(c_{n_o}) - m_{\delta_1}(c_{n_o}) \geq \frac{1}{n}$, and this forces $M_\delta(d) - m_\delta(d) = \sup f(x) - \inf f(x)$ on $(d-\delta, d+\delta)$ to be $\geq \frac{1}{n}$ since we are taking this difference over a larger interval. So $d \in D_n$, which makes D_n closed. Since D_n is a closed subset of a compact set $[a,b]$, D_n is compact. ∎

Lemma 8.41 *If, for a fixed n, $[x_{i-1}, x_i]$ contains no point of D_n, then we can subdivide the interval $[x_{i-1}, x_i]$ further, into subintervals where $\sup f(x) - \inf f(x) < \frac{1}{n}$ for each interval.*

Proof. It follows from the definition of D_n, that for each point c in $[x_{i-1}, x_i]$ which is not in D_n there is a $\delta > 0$, and an interval $(c-\delta, c+\delta)$, where $M_\delta(c) - m_\delta(c) < \frac{1}{n}$. Of course, this inequality will hold on the smaller interval $I_{\delta/2}(c) = (c-\delta/2, c+\delta/2)$. Consider the set of intervals of the form $I_{\delta/2}(c) = (c-\delta/2, c+\delta/2)$, for each such c. These cover the interval $[x_{i-1}, x_i]$, and by compactness of $[x_{i-1}, x_i]$ there is a finite subcover of $[x_{i-1}, x_i]$, $I_{\delta_1/2}(c_1), I_{\delta_2/2}(c_2),...,I_{\delta_k/2}(c_n)$. Now, let $\delta = $ the minimum of the $\delta_k/2$. Now, if we divide $[x_{i-1}, x_i]$ further so that the norm of the partition is less than δ, then if $|x-y| < \delta$ (where x and y are in $[x_{i-1}, x_i]$), then x and y will be in the same interval within that finite subcover, and it follows that $|f(x) - f(y)| < \frac{1}{n}$. (That x and y are in the same interval follows from the fact that if $x \in I_{\delta_j/2}$, then $|x - c| < \delta_j/2$. So if $|x-y| < \delta$, then $|y-c| \leq |y-x| + |x-c| < \delta + \delta_j/2 \leq \delta_j/2 + \delta_j/2 = \delta_j$. So $y \in I_{\delta_j/2}$ also.) ∎

Theorem 8.42 *(The second half of Lebesgue's theorem.) If D has measure 0, then for any $\varepsilon > 0$ we can find a partition of $[a,b]$ where $U(P,f) - L(P,f) < \varepsilon$, and thus f is Riemann integrable.*

Proof. Since D_n has measure 0, being a subset of D which has measure 0, we can cover D_n by a countable collection of open intervals, the sum of whose lengths is $< \frac{1}{n}$. This is true for any n. Since D_n is compact, there are a finite number of these open

8.2. THE MODERN APPROACH TO THE LEBESGUE INTEGRAL

intervals which cover D_n and of course, the sum of the lengths of these intervals is also $< \frac{1}{n}$. Let U be the union of these open intervals.(Figure $A.1$ shows the case where the union consists of two open intervals. The dots represent points of discontinuity of f.)

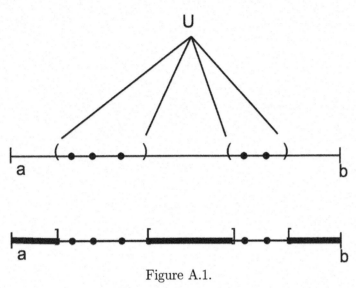

Figure A.1.

Now, $[a,b] \cap U^c$ is a set, S, of closed interval in $[a,b]$ containing no points of D_n. (See the bottom picture in Figure $A.1$. These closed intervals are bolded.) By the previous lemma, each of these closed subintervals can be further divided into subintervals where $\sup f(x) - \inf f(x) < \frac{1}{n}$. The set of subdivision points together with a and b, form a partition, P, of $[a,b]$. Consider $U(P,f) - L(P,f)$. The terms of this sum fall into two categories: those terms where $[x_{i-1}, x_i]$ contains at least one point of D_n, and the remaining terms. Suppose A is the set of intervals from the partition containing points of D_n and let B be the remaining intervals. Then $\sum_A (M_i - m_i) \Delta x_i \leq \sum_A (M - m) \Delta x_i \leq (M - m)$ (sum of the lengths of the intervals from A) $\leq (M - m) \cdot \frac{1}{n} = \frac{M-m}{n}$. Furthermore, $\sum_B (M_i - m_i) \Delta x_i < \frac{1}{n}$(sum of the lengths of the intervals)$< \frac{1}{n}(b-a) = \frac{b-a}{n}$, since we have subdivided the intervals in B so that $\sup f(x) - \inf f(x) < \frac{1}{n}$. Now, $U(P,f) - L(P,f) = \sum_{i=1}^n (M_i - m_i) \Delta x_i = \sum_A (M_i - m_i) \Delta x_i + \sum_B (M_i - m_i) \Delta x_i \leq \frac{M-m}{n} + \frac{b-a}{n} = \frac{M-m+b-a}{n} < \varepsilon$, when

$$n > \frac{M-m+b-a}{\varepsilon}.$$

For any $\varepsilon > 0$, we have found a partition P of $[a,b]$ such that $U(P,f) - L(P,f) < \varepsilon$, which implies f is Riemann integrable on $[a,b]$. ∎

Bibliography

1. Apostol, T., *Mathematical Analysis*, 2nd ed. Addison Wesley, Reading Massachusetts, 1974.

2. Ash, R. Real Variables With Metric Space Topology, Dover Publications, New York, 2007.

3. Bartle, R.G. and Sherbert, D.R., Introduction to Real Analysis, 3rd ed. John Wiley and Sons, New York, 1999.

4. Johnsonbough, R.W. and Pfaffenberger, W.E., *Foundations of Mathematical Analysis*, Dover Publications, New York, 2002.

5. Rudin, W. , *Principles of Mathematical Analysis*, 3rd ed McGraw Hill, New York, 1976.

Solutions to Most Exercises

Section 1.1

1.
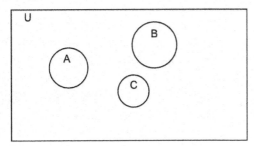

2.
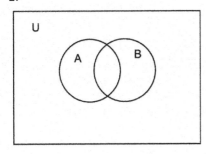

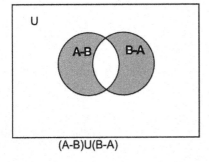
(A-B)U(B-A)

3. $A \times A = \{(1,1), (1,2), (2,1), (2,2)\}$, $B \times A = \{(2,1), (2,2), (4,1), (4,2), (5,1), (5,2)\}$

4. $\cup O_x = (-.5, 1.5)$. $\cap O_x = \emptyset$. If $O_x = [x - .5, x + .5]$, $\cap O_x = \{\frac{1}{2}\}$.

5. $\{2\}$

6. If $x \in A - B$, then $x \in A$ and not in B, so $x \in A \cap B^c$. So $A - B \subset A \cap B^c$. All the steps are reversible, so $A \cap B^c \subset A - B$, and we have that $A - B = A \cap B^c$.

7. If $x \in B^c$, then x is not in B. So x cannot be in A, since $A \subset B$. We have shown that everything not in B is not in A, so $B^c \subset A^c$.

8. $((A \cap B) - C) \cup ((A \cap C) - B) \cup ((B \cap C) - A)$

9.   a. If $f(a_1) = f(a_2)$, then $a_1 = a_2$.

    b. We will show that if $f(a_1) = f(a_2)$, then $a_1 = a_2$. But if $f(a_1) = f(a_2)$, $ma_1 + b = ma_2 + b$. Subtracting b from both sides and dividing by m, which is not 0, we get that $a_1 = a_2$.

    c. If $(g \circ f)(a_2) = (g \circ f)(a_1)$, $g(f(a_2)) = g(f(a_1))$, and since g is $1-1$, it follows that $f(a_2) = f(a_1)$, and since f is $1-1$, it follows that $a_2 = a_1$. So, $(g \circ f)(a_2) = (g \circ f)(a_1)$ implies that $a_2 = a_1$, and therefore $g \circ f$ is $1-1$.

    d. If $x \in \cap f(E_\alpha)$ (if x is the image of something in each $f(E_\alpha)$) then it need not be true that x is the image of something that occurs simultaneously in each E_α. So x need not be in $f(\cap E_\alpha)$.

    e. True. We know that $f(\cap E_\alpha) \subset \cap f(E_\alpha)$. To prove the inclusion the other way, suppose that $p \in \cap f(E_\alpha)$. Then $p \in f(E_\alpha)$ for each α. p cannot be the image of several different things since f is $1-1$. So p is the image of only one thing, z where $z \in \cap E_\alpha$, and thus $p \in f(\cap E_\alpha)$. So $\cap f(E_\alpha) \subset f(\cap E_\alpha)$. This together with the reverse inclusion proves $f(\cap E_\alpha) = \cap f(E_\alpha)$.

10. a. If b is in the range of a function f, then $b = f(a)$, for some a in the domain of f by definition of the range. So any function is onto its range.

    b. For any c in the range, $c = f(\frac{c-b}{m})$.

    c. $f(x) = \tan\left(\frac{\pi}{2}x\right)$

    d. True. If $c \in C$, then $c = g(b)$ for some $b \in B$ since g is onto C. But since f is onto B, $b = f(a)$ for some $a \in A$. It follows that $c = g(b) = g(f(a))$, so $g \circ f$ is onto C.

    e. True. We already know that $E \subset f^{-1}(f(E))$. To prove the inclusion the other way, suppose that $p \in f^{-1}(f(E))$. Then $f(p) \in f(E)$, which means that $f(p) = f(z)$ for some $z \in E$. Since f is $1-1$, $p = z \in E$, so $p \in E$. We showed that $p \in f^{-1}(f(E))$ implied that $p \in E$, so $f^{-1}(f(E)) \subset E$. This, together with the reverse inclusion, which is always true, shows that $E = f^{-1}(f(E))$ when f is $1-1$.

f. False. Let $f(x) = 1$ for all real numbers. Then f is onto $\{1\}$. Let $E = \{1\}$. Then $f(E) = 1$, and $f^{-1}(\{1\}) = R$. So $f^{-1}(f(E)) = R$, which does not equal E.

11. f is $1-1$.

Section 1.2

1. If $\dfrac{p}{q}$ and $\dfrac{r}{s}$ are rational numbers, then $\dfrac{p}{q} + \dfrac{r}{s} = \dfrac{ps + rq}{qs}$ is rational since the numerator and denominator are integers. The others are proved similarly.

2. Suppose that there were only finitely many rationals between the two real numbers, r and s, where $r < s$. Let d be the half the smallest distance between these points and r and let $r_0 = r + d$. By hypothesis, there are no rationals between r and r_0 which contradicts Theorem 1.47.

3. Counterexample: $\sqrt{2} + (1 - \sqrt{2}) = 1$

4. If $\varnothing_1$ and $\varnothing_2$ are two null sets, then since the null set is a subset of every set, $\varnothing_1 \subset \varnothing_2$, and $\varnothing_2 \subset \varnothing_1$, implying that $\varnothing_1 = \varnothing_2$.

5. If $x \in f^{-1}(S)$, then $f(x) \in S$, which means that $f(x) \in T$ since $S \subset T$. But if $f(x) \in T$, then $x \in f^{-1}(T)$. We started with an arbitrary $x \in f^{-1}(S)$ and showed it was in $f^{-1}(T)$, so $f^{-1}(S) \subset f^{-1}(T)$.

6. $|a| = |a - b + b| \leq |a - b| + |b|$, from which the result follows.

7. We show that $|b| - |a| \leq |a - b|$ by switching the roles of a and b in the inequality from question 6. Calling $N = |a| - |b|$, our two inequalities, $|b| - |a| \leq |a - b|$ and $|a| - |b| \leq |a - b|$, tell us that $-N \leq |a - b|$ (or equivalently $N \geq -|a - b|$) and $N \leq |a - b|$ from which it follows that $|N| \leq |a - b|$ (Proposition 1.50 part 5), or said differently, $||a| - |b|| \leq |a - b|$.

8a. There are $d, e > 0$ such that $a + d = b$, and $b + e = c$. Substituting for $a + d$ for b in the second equation we get $a + d + e = c$. Since $d + e > 0$, $a < c$.

b. There is a d such that $a + d = b$. Multiplying by c we get $ca + cd = cb$. Since cd is positive, $ca < cb$.

c. Since $-c$ is positive, $-ca < -cb$. This implies that there is a $d > 0$ such that $-ca + d = -cb$. Multiplying by -1, we get that $ca - d = bc$, or that $cb + d = ca$. So, $cb < ca$.

d. If we multiply $a < b$ by a, we get $a^2 < ab$ since $a > 0$. If we multiply $a < b$ by b, we get $ab < b^2$ since $b > 0$. Using these two inequalities and part (a), we get that $a^2 < b^2$. If $a = -3$ and $b = 2$, then $a < b$, but a^2 is not less than b^2.

9. a. When $a \geq 0$, $|a| = a \geq 0$. When $a < 0$, $|a| = -a > 0$.

b. If a and b are ≥ 0, so is ab. So $|ab| = |a||b|$. If $a > 0$ and $b < 0$, then ab is negative and $|ab| = -ab = a(-b) = |a||b|$, and so on.

c. If $|a| < c$, then the distance a is from the origin is less than c, which means $-c < a < c$ and $|a| < c$ are equivalent.

Section 1.3.

1. There is an x such that $|x| < 2$, but $|f(x)| \geq 4$.

2. There is an $x \in E$ such that for all r, $B_r \not\subset E$.

3. For some $\varepsilon > 0$, and for all $\delta > 0$, there is an $x \in E$ where $|x| < \delta$, but $|f(x)| \geq \varepsilon$.

4. For all $M > 0$, there exists an n such that $|a_n| \geq M$.

Section 2.1.

1. If r and s are suprema, then since a supremum is less than or equal to any other upper bound, we have $r \leq s$. Similarly, $s \leq r$, and these together imply that $r = s$.

2a. $GLB(A) = -1$, $LUB(A) = 1$. A is bounded.

   b. $GLB(B) = 0$, $LUB(B) = \frac{1}{2}$. B is bounded.
   c. $GLB(C) = \frac{1}{2}$, $LUB(C) = \infty$. C is unbounded.
   d. $GLB(D) = -\infty$, $LUB(D) = \infty$. D is unbounded.
   e. $GLB(E) = \frac{2}{3}$, $LUB(E) = 1$. E is bounded.
   f. $GLB(F) = 0$, $LUB(E) = 1$. F is bounded.
   g. $GLB(F) = 0$, $LUB(E) = \sqrt{2} - 1$. F is bounded.

3. If B is bounded above, then since everything in A is in B, every upper bound for B is an upper bound for A. So A is bounded above. Similarly, every lower bound for B is less than or equal to everything in B and hence less than or equal to everything in A, since $A \subset B$. So A bounded below. Being bounded above and below, A is bounded.

4. Since $a \leq b$ for all $b \in B$, $a \leq \inf B$ by Remark 2.15. So $\inf B$ is an upper bound for A, therefore $\sup A \leq \inf B$ again, by Remark 2.15.

5. Yes. 3 is clearly an upper bound for this set. Suppose that the least upper bound, a, for this set satisfies $1 < a < 3$. Then there is an irrational number i such that $1 < a < i < 3$, since between any two real numbers there is an irrational number (Theorem 1.49). This contradicts the fact that a is the least upper bound for the set of irrationals between 1 and 3.

6. $\sqrt{2}$ is an upper bound for the set. Suppose that a is the least upper bound for this set and that a is less than $\sqrt{2}$. Then there is a rational number between a and $\sqrt{2}$ (Theorem 1.47), contradicting that a is the least upper bound of the set.

7. Suppose that B is bounded below by a. Then $a \leq b$ for all b in B. Multiplying this inequality by -1, we get that $-a \geq -b$ for all $b \in B$, which says that $-a$ is an upper bound for $-B = \{-b | b \in B\}$. Since we are assuming that $-B$ has a least upper bound l, we have $l \leq -a$, from which it follows that $-l \geq a$. So $-l$ is greater than or equal to a, any lower bound for B. We need only show that $-l$ is a lower bound for B. Since l is an upper bound for $-B$, $l \geq -b$, for all $b \in B$, from which it follows that $-l \leq b$ for all $b \in B$, which shows that $-l$ is a lower bound for B. So $-l$ is a lower bound for B greater than or equal to any other lower bound for B, so $-l$ is the greatest lower bound of B.

Section 2.2.

1. If $\sup A \in A$, then $\sup A$ is in the set. But by definition of the supremum, $\sup A$ is an upper bound for the set and hence greater than or equal to everything in the set. These two observations make $\sup A$ the maximum element of the set.

2. This follows from Theorem 2.26 by taking $A = \{a\}$.

3. $\inf A \leq a$ for all $a \in A$. So $c \inf A \geq ca$ for all $a \in A$, from which we have (1): $c \inf A \geq \sup cA$ (by Remark 2.15). Also, $\sup cA \geq ca$ for all $a \in A$, so $\dfrac{\sup cA}{c} \leq a$, which implies that $\dfrac{\sup cA}{c} \leq \inf A$, again, by Remark 2.15. Multiplying by c we get $\sup cA \geq c \inf A$. This, coupled with $c \inf A \geq \sup cA$, which we already proved, shows that $\sup cA = c \inf A$.

4. True.

Section 2.3.

1. For any $\varepsilon > 0$, take $N = 1$.

2. False. $a_n = \dfrac{-1}{n} < \dfrac{1}{n} = b_n$, but $\lim a_n = \lim b_n = 0$. The correct conclusion is that $\lim a_n \leq \lim b_n$.

3. $1, 2, 1, 2, 1, 2, \ldots$

4. No such example.

5. $a_n \leq \sup\{a_n\}$ and $b_n \leq \sup\{b_n\}$. Adding, we get $a_n + b_n \leq \sup\{a_n\} + \sup\{b_n\}$. Thus the fixed number $\sup\{a_n\} + \sup\{b_n\}$ is greater than or equal to $a_n + b_n$ for all n, so $\sup\{a_n + b_n\} \leq \sup\{a_n\} + \sup\{b_n\}$, by Remark 2.15. This does not contradict Theorem 2.26 since that theorem refers to taking all possible sums from the sets, not just the sums of the corresponding terms of the sequences. To see that they are not equal, let $a_n = (-1)^n$ and $b_n = (-1)^{n+1}$. Then $\sup\{a_n\} = \sup\{b_n\} = 1$, but $\sup\{a_n + b_n\} = 0$.

6a. The limit is 0. $\left|\left(\dfrac{1}{2}\right)^n - 0\right| = \left|\left(\dfrac{1}{2}\right)^n\right| = \left(\dfrac{1}{2}\right)^n < \varepsilon$ when $n > N = \dfrac{\log \varepsilon}{\log(1/2)}$.

b. The limit is 0. $\left|\dfrac{-2}{3+n^2}\right| = \dfrac{2}{3+n^2} < \dfrac{2}{n^2} < \varepsilon$ when $n > N = \sqrt{\dfrac{2}{\varepsilon}}$.

c. The limit is $\dfrac{2}{3}$. $\left|\dfrac{2n+3}{3n+2} - \dfrac{2}{3}\right| = \dfrac{5}{9n+6} < \dfrac{5}{9n} < \varepsilon$ when $n > N = \dfrac{5}{9\varepsilon}$.

d. The limit is 0. $\left|\dfrac{3n^2 + 4n + 1}{4n^3 + 2n + 3} - 0\right| < \dfrac{3n^2 + 4n^2 + n^2}{4n^3} = \dfrac{10}{4n} < \varepsilon$ when $n > N = \dfrac{10}{4\varepsilon}$.

e. The limit is 0. Rationalizing the numerator, we get $\left|\sqrt{n+1} - \sqrt{n}\right| = \left|\dfrac{1}{\sqrt{n+1} + \sqrt{n}}\right| < \dfrac{1}{2\sqrt{n}} < \varepsilon$ when $n > N = \left(\dfrac{1}{2\varepsilon}\right)^2$.

f. The limit is $\dfrac{1}{2}$. $\left|\sqrt{n^2 + n} - n - \dfrac{1}{2}\right| = \left|\dfrac{2(\sqrt{n^2+n}) - 1 - 2n}{2}\right|$
$= \left|\dfrac{2\sqrt{n^2+n} - (2n+1)}{2}\right| = \left|\dfrac{-1}{2(2\sqrt{n^2+n} + (2n+1))}\right|$

$$= \frac{1}{2(2\sqrt{n^2+n}+(2n+1))} =$$
$$\frac{1}{4\sqrt{n^2+n}+(4n+2)} < \frac{1}{4\sqrt{n^2}} = \frac{1}{4n} < \varepsilon \text{ when } n > N = \left(\frac{1}{4\varepsilon}\right).$$

7. $\lim \frac{3n}{n+1} = \lim \frac{3}{1+\frac{1}{n}} = \frac{\lim 3}{\lim 1 + \frac{1}{n}} = \frac{3}{1} = 3$. The other part is done similarly by dividing the numerator and denominator by x^n, where n is the degree of the polynomials.

8. $a_n = (-1)^n$

9. Since $\{a_n\}$ converges to 3, for any $\varepsilon > 0$, there is an N such that $n > N$ implies that $|a_n - 3| < \varepsilon\sqrt{3}$. Multiplying $\sqrt{a_n} - \sqrt{3}$ by $\frac{\sqrt{a_n}+\sqrt{3}}{\sqrt{a_n}+\sqrt{3}}$, which equals 1, we get $\left|\sqrt{a_n} - \sqrt{3}\right| = \left|\sqrt{a_n} - \sqrt{3} \cdot \frac{\sqrt{a_n}+\sqrt{3}}{\sqrt{a_n}+\sqrt{3}}\right| = \frac{|a_n-3|}{\sqrt{a_n}+\sqrt{3}} \leq \frac{|a_n-3|}{\sqrt{3}} < \varepsilon$. The analogous result is that if $\{a_n\}$ converges to L, then $\sqrt{a_n}$ converges to $\sqrt{L}$. Thus $\lim \sqrt{a_n} = \sqrt{\lim a_n}$, so we can put the word "limit" inside the square root.

10. By definition, if $\lim a_n = 0$ then for any $\varepsilon > 0$, we can find an N so that $|a_n| < \varepsilon$ for $n > N$. This can be written as $||a_n|| < \varepsilon$, which is the definition of $\lim |a_n| = 0$.

11. $n! = n(n-1)(n-2)...1 \geq 2 \cdot 2 \cdot 2 \cdot ... \cdot 2 \cdot 1 = 2^{n-1}$. Since $n! \geq 2^{n-1}$, $\frac{1}{n!} \leq \frac{1}{2^{n-1}}$. Now, consider $\left|\frac{c}{n!}\right| = \frac{|c|}{n!} \leq \frac{|c|}{2^{n-1}} < \varepsilon$ if $n > 1 + \log_2 \frac{|c|}{\varepsilon}$.

12. Since $-1 \leq \sin 3n \leq 1$, we have that $-\frac{1}{n} \leq \frac{\sin 3n}{n} \leq \frac{1}{n}$. Since $\lim \frac{-1}{n} = \lim \frac{1}{n} = 0$, by the squeeze theorem, $\lim \frac{\sin 3n}{n} = 0$.

13. Since $0 < a_n < 1$, we can multiply the inequalities by a_n to get $0 < a_n^2 < a_n$, and by induction we get $0 < a_n^k \leq a_n^{k-1}$. So the sequence is decreasing and bounded below by 0. So the sequence converges by the Completeness Theorem (Theorem (2.16)). Calling the limit of $\{a_n\}$, L, we get from $a^n = a(a^{n-1})$, that $\lim a^n = \lim a \lim(a^{n-1})$, or that $L = aL$. Now, if $L \neq 0$, we get $a = 1$, which is a contradiction, so $L = 0$.

14. We can make $|a^n - 0| = |a|^n < \varepsilon$ when $n > \frac{\log \varepsilon}{|a|}$.

15. Since after a certain point $n! > 3^n$, we have that $0 < \dfrac{2^n}{n!} < \dfrac{2^n}{3^n}$. Since $\dfrac{2^n}{3^n} = \left(\dfrac{2}{3}\right)^n$ converges to 0, by the previous exercise, we have by the squeeze theorem that $\dfrac{2^n}{n!}$ converges to 0.

16. We can find an N such that $n > N$ implies that $|a_n| < \dfrac{\varepsilon}{C}$, since $\lim a_n = 0$. This implies that $|x_n - x| < \varepsilon$ for $n > N$ which implies that $x_n \to x$.

17. Pick $r - \frac{1}{n} < r_n < r$. Since there are infinitely many rationals between two points, we can take the r_n to all be different and $r_n > r_{n-1}$. The sequence $\{r_n\}$ converges to r.

18. For the first part, there is an N_1 such that $n > N_1$ implies that $a_n > \frac{M}{2}$, and there is an N_2 such that $n > N_2$ implies that $b_n > \dfrac{M}{2}$. If $N = \max\{N_1, N_2\}$ then $n > N$ implies that $a_n + b_n > \dfrac{M}{2} + \dfrac{M}{2} > M$. For the second part, there is an N_1 such that $n > N_1$ implies that $a_n > \sqrt{M}$, and there is an N_2 such that $n > N_2$ implies that $b_n > \sqrt{M}$. If $N = \max\{N_1, N_2\}$ then $n > N$ implies that $a_n \cdot b_n > \sqrt{M} \cdot \sqrt{M} > M$. For the third part, if we can make $a_n > M$ after a certain point, then we can do the same for b_n since $b_n > a_n$.

19. Since $\lim a_n$ exists, $\{a_n\}$ is bounded. So $|a_n| \leq M$ for some positive M. Since $\lim b_n = \infty$, there is an N such that $n > N$ implies that $b_n > \dfrac{M}{\varepsilon} > 0$, so that $\dfrac{1}{b_n} < \dfrac{\varepsilon}{M}$, and $\left|\dfrac{a_n}{b_n}\right| = \dfrac{|a_n|}{|b_n|} = \dfrac{|a_n|}{b_n} = |a_n| \cdot \dfrac{1}{b_n} < M\left(\dfrac{\varepsilon}{M}\right) = \varepsilon$.

20. By the binomial theorem, $c^n = (1+b)^n = 1 + nb + \ldots +$ (positive terms) $> 1 + nb > M$, for any number M as long as $n > \dfrac{M-1}{b}$. So, $\lim c^n = \infty$.

21. For any positive number M, there is an N such that $n > N$ implies that $a_n < -M$.

22. $a_n = \left(1 + \frac{1}{n}\right)^n$

$= 1 + n \cdot \dfrac{1}{n} + \dfrac{n(n-1)}{2!} \cdot \dfrac{1}{n^2} + \dfrac{n(n-1)(n-2)}{3!} \cdot \dfrac{1}{n^3} + \ldots + \dfrac{n(n-1)(n-2)\ldots 1}{n!} \cdot \dfrac{1}{n^n}$

$= 1 + 1 + \dfrac{1}{2!} \cdot \dfrac{n(n-1)}{n^2} + \dfrac{1}{3!} \cdot \dfrac{n(n-1)(n-2)}{n^3} + \ldots + \dfrac{1}{n!} \cdot \dfrac{n(n-1)(n-2)\ldots 1}{n^n}$

$$\leq 1+1+\frac{1}{2!}+\frac{1}{3!}+...+\frac{1}{n!} < 1+1+\frac{1}{2}+\frac{1}{2^2}+...+\frac{1}{2^{n-1}}$$
$$< 1+1+\frac{1}{2}+\frac{1}{2^2}+...+\frac{1}{2^{n-2}}+... \leq 1+1+\frac{1/2}{1-1/2} = 3.$$

This shows the sequence is bounded.

23. $a_n - a_{n+1} = a_n - (2 - \frac{1}{a_n}) = (\frac{1}{a_n} + a_n) - 2 \geq 0$, since $\frac{1}{a_n} + a_n \geq 2$. So $a_n \geq a_{n+1}$ for all n, which means that $\{a_n\}$ is decreasing. The sequence is bounded below by 0 since each term is 2 minus a term strictly less than 2. So we know that the sequence, being bounded and monotonic, must converge. Taking limits as $n \to \infty$, we get that $\lim a_{n+1} = \lim(2 - \frac{1}{a_n})$ or $L = 2 - \frac{1}{L}$, which when solved for L gives $L = 1$.

24. We show by induction that a_n is bounded above by 2, and it is clear from the definition of a_n that $a_{n+1} \geq a_n$. So the limit, L, exists. Taking the limit of $a_{n+1} = \sqrt{2 + a_n}$ as $n \to \infty$, we get $L = \sqrt{2 + L}$. Squaring, we get $L^2 = 2 + L$, or $L^2 - L - 2 = 0$, from which we get that $L = 2$, or -1. Since $L \geq 0$ by virtue of each a_n being greater than 0, we have $L = 2$.

25. It is clearly monotonic since we are adding a positive number to each term of the sequence to get to the next term. It is bounded since $x_n = \frac{1}{1^2}+\frac{1}{2^2}+\frac{1}{3^2}+...+\frac{1}{n^2} < 1 + (1 - 1/2) + (\frac{1}{2}-\frac{1}{3}) + (\frac{1}{3}-\frac{1}{4}) + (\frac{1}{4}-\frac{1}{5}) + ... + (\frac{1}{n-1}-\frac{1}{n}) =$ (by cancellations) $2 - \frac{1}{n} < 2$. So the sequence is bounded. Since it is bounded and monotonic, it converges.

Section 2.4.

1. $1, 2, 1, 3, 1, 4, ...$

2. $1, 2, 3, 1, 2, 3, 1, 2, 3, ...$

3. If it converged then the subsequence consisting of odd terms would converge to the same limit as the subsequence consisting of the even numbered terms, and they don't.

4. If the sequence $\{(-1)^n a_n\}$ converged, then the subsequence of even numbered terms would have to converge to the same limit as the subsequence of odd numbered terms. But the first subsequence converges to c and the second to

$-c$. So $\{(-1)^n a_n\}$ cannot converge. On the other hand, if $\lim a_n = 0$, then we can find an N such that when $n > N$, $|a_n| < \varepsilon$. But then $|(-1)^n a_n| = |a_n| < \varepsilon$ when $n > N$, so $\lim(-1)^n a_n = 0$.

5. If $\{a_n\}$ is unbounded, there is a term a_{n_1} where $|a_{n_1}| > 1$, and a term $|a_{n_2}| > 2$, etc., giving terms a_{n_k}, where $|a_{n_k}| > k$. So $\lim \dfrac{1}{a_{n_k}} = 0$ since $|\dfrac{1}{a_{n_k}} - 0| < \frac{1}{k}$ which can be made $< \varepsilon$ for any $\varepsilon > 0$.

6. True

7. $a_n = 1, 2, 1/2, 2, 1/3, 2, ...$, $b_n = 1, 3, -1, 3, 1, 3, -1, 3, ...$ The sequence a_{n_k} of odd numbered terms from a_n converges, but the sequence of odd numbered terms of b_n does not. But this $b_{n_k} : -1, 1, -1, 1, ...$, has a convergent subsequence, and if we take a look at the corresponding terms of a_{n_k}, we get a subsequence of a_{n_k}, which must converge since a_{n_k} already does.

8. $L = \lim c^{\frac{1}{n}} = \lim c^{1/2n} = \lim(c^{1/n})^{1/2} = L^{1/2}$. Since $L = L^{1/2}$, $L = 0$ or 1. Since each term is > 1, the limit is ≥ 1, so $L = 1$.

Section 2.5.

1. Any subsequence of $\{a_{n_k}\}$ is a subsequence of $\{a_n\}$, so any subsequential limit of $\{a_{n_k}\}$ is a subsequential limit of $\{a_n\}$. Therefore the largest subsequential limit of $\{a_{n_k}\}$ is less than or equal to the largest subsequential limit of $\{a_n\}$. That is, $\overline{\lim}\{a_{n_k}\} \leq \overline{\lim}\{a_n\}$.

2. If A represents the set of subsequential limits of $\{a_n\}$ then $-A$ is the set of subsequential limits of $\{-a_n\}$. We know that $\overline{\lim}\{a_n\} = \sup A$, $\underline{\lim}\{-a_n\} = \inf(-A)$ and since for any set, A, of real numbers $\sup A = -\inf(-A)$, we have $\overline{\lim}\{a_n\} = -\underline{\lim}\{-a_n\}$.

3. Observe that $A_n \leq B_n$ for all n, so $\overline{\lim}\{a_n\} = \lim A_n \leq \lim B_n = \overline{\lim}\{b_n\}$

4. $a_n \leq A_n$ for each n and $b_n \leq B_n$ for each n. Add to get $a_n + b_n \leq A_n + B_n$. Now, use the previous exercise to get that $\overline{\lim}\{a_n + b_n\} \leq \overline{\lim}\{A_n + B_n\} = \lim(A_n + B_n) = \lim A_n + \lim B_n$, since the limits on the right of the inequality exist. But $\lim A_n = \overline{\lim}\{a_n\}$ and $\lim B_n = \overline{\lim}\{b_n\}$, so we have $\overline{\lim}\{a_n + b_n\} \leq \overline{\lim}\{a_n\} + \overline{\lim}\{b_n\}$.

5. $\overline{\lim}\, ca_n = \lim\left(\sup_{k\geq n}\{ca_k\}\right) = \lim\left(c\sup_{k\geq n}\{a_k\}\right) = c\lim\left(\sup_{k\geq n}\{a_k\}\right) = c\,\overline{\lim}\, a_n$.

6. Let $a_n = (-1)$ if n is odd and 2 if n is even, and let $b_n = -2$ if n is odd, and -1 if n is even. Then $a_n b_n = 2$ if n is odd and -2 if n is even. So $\overline{\lim}\{a_n b_n\} = 2$, while $\overline{\lim}\{a_n\} = 2$ and $\overline{\lim}\{b_n\} = -1$.

7. $a_n \leq \sup_{k \geq n}\{a_k\} = A_n$ and $b_n \leq \sup_{k \geq n}\{b_k\} = B_n$, so $a_n b_n \leq A_n B_n$. So, by Exercise 3, with $a_n b_n$ in place of a_n and $A_n B_n$ in place of b_n, we have $\overline{\lim}\{a_n b_n\} \leq \overline{\lim} A_n B_n = \lim(A_n B_n)$ (since the limits of each exist)$=\lim A_n \lim B_n = \overline{\lim}\{a_n\}\overline{\lim}\{b_n\}$.

8. If a_n is any subsequence of the rationals in $[0,1]$ once they have been sequenced, then all terms are ≤ 1, so any subsequential limit is ≤ 1. To show that 1 is a subsequential limit, follow the argument in Example 2.72 in the text.

Section 2.6.

1. $\{n\}$

2. We know that for any $\varepsilon > 0$, there is an N such that $|x_p - x_q| < \varepsilon$ when $p, q > N$. But for any subsequence $\{x_{n_k}\}$, $n_p \geq p$ and $n_q \geq q$, so $|x_{n_p} - x_{n_q}| < \varepsilon$ when $p, q > N$, and the subsequence $\{x_{n_k}\}$ is Cauchy.

3. We know that there are numbers N_1 and N_2 such that $|a_n - a_m| < \frac{\varepsilon}{2}$ and $|b_n - b_m| < \frac{\varepsilon}{2}$ respectively when $m, n > N_1, N_2$. Now, if $n, m > \max\{N_1, N_2\}$ we have $|a_n + b_n - (a_m + b_m)| = |a_n - a_m + (b_n - b_m)| \leq |a_n - a_m| + |b_n - b_m| < \frac{\varepsilon}{2} + \frac{\varepsilon}{2} = \varepsilon$. The sequences $\{a_n\}$ and $\{b_n\}$ converge because they are both Cauchy, so by Theorem 2.45 the sequences $\{a_n^2 b_n^3\}$ and $\left\{\frac{a_n}{b_n}\right\}$ converge, and hence they are Cauchy also.

4. $a_{2n} - a_n = \frac{1}{n+1} + \frac{1}{n+2} + \ldots + \frac{1}{2n} > \frac{1}{2n} + \frac{1}{2n} + \ldots + \frac{1}{2n}$ (n times)$=\frac{1}{2}$

5. Suppose that $m > n$. $|x_m - x_n| = \left|\frac{(-1)^{n+2}}{(2n+1)!} + \frac{(-1)^{n+3}}{(2n+3)!} + \ldots + \frac{(-1)^{m+1}}{(2m-1)!}\right| \leq$
$\frac{1}{(2n+1)!} + \frac{1}{(2n+3)!} + \ldots + \frac{1}{(2m-1)!}$ (Triangle inequality)
$\leq \frac{1}{2^{2n}} + \frac{1}{2^{2n+2}} + \ldots + \frac{1}{2^{2m-2}} \leq \frac{1}{2^{2n}} + \frac{1}{2^{2n+2}} + \ldots + \frac{1}{2^{2m-2}} + \ldots = \left(\frac{1}{2^{2n}}\right)/(1 - 1/4)$
$= \frac{4}{3 \cdot 2^{2n}}$ which can be made less than any $\varepsilon > 0$ by taking $n > \frac{1}{2}\log_2\left(\frac{3}{4\varepsilon}\right)$. A similar argument is done if $n > m$.

Section 3.1.

1. a. If $p = (0, 1)$ and $q = (0, 2)$, then $d(p, q) = 0$, but $p \neq q$.
   b. $d(p, q) = 1$ when $p = q$.
   c. Clearly $d(p, q) = d(q, p)$ and $d(p, q) \geq 0$. If $d(p, q) = 0$, then $x = a$ and $y = b$, so $p = q$. Let $r = (c, d)$. Then $d(p, r) = |x-c|+|y-d| \leq |x-a|+|a-c|+|y-b|+|b-d|$ (by the ordinary triangle inequality)$= d(p, q) + d(q, r)$.

2. Yes.

3. Yes.

4. By the triangle inequality, $d(x, y) \leq d(x, y_0) + d(y_0, y)$, which implies (1): $d(x, y) - d(x, y_0) \leq d(y_0, y)$. Similarly, $d(x, y_0) \leq d(x, y) + d(y, y_0)$ which implies (2): $d(x, y_0) - d(x, y) \leq d(y, y_0)$. Calling $d(x, y) - d(x, y_0) = P$, we have from (1) and (2): $P \leq d(y_0, y)$ and $-P \leq d(y_0, y)$, so that $|P| \leq d(y_0, y)$, which means that $|d(x, y) - d(x, y_0)| \leq d(y_0, y)$.

5. We only verify the triangle inequality. Let $p = (x_1, y_1), q = (x_2, y_2)$ and $r = (x_3, y_3)$. Then $d(p, q)+d(q, r) = \sup(|x_1-x_2|, |y_1-y_2|)+\sup(|x_2-x_3|, |y_2-y_3|) \geq |x_1 - x_2| + |x_2 - x_3| \geq |x_1 - x_3|$. So $d(p, q) + d(q, r) \geq |x_1 - x_3|$. Similarly, $d(p, q)+d(q, r) \geq |y_1-y_3|$. So, $d(p, q)+d(q, r) \geq \sup(|x_1-x_3|, |y_1-y_3|) = d(p, r)$.

Section 3.2.

1. If $\lim p_n = p$, then $d(p_n, p) < \frac{1}{2}$ for $n > N$, for some N, which implies that $p_n = p$ for $n > N$.

2. For Theorems 3.23 and 3.24, mimic the proofs on the real line. For Theorem 3.25, if $a_n \to L$, then for any $\varepsilon = 1$, there is an integer N such that $n \geq N$ implies that $d(a_n, L) < 1$. Let $M' = \max\{d(a_i, L)|\ i = 1, 2, 3, ..., N\}$. Now, $d(a_n, a_m) \leq d(a_n, L) + d(L, a_m)$. The right side of this inequality is ≤ 2 if $m, n \geq N$, and less than or equal to $2M'$ if $m, n < N$, and less than $1 + M'$ if only one of m or n is $< N$. In any event, $d(a_m, a_n) \leq \max\{2, 2M', 1 + M'\} = M$, so the sequence is bounded. For Theorem 3.26 there is an integer N such that $d(a_n, a_m) < 1$ when $m, n \geq N$. In particular, $d(a_N, a_m) < 1$ when $m \geq N$. Let $M' = \max \{d(a_i, a_N)\}$ where $1 \leq i \leq N$. We have $d(a_m, a_n) \leq d(a_m, a_N)+d(a_N, a_n)$, which is ≤ 2 when $m, n \geq N$, and $\leq 2M'$ when $m, n < N$ and $\leq 1+M'$ when only one of m or n is $< N$. So $d(a_m, a_n) \leq \max\{1, 2M', M' + 1\} = M$ and the sequence is bounded.

3. $(0, 3, 1, 1)$

Section 3.3.

1. $[0, 1) \cup (1, 2]$

2. A finite set, F, in a metric space cannot have limit points (or else how would an open ball surrounding the limit point have infinitely many points of the set?). Thus, $F' = \emptyset$, and since $F' \subset F$, F is closed.

3. It is a square with center $(0, 0)$ and vertices at $(1, 0), (0, 1), (0, -1)$, and $(-1, 0)$, and so looks like a diamond. The open ball of radius r and center (a, b) is also diamond shaped with center (a, b) and diagonals having length $2r$.

4. $B_1(0, 0)$ is a square with center $(0, 0)$ and sides parallel to the coordinate axes. The vertices of the square are $(1, 1), (1, -1), (-1, 1)$, and $(-1, -1)$. $B_r(a, b)$ is also a square with center (a, b) and sides having length $2r$, parallel to the coordinate axes.

5. a. The triangle inequality follows from the ordinary triangle inequality for real numbers. The rest of the conditions for metric space are clear.

   b. Choose K so that $\frac{1}{K} < \frac{\varepsilon}{2}$. Now, choose $m, n > K$. Then $d(m, n) = \left|\frac{1}{m} - \frac{1}{n}\right| \leq \left|\frac{1}{m}\right| + \left|\frac{1}{n}\right| = \frac{1}{m} + \frac{1}{n} < \frac{\varepsilon}{2} + \frac{\varepsilon}{2} = \varepsilon.$

   c. It is not complete since the Cauchy sequence from (b) can only converge to 0 which is not in Z^+.

6. If $m, n \in Z^+$, then if $d(m, n) = |m - n| < \frac{1}{2}$, we have that $m = n$. So any Cauchy sequence is eventually constant, which makes it convergent.

7. Look at the complement. We will show it is open. Pick a point t in S^c. Choose a number $N < d(p, t) - r$. We claim that $B_N(t) \subset S^c$ which will show that S^c is open and hence that S is closed. Now, pick $v \in B_N(t)$. If $d(v, p) < r$, then $d(p, t) < d(p, v) + d(v, t) < r + N < r + d(p, t) - r = d(p, t)$. So, we have that $d(p, t) < d(p, t)$. This contradiction arose from assuming that $d(p, v) < r$, so $d(p, v) \geq r$ and no point v of $B_N(t)$ is in S. Thus, $B_N(t) \subset S^c$.

8.

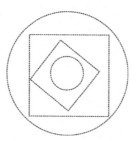

9. It contradicts nothing. Open balls in R are intervals, and open balls in R^2 are interiors of circles. So the definition of open in R and in R^2 requires checking different things.

10. The first two definitions of bounded were shown to be equivalent in the text. If $|a| \leq M$ for all $a \in S$ and some M, then $a \in B_{M+1}(0)$ for all $a \in S$ which means that $S \subset B_r(0)$ where $r = M+1$ and S is bounded according to the third definition. Conversely, if $S \subset B_{M'}(p)$ for some M', then each $a \in S$ is in $B_{M'}(p)$, so $|a-p| < M'$. By the reverse triangle inequality, $||a|-|p||$ is also $< M'$, which implies that $|a| < |p|+M' = M$ which implies $|a| \leq M$, and the set is bounded by the first definition. Therefore, second and third definitions are equivalent, and we now have the first 3 definitions of bounded are equivalent. Now, if a, b are in S, and $S \subset B_M(p)$, then by the previous statement, $|a-p| < M$ and $|b-p| < M$ so $|a-b| < |a-p|+|p-b| < M+M = 2M = M'$, and S is bounded by the fourth definition. So the third definition implies the fourth. Now, suppose that (1): $|a-b| \leq M'$, for all $a, b \in S$, and it is not true that $|a| \leq M$ for some M and all $a \in S$. Then for each $M > 0$ you choose, there is an $p \in S$ where $|p| > M$. Now, pick any element $b \in S$. Then by what we just said, there is an element a of S such that $|a| > M = |b| + M'$ which implies that (2): $|a| - |b| > M'$. But by the reverse triangle inequality, $|a| - |b| \leq ||a| - |b|| \leq |a - b| < M'$ which contradicts (2). This contradiction, which came from assuming that there was no M such that $|a| \leq M$ for all $a \in S$, shows that there is a number M such that $|a| \leq M$ for all $a \in S$. So the fourth definition implies the first implies the third, and since the third implies the fourth definition, all the definitions are equivalent.

11.   a. closed

     b. open

     c. open

     d. closed

e. neither

12. a. $[1, \infty)$
    b. $[-3, 2] \cup \{0\}$
    c. $\{-1, 1\}$
    d. R
    e. R
    f. All numbers of the form $3^{-n}, 5^{-n}$ and 0.

13. a. open
    b. closed
    c. closed
    d. neither
    e. neither

14. If O is open in S then for each point $p \in O$, there is an open ball, B_p, in S totally contained within O. But $B_p = S \cap C_p$ where C_p is an open ball in M. Now, $O = \bigcup_{p \in O} B_p = \bigcup_{p \in O} (S \cap C_p) = S \cap (\cup C_p) = S \cap C$ where $C = \bigcup_{p \in O} C_p$ is open since each C_p is.

15. $[3, \infty) = E \cap (2.5, \infty)$ and so is open in E being the intersection of E with an open ball in R. It complement in E is $(-\infty, 2] = E \cap (-\infty, 2, 5)$ is also open in E, so $[3, \infty)$ is closed.

16. Since every open ball centered at p has infinitely many points of the sequence (all of which are in S), every open ball centered at p has a point of S other than p, so p is a limit point of S.

17. a. We verify the triangle inequality. Let $p = (a, b)$, $q = (c, t)$ and $r = (e, f)$. Now $d^*(p, r) = d(a, e) + d(b, f) \leq d(a, c) + d(c, e) + d(b, t) + d(t, f) = d^*(p, q) + d^*(q, r)$.
    b. $d^*(p_n, p) = d(a_n, a) + d(b_n, b) \geq d(a_n, a)$ and $d(b_n, b)$. So if $d^*(p_n, p)$ can be made $< \varepsilon$ for $n > N$, then the same is true for $d(a_n, a)$ and $d(bn, b)$, which says that if p_n converges to p, in (A, d^*), a_n converges to a and b_n converges to b in (M, d). For the converse, make $d(a_n, a)$ and $d(b_n, b) < \frac{\varepsilon}{2}$ for $n > N$, and it will follow that $d^*(p_n, p) < \varepsilon$, which means that if $a_n \to a$, and $b_n \to b$ in (M, d) then $p_n \to p$ in (A, d^*). The proof for Cauchy is similar, just replace p by $p_m = (a_m, b_m)$.

c. We need to show that if a convergent sequence $p_n = (a_n, b_n) \in C \times C$, converges to a point $p = (a, b)$, p must be in $C \times C$. But if p_n converges to p, in $(C \times C, d^*)$ then $a_n \to a$ and $b_n \to b$, in (M, d), and both a and b must be in since C is closed. So $(a, b) \in C \times C$ and we are done. The sequence p_n converges to a point in $C \times C$.

d. If $p_n = (a_n, b_n)$ is Cauchy in $(C \times C, d^*)$ then $\{a_n\}$ and $\{b_n\}$ are Cauchy in (C, d) and hence converge to a and b respectively, both of which are in C since C is complete. So (a, b) is in $C \times C$ and we have shown that any Cauchy sequence p_n in $C \times C$ converges to a point in C.

18.

a. $int(R) = R, int(Q) = \emptyset$.

b. (a, b)

c. $\{(x, y) |\ y < x\}$

d. If U is open and $p \in U$, then there is a ball $B_r(p) \subset U$, so $p \in intU$. It follows from this that $U \subset intU$. Conversely, if $p \in intU$, then $p \in U$ by definition. So $U \subset intU$. The two set inclusions show that $U = intU$

e. If $p \in int(A) \cup int(B)$, then $p \in int(A)$ or $p \in int(B)$. If $p \in intA$, there is a ball, $B_r(p)$ such that $B_r(p) \subset A$. Since $A \subset A \cup B$, and $B_r(p) \subset A$, $B_r(p) \subset A \cup B$, which means that $p \in int\ (A \cup B)$. So, $int(A) \cup int(B) \subset int(A \cup B)$. To see that the set inclusion doesn't go the other way, $int(Q \cup (R - Q)) = int(R) = R$, but $int(Q) \cup int(R - Q) = \emptyset$.

19.

a. If $p \in \overline{A}$, the $p \in A$ or $p \in A'$. If $p \in A$, then $p \in B$ since $A \subset B$. If $p \in A'$, then every $B_r(p)$ contains a point $q \in A$ other than p, and $q \in B$ too since $A \subset B$. So $p \in B'$. Either way, $p \in B$ or $p \in B'$ which means that $p \in \overline{B}$. We took an arbitrary point in $\overline{A}$ and showed it was in $\overline{B}$, so $\overline{A} \subset \overline{B}$.

b. True. Since $A, B \subset A \cup B$, $\overline{A} \subset \overline{A \cup B}$ and $\overline{B} \subset \overline{A \cup B}$. So $\overline{A} \cup \overline{B} \subset \overline{A \cup B}$.by part a. To get the set inclusion the other way, suppose that (1): $p \in \overline{A \cup B}$ but $p \notin \overline{A}$ and $p \notin \overline{B}$. Since $p \notin \overline{A}$ and $\overline{A}$ is closed, $p \in \overline{A}^c$ which is open. So, there is an open ball with center p whose intersection with A is empty. Similarly, there is an open ball with center p whose intersection with B is empty. The intersection of these open balls has nothing in common with $A \cup B$, and so $p \notin A \cup B$ and $p \notin (A \cup B)'$. So $p \notin \overline{A \cup B}$ and this contradicts (1). So, if $p \in \overline{A \cup B}$ then p must be in

$\overline{A}$ or $\overline{B}$ and hence in $\overline{A} \cup \overline{B}$. Therefore, $\overline{A \cup B} \subset \overline{A} \cup \overline{B}$. Since we have already shown the inclusion the other way, we have that $\overline{A \cup B} = \overline{A} \cup \overline{B}$.

20. We need to show $\overline{E}$ contains all its limit points. Suppose $p \in \overline{E}'$, and suppose that $p \notin \overline{E}$. Then $p \notin E$ and $p \notin E'$. Since $p \notin E'$, there is a $B_r(p)$ whose intersection with E is either $\{p\}$ or empty. If the intersection with E is $\{p\}$, then $p \in E$ which is a contradiction. So, $B_r(p) \cap E = \emptyset$. We claim that $B_r(p) \cap \overline{E} = \emptyset$. For if $z \in B_r(p) \cap \overline{E}$, then $z \in B_r(p)$ and $z \in \overline{E}$. By virtue of $B_r(p)$ being open, there is a $B_u(z) \subset B_r(p)$, and this ball can't intersect E since $B_r(p)$ doesn't. This contradicts that $z \in \overline{E}$. To summarize, we took a point $p \in \overline{E}'$ and showed that it had to lie in $\overline{E}$, which proves that $\overline{E}$ is closed.

21. Since $E \subset \overline{E}$, need only show that $\overline{E} \subset E$, when E is closed. If not, then there is a point p of $\overline{E}$ which is not in E, and since we are assuming that E is closed, the complement is open. So, there is a neighborhood of p whose intersection with $\overline{E}$ is empty, which shows that p can't be in E or E' and therefore can't be in $\overline{E}$. We have our contradiction which arose from assuming that there was a point of $\overline{E}$ which is not in E. So $\overline{E} \subset E$. Since we have the inclusion the other way, $E = \overline{E}$.

22. A is certainly contained in I, the intersection of all closed sets containing A, and since I is closed, $\overline{A} \subset \overline{I} = I$ (by the previous exercise). We need to show that $I \subset \overline{A}$. Well, $\overline{A}$ being one of the sets whose intersection forms I, shows that $I \subset \overline{A}$. The two set inclusions show that $I = \overline{A}$.

Section 4.1.

1. $|2x + 1 - 5| = |2x - 4| = 2|x - 2| < \varepsilon$ when $0 < |x - 2| < \varepsilon/2 = \delta$.

2. Let $|x - 3| < 1$ so that x is between 2 and 4. Now, $|x^2 - 9| = |x - 3||x + 3| \leq (|x| + |3|)|x - 3| \leq (4 + 3)|x - 3| = 7|x - 3| < \varepsilon$ if $\delta < \min(1, \varepsilon/7)$.

3. $|\sqrt{x} - \sqrt{a}| = \left| \dfrac{x - a}{\sqrt{x} + \sqrt{a}} \right| \leq \dfrac{|x - a|}{\sqrt{a}} < \varepsilon$ if $0 < |x - a| < \varepsilon\sqrt{a} = \delta$.

4. The limit is concerned with what happens when $x \neq 1$. So, the cancellation is valid.

5. Take the limit of the numerator divided by the limit of the denominator, both of which are found by substituting 1 for the variable since the functions are polynomials.

6. By the reverse triangle inequality, $||f(x)| - |L|| \leq |f(x) - L|$. So if the right side of this inequality can be made less than ε for $0 < |x - a| < \delta$, then so can the left side.

7. Let $0 < |x - 0| < \sqrt{\frac{1}{N}}$, then $|x|^2 = x^2 < \frac{1}{N}$ so $f(x) = \frac{1}{x^2} > N$. For $\lim_{x \to 0} \frac{1}{x}$, we observe that we can't make $f(x) > N$ for $N > 0$ in a neighborhood of 0, since the function takes on negative values to the left of 0.

Section 4.2

1. $x_n = \dfrac{1}{\sqrt{n}}$

2. We need to show that if the distance between (x, y, z) and $(0, 0, 0)$ (which is $\sqrt{x^2 + y^2 + z^2}$) is less than δ, then the distance between $f(x, y, z)$ and 0 is less than ε. Since $x^2 \leq x^2 + y^2 + z^2$, we have that $|f(x, y, z) - 0| = \left|\dfrac{x^2}{\sqrt{x^2 + y^2 + z^2}}\right| = \dfrac{x^2}{\sqrt{x^2 + y^2 + z^2}} \leq \dfrac{x^2 + y^2 + z^2}{\sqrt{x^2 + y^2 + z^2}} = \sqrt{x^2 + y^2 + z^2} < \varepsilon$ when $\sqrt{x^2 + y^2 + z^2} < \delta = \varepsilon$. So our δ is the same as our ε.

3. If we can find two different sequences that converge to $(0, 0, 0)$ and we get different limits for the function value on these sequences, then the limit of $f(x, y, z)$ as $(x, y, z) \to (0, 0, 0)$ does not exist. Consider the sequence of points $p_n = (\frac{1}{n}, 0, 0)$. Now, $p_n \to (0, 0, 0)$ and $f(p_n) \to 0$. But the sequence $q_n = (\frac{1}{n}, \frac{1}{n}, 0)$ also converges to $(0, 0, 0)$ but $f(q_n) \to \frac{1}{2}$.

4. Since $\lim_{(x,y,z) \to (a,b,c)} f(x, y, z) = p$, we know that we can make $|f(x, y, z) - p| < \varepsilon$ by making $\sqrt{(x-a)^2 + (y-b)^2 + (z-c)^2} < \delta$. Now, consider $|\sqrt{f(x, y, z)} - \sqrt{p}| = \left|\dfrac{f(x, y, z) - p}{\sqrt{f(x, y, z)} + \sqrt{p}}\right| = \dfrac{|f(x, y, z) - p|}{\sqrt{f(x, y, z)} + \sqrt{p}} < \dfrac{|f(x, y, z) - p|}{\sqrt{p}}$. We can make this $< \varepsilon$ by taking $|f(x, y, z) - p| < \varepsilon \sqrt{p}$. So our δ is $\varepsilon \sqrt{p}$.

5. If $x_n \to c$, where $x_n \neq c$, and the x_n come from A, then $\lim f(x_n)$ converges to L. Since $a \leq f(x_n) \leq b$, we have by our limit theorems for sequences, that $a \leq \lim f(x_n) \leq b$, or that $a \leq L \leq b$.

6. Note that c is a limit point of I. If $x_n \to c$, then $f(x_n) \leq g(x_n) \leq h(x_n)$. Using the Squeeze Theorem for sequences, we have, since $\lim f(x_n) = \lim h(x_n) = L$, that $\lim g(x_n) = L$. So for any sequence $\{x_n\}$ converging to c, $\lim g(x_n) = L$, and so $\lim_{x \to c} g(x) = L$.

Section 4.3.

1. To find the limit as $x \to 0$, we must rationalize the numerator. $\lim\limits_{x\to 0} f(x) =$
$\lim\limits_{x\to 0} \dfrac{\sqrt{1+x}-1}{x} = \lim\limits_{x\to 0} \dfrac{\sqrt{1+x}-1}{x} \cdot \dfrac{\sqrt{1+x}+1}{\sqrt{1+x}+1} = \lim\limits_{x\to 0} \dfrac{x}{x(\sqrt{1+x}+1)} =$
$\lim\limits_{x\to 0} \dfrac{1}{(\sqrt{1+x}+1)} = \dfrac{1}{2}$, and $f(0) = \dfrac{1}{2}$. Since $\lim\limits_{x\to 0} f(x) = f(0)$, this function is continuous at $a = 0$.

2. $\lim\limits_{(x,y)\to(2,2)} f(x,y) = \lim\limits_{(x,y)\to(2,2)} \dfrac{x^2-y^2}{x^2-xy} = \lim\limits_{(x,y)\to(2,2)} \dfrac{(x-y)(x+y)}{x(x-y)} = \lim\limits_{(x,y)\to(2,2)} \dfrac{x+y}{x} =$
2. Since this is not equal to $f(2,2) = 3$, this function is not continuous at $(2,2)$.

3. True. If $x \neq 0$ we can take a sequence, $\{i_n\}$ of irrational numbers converging to x, but $f(i_n) \not\to f(x)$ since $f(i_n) = 0$ and $f(x) = x \neq 0$. So, f is not continuous at any nonzero x. But if $x = 0$, then $|f(x) - f(0)| = |x - 0| < \epsilon$ when $|x| < \delta = \varepsilon$.

4. Only at $c = 1$. At any point $c \neq 1$, we can take a sequence $\{r_n\}$ of rational numbers converging to c, and a sequence $\{i_n\}$ of irrational numbers converging to c, and $f(r_n)$ converges to $2c$, while $f(i_n)$ converges to $c + 1$. These are not equal. But if c is 1, these are equal, so the only possible point of continuity is at $c = 1$. To show that the function is continuous at $c = 1$, take any sequence $\{x_n\}$ converging to 1. Then $f(x)$ is either $2x_n$ or $x_n + 1$ both of which converge to 2. So $\lim f(x_n) = 2$ for each such sequence $\{x_n\}$ converging to 1, and it follows that $\lim\limits_{x\to 1} f(x) = 2$. Since $f(1) = 2$, $\lim\limits_{x\to 1} f(x) = f(1)$, $f(x)$ is continuous at 1.

5. If p is an isolated point of the domain, then f is and $|f|$ are automatically continuous. If p is a limit point of the domain, then we just have to show that $\lim\limits_{x\to a} |f(x)| = |f(a)|$, but this follows from the solution to question 6 in section 4.1 where we need only take $\delta = \varepsilon$.

6. Take any irrational number, i, and take a sequence, $\{r_n\}$ of rational numbers converging to i. By the continuity, $f(r_n) \to f(i)$. Since $f(r_n) = 0$, $f(i) = 0$. So $f(x)$ is 0 for both rational and irrational x, and hence for all x.

7.   a. Since $f(0+0) = f(0)$, we get that $f(0) + f(0) = f(0)$, which implies that $f(0) = 0$.

      b. By induction we get that $f(x_1 + x_2 + x_3 \ldots + x_n) = f(x_1) + f(x_2) + f(x_3) + \ldots + f(x_n)$. Letting all the x_i in this statement be equal to x, we get $f(nx) = n(f(x))$. In particular, when $x = 1$, we have $f(n) = n$ for any integer n.

c. From part (b), we have for any rational number $\frac{p}{q}$, that $f(\underbrace{\frac{p}{q} + \frac{p}{q} + \ldots \frac{p}{q}}_{q \text{ times}}) = qf\left(\frac{p}{q}\right)$ or just $f(p) = qf\left(\frac{p}{q}\right)$ from which it follows that $\frac{f(p)}{q} = f\left(\frac{p}{q}\right)$. Since we know that $f(p) = p$, this reduces to $\frac{p}{q} = f\left(\frac{p}{q}\right)$. So, for any rational number, r, $f(r) = r$.

d. Suppose that x is irrational. We know there is a sequence of rationals numbers $r_n \to x$. By the continuity of f, $f(r_n) = r_n \to f(x)$. So both r_n and $f(r_n)$ converge to x and since a sequence can only have one limit, $f(x) = x$. Since $f(x) = x$ for both rational and irrational numbers, we are done.

8. We can make $|f(x) - f(c)| < \frac{f(c)}{2}$ in some open interval, I, centered at c by the definition of continuity at c. So $\frac{-f(c)}{2} < f(x) - f(c) < \frac{f(c)}{2}$. Adding $f(c)$ to each part of the inequality, we get that $\frac{f(c)}{2} < f(x) < \frac{3f(c)}{2}$ which implies that $f(x) > \frac{f(c)}{2} > 0$ on I.

9. a. By the triangle inequality, $d(x,y) \leq d(x,y_0) + d(y_0,y)$, which implies that $d(x,y) - d(x,y_0) \leq d(y_0,y)$. Similarly, $d(x,y_0) \leq d(x,y) + d(y,y_0)$ which implies that $d(x,y_0) - d(x,y) \leq d(y,y_0)$. Calling $d(x,y) - d(x,y_0) = P$, we have $P \leq d(y_0,y)$ and $-P \leq d(y_0,y)$, so that $|P| \leq d(y_0,y)$, which means that $|d(x,y) - d(x,y_0)| \leq d(y_0,y)$.

b. We need to show that if $\varepsilon > 0$, $|f_x(y) - f_x(y_0)|$ can be made less than ε by making $d(y,y_0) < \delta$ (for some suitable δ). But $|f_x(y) - f_x(y_0)| = |d(x,y) - d(x,y_0)| < d(y,y_0)$ by part (a). So we just take $\delta = \varepsilon$.

c. Since $x_n \to x$ and $y_n \to y$ we can find an N large enough so that $d(x_n,x) < \frac{\varepsilon}{2}$ and $d(y_n,y) < \frac{\varepsilon}{2}$ for $n > N$. So, for $n > N$, we have that $|d(x,y) - d(x_n,y_n)| = |d(x,y) - d(x_n,y) + d(x_n,y) - d(x_n,y_n)| \leq |d(x,y) - d(x_n,y)| + |d(x_n,y) - d(x_n,y_n)| \leq d(x_n,x) + d(y_n,y)$ [part (a)] $< \frac{\varepsilon}{2} + \frac{\varepsilon}{2} = \varepsilon$.

10. f is not continuous, but g is. g is continuous since the preimage of every open set (in fact every set) is open. f is not continuous since the preimage of the open set $\{1\}$ in the discrete metric space is not open in M_1.

11. It is a polynomial which is continuous everywhere.

12. The functions $f(t) = \sin t$ and $g(x,y) = x^2 y$ are both continuous everywhere their composition $f \circ g = \sin(x^2 y)$ is continuous everywhere.

13. If f is continuous, then of course the preimage of every open ball is open, since the preimage of any open set is open and an open ball is open. To show the proof the other way, if O is open in (M_2, d_2), then for each point p in O, there is an open ball O_p containing p totally contained within O. It follows that $O = \bigcup_{p \in O} O_p$. So $f^{-1}(O) = \bigcup_{p \in O} f^{-1}(O_p)$ which, being a union of open sets, is open. The preimage of every open ball need not be an open ball. For example the function $f(x) = x^2$ is continuous, but $f^{-1}((1,4)) = (1,2) \cup (-2,-1)$.

14. Suppose that O is open in R. Consider $f^{-1}(O)$. Since f restricted to A, which we call f_A, is continuous, $f_A^{-1}(O)$ is open in A, and so $f_A^{-1}(O) = A \cap G_1$ where G_1 is open in R. But since A is open in R, $f_A^{-1}(O) = A \cap G_1$ is open in R, being the intersection of two open sets. Similarly $f_B^{-1}(O) = B \cap G_2$ is open in R. So $f^{-1}(O) = f_A^{-1}(O) \cup f_B^{-1}(O)$ is open in R and f is continuous. A similar proof works when A and B are closed sets since the preimage of every closed set under a continuous function is closed.

Section 4.4

1. The sets are expanding. If there were a finite cover there would be a largest n in that subcover, which we call N. This would also be the largest interval in the subcover. But then there would still be numbers between $1 - \frac{1}{N}$ and 1, and this contradicts the assumed fact, that this was a finite cover of $(0,1)$.

2. The union of each collection of sets contains $[2, \infty)$.

3. For $n = 1, 2, 3, ...$, let $O_n = (n - \frac{1}{2}, n + \frac{1}{2})$. $\{O_n\}$ is an open cover with no finite subcover. An open cover with a finite subcover would be $\{O_n\}$ where $O_n = (n, \infty)$.

4. Enclose each point, $\frac{1}{n}$, by an open ball of radius r_n where $r_n = \frac{1}{2}\left(\frac{1}{n} - \frac{1}{n+1}\right)$.

5. R

6. No such example since a compact subset of R is closed, and a closed subset of a complete metric space is complete.

7. The range has to be compact which $[c, d)$ isn't.

8. v must be in $[a,b]$ since $[a,b]$ is closed. By the continuity $f(p_{n_k}) \to f(v)$, which means that $f(p_n)$ is bounded, and this is our contradiction.

9. $f(x_{n_k}) \to f(d)$, and also to s, since $f(x_n) \to s$. So by the uniqueness of the limit, $f(d) = s$.

10. Since K is compact, it is closed, and the preimage of a closed set under a continuous function is closed.

Section 4.5.

1. Suppose that $\{O_\alpha\}$ is an open cover of $\bigcup_{i=1}^{n} A_i$. Then $A_i \subset \cup O_\alpha$ for each i, which means that each A_i has a finite subcover. The union of these finite subcovers is a finite subcover of $\bigcup_{i=1}^{n} A_i$. For the second part, observe (since every compact set is closed) that $\bigcap_{i=1}^{n} A_i$ is a closed subset of A_1, hence is compact.

2.   a. It is not bounded.

   b. It is not closed.

   c. It is neither closed nor bounded.

3. Let $O_n = \{(x,y)|\ x^2 + y^2 < 1 - \frac{1}{n}\}$.

4. Since K is bounded, the sequence has a convergent subsequence which converges to some point p by the Bolzano-Weierstrass theorem. We know that p must be in K by Corollary 3.65.

5. Since $p \in [a_n, b_n]$ for all n, $|p - a_n| \leq b_n - a_n = \dfrac{b-a}{2^n} < \varepsilon$, for $n >$ some N with a similar statement for $|p - b_n|$. So $a_n \to p$ and so does b_n. Therefore $f(a_n) \to f(p)$ and $f(p_n) \to f(p)$. Since $f(a_n) > 0$, we have (1): $f(p) \geq 0$ (Corollary 2.49 with $M = 0$). Similarly, since $f(b_n) < 0$, we have (2): $f(p) \leq 0$. (1) and (2) imply that $f(p) = 0$. Clearly, p can't be a or b since $f(a)$ and $f(b)$ are both nonzero, so $p \in (a,b)$.

6. $h(a) > 0$ and $h(b) < 0$, so by the previous result, there is a $d \in (a,b)$ such that $h(d) = 0$. But, $h(d) = 0$ is equivalent to $f(d) - c = 0$, and so $f(d) = c$.

7. If M and m are the maximum and minimum respectively of $f(x)$ on $[a,b]$, we know there are points c and d such that $f(c) = m$ and $f(d) = M$. If k is such that $f(c) \leq k \leq f(d)$, then by the previous theorem, and (allowing k to be $f(c)$ or $f(d)$) we can find an $e \in [a,b]$ such that $f(e) = k$. So the function takes on all values between m and M.

8. We leave the proof that it is a metric to you. It is not bounded since $d(a_n, b_n) = n$ where a_n is the point $a_n = (n, 0, 0, 0, ...)$ and $b_n = (0, 0, 0, ...)$ can be made as large as we want, by taking n sufficiently large.

9. We have to show that the preimage of any closed set, F, in K under f^{-1} is closed. But the preimage under f^{-1} of F is $(f^{-1})^{-1}(F) = f(F)$. So we must show that $(f^{-1})^{-1}(F) = f(F)$ is closed. But if F is closed in K, then F is compact since K is (Theorem 4.40), and $(f^{-1})^{-1}(F) = f(K)$ is compact, (Theorem 4.50) hence closed (Theorem 4.46). So f^{-1} is continuous.

10. Pick a sequence, $p_n = (x_n, f(x_n))$ of point from K. Since the sequence $\{x_n\}$ comes from the compact set $[a, b]$, there is a subsequence x_{n_k} converging to $p \in [a, b]$. By the continuity, $f(x_{n_k}) \to f(p)$. Since x_{n_k} converges to p and $f(x_{n_k})$ converges to $f(p)$, the sequence $p_{n_k} = (x_{n_k}, f(x_{n_k}))$ converges to $(p, f(p)) \in K$. We have shown that every sequence in K has a convergent subsequence with limit in K. Thus K is compact.

Section 4.6.

1. The δ that "works" for a given $\varepsilon > 0$, works for all points in M_1 and so works for all points in S.

2. False: The sequence $\{1/n\}$ is Cauchy in R with the usual metric, but if $f(x) = 1/x$, then $f(1/n) = n$ is not Cauchy in R with the same metric.

3. True. If f is uniformly continuous, then there is a $\delta > 0$, such that if $d_1(x, y) < \delta$, then $d_2(f(x), f(y)) < \varepsilon$. If $\{a_n\}$ is Cauchy in M_1, then $d_1(a_n, a_m) < \delta$ for $n, m > N$ some N. So we have that $d_2(f(a_m), f(a_n)) < \varepsilon$, if $m, n > N$, which says that the image of the Cauchy sequence $\{a_n\}$ is Cauchy.

4. The derivative is bounded.

5. Look at the function on $[0, p]$. It is uniformly continuous since $[0, p]$ is compact (Corollary 4.74). Since the function repeats on every interval of length p. The δ that works for this interval, works for all of them.

6. The derivative is bounded.

7. Consider $f(x) = \sqrt{x}$ on $[0, 2]$. We know it is uniformly continuous by Corollary 4.74. So there is a δ_1 that "works" on $[0, 2]$. Also, $f(x)$ is uniformly continuous on $[1, \infty)$ since the derivative is bounded. So there is a δ_2 that works on $[1, \infty)$. The only case to consider is when one of $x, y \in [0, 1)$ and the other is in $[1, \infty)$ Let $\delta = \min(1, \delta_1, \delta_2)$. If $x \in [0, 1]$ and $y \in [1, \infty)$ then since $\delta < 1$, if $|x - y| < \delta$, then $x, y \in [0, 2]$, and δ_1 will work for these points. So δ works for all points.

8. From the inequality we have that $|\sqrt{x} - \sqrt{y}| \leq \sqrt{|x-y|}$. So we need only make $|x - y| < \delta = \varepsilon^2$

9. If the range is not bounded, then for each positive integer n, there are points p_n in the domain such that $|f(p_n)| > n$. Since the domain is bounded, by the Bolzano-Weierstrass theorem, $\{p_n\}$ has a convergent subsequence $\{p_{n_k}\}$, which must be Cauchy. So $\{f(p_{n_k})\}$ must be Cauchy, by Exercise 3. This implies that $\{f(p_{n_k})\}$ is bounded, which is a contradiction since $|f(p_{n_k})| > n_k \to \infty$. Since the assumption that the range was unbounded led to a contradiction, the range must be bounded. An alternative proof would be to extend f to the closure, which is compact, and then since the range of the extended function is bounded, so is the range of the original function.

10. The first has no limit as $x \to 0$, so the first can't be extended to a continuous function on the closure of $(0, 1]$ which $[0, 1]$, but the second can be extended to a continuous function on the closure since the limit of the second as $x \to 0$ is 0.

Section 5.1.

1. Saying that $\{f_n\}$ converges to f means that for any $\varepsilon > 0$, we can find an N such that $n > N$ implies that $|f_n(x) - f(x)| < \epsilon$ for all $x \in X$. If you fix x, you get the definition of pointwise convergence at x.

2. If $|f(x)| \leq M$, then $|f_n(x)| = \left|\dfrac{f(x)}{n}\right| = \dfrac{|f(x)|}{n} \leq \dfrac{M}{n} < \varepsilon$ when $n > \dfrac{M}{\varepsilon}$.

3. $f_n(n) = 1$, so we can't make $|f_n(n) - f(n)| < 1/2$ for example. Each curve has a hump with the same height, and the hump moves over to the right as n gets larger and larger, so you can't get uniformly close to 0.

4. $|f_n(x) - 0| = \left|\dfrac{\sin(x^2 + x + 1)}{\sqrt{n+5}}\right| = \dfrac{|\sin(x^2+x+1)|}{\sqrt{n+5}} \leq \dfrac{1}{\sqrt{n+1}} < \varepsilon$ for all x when $n > (1/\varepsilon^2) - 1$.

5. Mimic the proof that the limit of the sum is the sum of the limits for sequences. (Theorem 2.45)

6. Mimic the proof that the limit of the product of two sequences is the product of the limits. (Theorem 2.45)

7. For real valued functions we know that we can make $|f(x) - f_n(x)| < 1$ for $n \geq N$, some integer N. Now, $|f(x)| \leq |f(x) - f_N(x)| + |f_N(x)| < 1 + M$, where M is a bound for the range of $|f_N|$. So $f(x)$ is bounded. To see the

$\{f_n\}$ are uniformly bounded, we have by the reverse triangle inequality that $|f_n(x)| - |f(x)| \leq ||f_n(x)| - |f(x)|| \leq |f_n(x) - f(x)| \leq 1$ for $n \geq N$. It follows that $|f_n(x)| \leq 1 + |f(x)| \leq 1 + 1 + M = 2 + M$. So the $f_n(x)$ are uniformly bounded after a certain point. We need only take $M' = $ the largest of a set of bounds for $|f_1|, |f_2|, |f_3| \ldots |f_{N-1}|$, and the number $2 + M$, and we have $|f_n(x)| \leq M'$ for all n and x.

8. Pick an $\varepsilon > 0$. Then since $f_n \to f$ uniformly, there is an integer $N > 0$ such that $d(f_n(x), f(x)) < \varepsilon/3$ for all $x \in X$ and all $n > N$. In particular, we can make
$$d(f_{N+1}(x), f(x)) < \varepsilon/3 \text{ for all } x \in X. \tag{8.6}$$
Since $f_{N+1}(x)$ is continuous, we know that we may find a $\delta > 0$ such that
$$d(f_{N+1}(x), f_{N+1}(x_0)) < \varepsilon/3 \text{ when } d(x, x_0) < \delta. \tag{8.7}$$
Now, let $d(x, x_0)$ be less than δ, and consider $d(f(x), f(x_0))$. We have
$$d(f(x), f(x_0))$$
$$\leq d(f(x), f_{N+1}(x)) + d(f_{N+1}(x), f_{N+1}(x_0)) + d(f_{N+1}(x_0), f(x_0))$$
(triangle inequality)
$$\leq \varepsilon/3 + \varepsilon/3 + \varepsilon/3 = \varepsilon \quad \text{(by (8.6), (8.7) and (8.6) respectively)}.$$

9. If for some x, $|f(x)| \leq 5$, then $f_n(x) = f(x)$ when $n \geq 5$. So clearly $\{f_n(x)\}$ converges pointwise to f at this x. A similar argument holds for each x.

10. For any $\varepsilon > 0$, we can find an N such that $|f_n(x) - f(x)| = \left|x + \dfrac{1}{n} - x\right| = \dfrac{1}{n} < \varepsilon$ for $n > N$ and all x. So f_n converges to f uniformly. Now, $|f_n(x)g_n(x) - f(x)g(x)| = \left|(x + \dfrac{1}{n})^2 - x^2\right| = \dfrac{2x}{n} + \dfrac{1}{n^2}$ which cannot be made less than a fixed ε for all $x \in R^+$. So the convergence is not uniform. If the set on which the convergence is taking place is bounded, then $|x| \leq M$ for some positive M and $\dfrac{2x}{n} + \dfrac{1}{n^2}$ can be made as small as we want independent of x by taking n sufficiently large. So the convergence will be uniform on such a set.

11. $f_n(x) = \dfrac{1}{n}$ if $x \in (0, 1]$ and $\dfrac{1}{2n}$ if $x = 0$. This sequence converges to $f(x) = 0$.

12. In the first case, $f(x) = 0$ for all x except $\pi/2$, where $f = 1$. The convergence is not uniform since the limit function is not continuous, For $f(x) = \sin(x/2)$ the limit function is $f(x) = 0$ for all x. The convergence is uniform by Dini's theorem for example since the sequence is decreasing.

13. $f_n(x) \leq \dfrac{x^n}{nx^n} = \dfrac{1}{n} < \varepsilon$ for all x if n is sufficiently large. In the second example, the functions converge to $f(x) = 1/3$ if $x = 1$, and $1/2$ if $x > 1$, and 0 if $x < 1$, which is not a continuous function, so the convergence is not uniform.

14. $f_n(x_n)$ converges to 0. But $f(1) = 1$.

15. Since f_n converges uniformly to f, there is an N_1 such that $n > N_1$, implies that $d_2(f(x), f_n(x)) < \varepsilon/2$ for all x. In particular, we have (1): $d_2(f(x_n), f_n(x_n)) < \varepsilon/2$. By the continuity of f_n, we can make $d_2(f(x_n), f(x)) < \varepsilon/2$ if $d_1(x_n, x) < \delta$ (for some δ). Since $x_n \to x$, there is an N_2 such that $n > N_2$ implies that $d(x_n, x) < \delta$. And from this it follows that (2): $d_2(f(x_n), f(x)) < \varepsilon/2$ Now, if we can show that $d_2(f_n(x_n), f(x))$ can be made less than any $\varepsilon > 0$ for $n > N$, some N, we will be done. Let $N = \max(N_1, N_2)$ Now, if $n > N$, then from (1) and (2) we have, $d_2(f_n(x_n), f(x)) \leq d_2(f(x), f(x_n)) + d_2(f(x_n), f_n(x_n)) < \varepsilon/2 + \varepsilon/2 = \varepsilon$.

16. Since $g_1(x) \geq g_2(x) \geq g_3(x)...$ for each x, if $g_2(x) \geq \varepsilon$, $g_1(x) \geq \varepsilon$ also. So $K_2 \subset K_1$. Similarly, $K_3 \subset K_2$ etc.

17. If f_n increases to f, then $-f_n$ decrease to $-f$, and by Dini's theorem, the convergence is uniform. That is, we can make $|-f_n(x) - (-f(x))| < \varepsilon$ for all $x \in K$ when $n > N$ some N. But this can be simplified to $|f(x) - f_n(x)| < \varepsilon$ for all $x \in K$, when $n > N$ which says the sequence of functions converges uniformly on K.

Section 6.1.

1. We can list the elements of $Z : 0, 1, -1, 2, -2...$ which makes the set countable. Each of sets $\{1\} \times Z, \{2\} \times Z, ...$ is clearly countable, so $Z \times Z$ being the union of this countable collection of countable sets is countable. Continue in the manner considering $\{1\} \times Z \times Z, \{2\} \times Z \times Z,...$ to show $Z \times Z \times Z$ is countable, or just use induction.

2. If B were countable, the so would A since a subset of a countable set is countable. Since $[0, 1]$ is uncountable, R is.

3. There are only finitely many subsets of S_n. The union of all these sets for each $n = 1, 2, 3, ...$ will give all finite subsets of Z^+ and this is a countable union of countable sets.

4. This set can be identified with $Z \times Z \times Z \times ..... \times Z$ (where we have $n+1$ factors of Z). The polynomial $a_0 + a_1 x + a_2 x^2 + ... + a_n x^n$ is thought of as the $n+1$ tuple $(a_0, a_1, a_2, ..., a_n)$. By the previous problem $Z \times Z \times Z \times ..... \times Z$ (where we have $n+1$ factors of Z) is countable.

5. Every polynomial of degree n with integer coefficients has at most n real roots. For any such polynomial p, let $R(p)$ be the set of real roots of p, and let P be the set of all such polynomials. The algebraic numbers is $\bigcup_{p \in P} R(p)$. But since P is countable, this union of sets being a countable union of countable sets is countable.

6. The union of the algebraic and transcendental numbers is the set of all real numbers. By the previous exercise, the former set is countable. If the latter set were countable, the set of real numbers would be, which is not true. So the number of transcendental numbers must be uncountable.

7. Pick one rational number in each interval. Since the intervals are disjoint, the rationals we picked are different and constitute a subset of the rational numbers which is countable. So the number of intervals is countable.

8. Once you show they are disjoint, they must be countable, by the previous exercise. To show disjointness, observe that if two open intervals I_x and I_y (both contained in I) overlap, we can unite them to get a larger interval contained in I contradicting that I_x is the largest open interval contained in I containing x. So they must be disjoint.

9. Since the intervals on the y-axis arising from the discontinuities are disjoint, we simply pick one rational in each, showing the that the number of such intervals, and hence the number of discontinuities is countable.

10. $\cup \{x\}$ where $x \in R$.

11. Suppose we could do this. Examine r_1 and r_2, and suppose that $r_1 < r_2$. There is a rational number, r_p such that $r_1 < r_p < r_2$ whose subscript must be between 1 and 2, which is impossible. We argue similarly if $r_2 < r_1$.

12. Every such sequence of 0s and 1s can be considered a binary representation of a number in $[0,1]$. We consider, for example, the sequence 101100000.... as the binary representation of the number .101100000... $= \frac{1}{2} + \frac{0}{2^2} + \frac{1}{2^3} + \frac{1}{2^4} + \frac{0}{2^4} + ...$ (which is a number in $[0,1]$). Since every number in $[0,1]$ has a binary representation, the number of binary sequences is "at least" the number of points in $[0,1]$ which is not countable.

13. Pick $I_n \subset I_{n-1}$ such that x_n is not in I_n. Now, $\cap I_n \neq \emptyset$ since the sequence of intervals is nested, and so there is a point $p \in \cap I_n$, and since $p \in I_n$, $p \in [0,1]$.

But p is not equal to any of the x_n by definition of I_n. This contradicts that assumption that $I = [0,1] = \{x_1 x_2, x_3, ...\}$.

14. The function f is 1-1 and onto. If $[a, b]$ were countable, we could sequence it as $a_1, a_2...$, which would imply that we could sequence $[0, 1]$, namely, $f(a_1)$, $f(a_2)$ etc., which is a contradiction since $[0, 1]$ is uncountable.

Section 6.2.

1. $A - B$ is a subset of A and so has measure 0. $B - A$ is a subset of B and so has measure 0. Their union therefore has measure 0.

2. The set of rationals numbers in $[0, 1]$. The closure of this set is $[0, 1]$.

Section 7.1.

1. The partition points are $1, 1.25, 1.50, 1.75$ and 2. The sup of the function on any subinterval occurs at the right endpoint, while the inf of the function occurs at the left endpoint. So the upper sum is $U(P, f) = f(1.25)(.25) + f(1.50)(.25) + f(1.75)(.25) + f(2)(.25) = 1.25^2 \cdot .25 + 1.50^2 \cdot .25 + 1.75^2 \cdot .25 + 2^2 \cdot .25 = 2.7188$. and $L(P, f) = f(1)(.25) + f(1.25) \cdot .25 + f(1.50)(.25) + f(1.75)(.25) = 1^2 \cdot .25 + 1.25^2 \cdot .25 + 1.50^2 \cdot .25 + 1.75^2 \cdot .25 = 1.9688$. $S(P, f) = f(1.125)(.25) + f(1.3750)(.25) + f(1.675)(.25) + f(1.875)(.25) = 1.125^2 \cdot .25 + 1.375^2 \cdot .25 + 1.6750^2 \cdot .25 + 1.875^2 \cdot .25 = 2.3694$. Notice that $L(P, f) \leq S(P, f) \leq U(P, f)$.

2. The partition points are $0, \dfrac{1}{n}, \dfrac{2}{n}, ... \dfrac{n}{n}$. Since the supremum of f occurs at the right endpoint, $U(P, f) = \sum_{i=1}^{n} f(\dfrac{i}{n}) \cdot \dfrac{1}{n} = \sum_{i=1}^{n} (\dfrac{i}{n}) \cdot \dfrac{1}{n} = \sum_{i=1}^{n} \dfrac{i}{n^2} = \dfrac{1}{n^2} \sum_{i=1}^{n} i = \dfrac{1}{n^2} \cdot \dfrac{n(n+1)}{2} = \dfrac{n+1}{2n}$. The upper sum equals $\inf \dfrac{n+1}{2n}$. But the sequence $\left\{\dfrac{n+1}{2n}\right\} = \left\{\dfrac{1}{2} + \dfrac{1}{2n}\right\}$ is decreasing and so the infimum of this sequence is its limit. So, $\overline{\int_0^1 x dx} = \lim \dfrac{n+1}{2n} = \dfrac{1}{2}$.

3. When $c \geq 0$, $\sup cf = c \sup f$ on $[a, b]$. If we let M'_i be $\sup cf$ on $[x_{i-1}, x_i]$, and $M_i = \sup f$ on $[x_{i-1}, x_i]$, then we have that $M'_i = cM_i$. So $\overline{\int_a^b cf(x)dx} = \inf U(P, f) = \inf \sum_{i=1}^{n} M'_i \Delta x_i = \inf \sum_{i=1}^{n} cM_i \Delta x_i = c \inf \sum_{i=1}^{n} M_i \Delta x_i = c\overline{\int_a^b f(x)dx}$.

4. We know that $\sup f(x) \geq f(x)$ and $\sup g(x) \geq g(x)$, so $\sup f(x) + \sup g(x) \geq f(x) + g(x)$ for each x in $[x_{i-1}, x_i]$ from which it follows that $\sup(f(x)+g(x)) \leq \sup f(x) + \sup g(x)$. If we let M_i, M_i' and M_i'' be $\sup(f(x) + g(x))$, $\sup f(x)$, $\sup g(x)$ on $[x_{i-1}, x_i]$ respectively, then as we have just shown, $M_i \leq M_i' + M_i''$. It follows from this that (1): $U(P, f+g) \leq U(P, f) + U(P, g)$ Now, because $\overline{\int_a^b f(x)dx} = \inf U(P,f)$, there is a partition P_1 of $[a,b]$ such that $U(P_1, f) < \overline{\int_a^b f(x)dx} + \frac{\varepsilon}{2}$. Because $\overline{\int_a^b g(x)dx} = \inf U(P,g)$, there is a partition P_2 of $[a,b]$ such that $U(P_2, g) < \overline{\int_a^b g(x)dx} + \frac{\varepsilon}{2}$. Let P be the common refinement of P_1 and P_2. Then because the upper sums decrease as we refine, we have $U(P, f) \leq U(P_1, f) < \overline{\int_a^b f(x)dx} + \frac{\varepsilon}{2}$ and $U(P, g) \leq U(P_2, f) \leq \overline{\int_a^b g(x)dx} + \frac{\varepsilon}{2}$. Adding these two we get that $U(P, f) + U(P, g) \leq \overline{\int_a^b f(x)dx} + \overline{\int_a^b g(x)dx} + \varepsilon$. But we know from (1) that $U(P, f) + U(P, g) \geq U(P, f+g) \geq \overline{\int_a^b f(x) + g(x)dx}$. So combining these inequalities we get that so that $\overline{\int_a^b f(x) + g(x)dx} \leq \overline{\int_a^b f(x)dx} + \overline{\int_a^b g(x)dx} + \varepsilon$ and since $\varepsilon > 0$ was arbitrary, $\overline{\int_a^b f(x) + g(x)dx} \leq \overline{\int_a^b f(x)dx} + \overline{\int_a^b g(x)dx}$.

5. Since $\overline{\int_a^c f(x)dx}$ and $\overline{\int_c^b f(x)dx}$ are the infima of upper sums, there are partitions, P_1 and P_2 of $[a,c]$ and $[c,b]$ respectively, such that $U(P_1, f) \leq \overline{\int_a^c f(x)dx} + \varepsilon/2$ and $U(P_2, f) \leq \overline{\int_c^b f(x)dx} + \varepsilon/2$. Adding these we get $U(P_1, f) + U(P_2, f) \leq \overline{\int_a^c f(x)dx} + \overline{\int_c^b g(x)dx} + \varepsilon$. Since P_1 and P_2 partition $[a,c]$ and $[c,b]$ respectively, $P_1 \cup P_2$ is a partition of $[a,b]$ so $U(P_1, f) + U(P_2, f) = U(P_1 \cup P_2, f)$. So our last inequality becomes, $U(P_1 \cup P_2, f) \leq \overline{\int_a^c f(x)dx} + \overline{\int_c^b g(x)dx} + \varepsilon$. And since this is true for any $\varepsilon > 0$, we have that $U(P_1 \cup P_2, f) \leq \overline{\int_a^c f(x)dx} + \overline{\int_c^b g(x)dx}$. But $\overline{\int_a^b f(x)dx} \leq U(P_1 \cup P_2, f)$. So substituting this into the previous inequality we get (1): $\overline{\int_a^b f(x)dx} \leq \overline{\int_a^c f(x)dx} + \overline{\int_c^b g(x)dx}$. To show the inequality the other way, we know that there is a partition, P' of $[a,b]$ such that $U(P', f) \leq \overline{\int_a^b f(x)dx} + \varepsilon$. Adjoin c to P'. Call the new partition P. Then we know the $U(P, f) \leq U(P', f)$. Let $P_1 = P \cap [a, c]$ and let $P_2 = P \cap [c, b]$. Then P_1 and P_2 are partitions of $[a, c]$ and $[c, b]$ respectively, so $\overline{\int_a^c f(x)dx} \leq U(P_1, f))$ and $\overline{\int_c^b g(x)dx} \leq U(P_2, f)$. Adding these we get that $\overline{\int_a^c f(x)dx} + \overline{\int_c^b g(x)dx} \leq U(P_1, f)) + U(P_2, f) = U(P, f) \leq U(P', f) \leq \overline{\int_a^b f(x)dx} + \varepsilon$. From this string we extract $\overline{\int_a^c f(x)dx} + \overline{\int_c^b g(x)dx} \leq \overline{\int_a^b f(x)dx} + \varepsilon$, and since $\varepsilon > 0$ is any we

get (2): $\overline{\int_a^c f(x)dx} + \overline{\int_c^b g(x)dx} \leq \overline{\int_a^b f(x)dx}$. Now combine (1) and (2) to get the result.

Section 7.2.

1. Suppose A_1 and A_2 are two possible values for the integral. If $\varepsilon > 0$ is given, there are $\delta_1, \delta_2 > 0$, such that if $||P|| < \delta_1$, $|S(P,f) - A_1| < \varepsilon/2$, and if $||P|| < \delta_2$, $|S(P,f) - A_2| < \varepsilon/2$. Now, $|A_1 - A_2| = |A_1 - S(P,f) + S(P,f) - A_2| \leq |A_1 - S(P,f)| + |S(P,f) - A_2| < \varepsilon$. Since $|A_1 - A_2|$ can be made less than any $\varepsilon > 0$, $A_1 = A_2$.

2. Any Riemann sum, $S(P,f)$ equals $c(b-a)$, so $|S(P,f) - c(b-a)| = 0 < \varepsilon$ for any $\varepsilon > 0$, and any partition which means that $\int_a^b f(x)dx = c(b-a)$. Or, use Theorem 7.27.

3. Any Riemann sum is between $m(b-a)$ and $M(b-a)$. Take the limit as the norm of the partition goes to 0 to get $m(b-a) \leq \int_a^b f(x)dx \leq M(b-a)$.

4. $f(x)$ is unbounded on this interval, so the Riemann integral of $f(x)$ does not exist.

5. Since the integral is $A_1 - A_2$ where A_1 is the area bounded by $f(x)$ and the x-axis in the first quadrant and A_2 is the area bounded by $f(x)$ and the x-axis, in the third quadrant, and since $A_1 = A_2$, the answer is 0. By Theorem 7.27, $\int_0^{2\pi} \sin x \, dx = -\cos x |_0^{2\pi} = -\cos(2\pi) - (-\cos(0)) = -1 - (-1) = 0$.

Section 7.3.

1. It is continuous everywhere except at $x = 0$, so it is continuous except on the set $\{0\}$ which is a set of measure 0.

2. $f(x) = -1$ if x is irrational, and 1 if x is rational.

3. The function may be unbounded or it may not be defined on [a,b]. For example, $f(x) = x$ is Riemann integrable on $[0,1]$ as is $g(x) = 1$, but $f(x)/g(x) = 1/x$ is neither bounded nor is it defined at all points of $[0,1]$.

4. Both are continuous everywhere except on a set of measure 0. But $g(f(x))$ is not.

5. Since the logarithm function is continuous and $f(x)$ is continuous except on a set of measure 0, their composition, $h(x)$, is continuous where $f(x)$ is, which is everywhere except on a set of measure 0. So $h(x)$ is Riemann integrable. m must be positive since the logarithm function is not defined at 0. Also, as we get close to 0, the logarithm function becomes unbounded and hence not Riemann integrable. So to keep the logarithm function bounded, we must stay away from 0.

6. Take the logarithm of M and use the previous exercise and the fact that the product of two functions each of which is continuous except on a set of measure 0, is continuous except on a set of measure 0. So $\ln M(x)$ is continuous except on a set of measure 0. Since $h(x) = e^x$ is continuous, $h(\ln(M(x))) = e^{\ln(M(x))} = M(x)$ is continuous except on a set of measure 0 and hence is Riemann integrable.

Section 7.4.

1. $f(x) = g(x)$ where $g(x) = x^2$ on $[a, b]$ except at one point. So $\int_1^2 f(x)dx = \int_1^2 g(x)dx = \int_1^2 x^2 dx = \frac{7}{3}$.

2. This is a simple induction proof

3. We know that $\int_a^b 0 dx = 0$. So this result follows from the chapter discussion, since $f(x)$ and $g(x) = 0$ agree except at b.

4. $\int_a^b f(x)dx = \int_{x_0}^{x_1} f(x)dx + \int_{x_1}^{x_2} f(x)dx + ... + \int_{x_{n-1}}^b f(x)dx$, and all the integrals on the right side of the equality are 0.

5. If we let $h(x) = f(x) - g(x)$, then $h(x) = 0$ except for a finite number of points, and so $\int_a^b h(x)dx = 0$. But $0 = \int_a^b h(x)dx = \int_a^b f(x)dx - \int_a^b g(x)dx$ from which it follows that $\int_a^b f(x)dx = \int_a^b g(x)dx$.

6. We know that $\int_a^b f(x)dx = \int_a^c f(x)dx + \int_c^d f(x)dx + \int_d^a f(x)dx \geq \int_c^d f(x)dx$ since the first and third integrals are nonnegative by virtue of f being nonnegative.

   To see that this is not true when $f(x)$ takes on negative values, let $f(x) = x^3$ on $[-1, 1]$. Then by the Fundamental Theorem of Calculus, $\int_{-1}^1 x^3 dx = 0$, but if we consider the integral of the function on the subinterval $[0, 1]$, we get $\int_0^1 x^3 dx = 1/4$.

7. If $f(x)$ is not 0, there is a point c where $f(c) > 0$. So there is an interval $(c - p, c + p)$ (where $p > 0$) containing c on which $f(x)$ is positive. (See Section Exercise 8 of Section 4.3 The function is also positive on $[c - \frac{p}{2}, c + \frac{p}{2}]$, and

since the function is continuous, it must take on its minimum value m on this interval, and m, must be > 0 since f is on this interval. So by the previous exercise, $\int_a^b f(x)dx \geq \int_{c-\frac{p}{2}}^{c+\frac{p}{2}} f(x)dx \geq m(c+\frac{p}{2}-(c-\frac{p}{2})) = mp > 0$, and this contradicts the given information. This contradiction tells us the $f(x)$ must be 0 everywhere on $[a, b]$

8. If $f(x)$ is never 0, then it is either positive or negative on $[a, b]$ (by the Intermediate Value theorem). Now use the previous exercise.

9. Obviously, if $f = g$ then $d(f, g) = 0$. If $d(f, g) = 0$, then $\int_a^b |f(x) - g(x)| dx = 0$. But we have already seen that the integral of a continuous nonnegative function being zero, implies that the function itself is 0. So $|f(x) - g(x)| = 0$ which implies that $f = g$. That $d(f, g) = d(g, f)$ is clear. The triangle inequality follows from the inequality we mentioned: $d(f, g) = \sqrt{\int_a^b |f(x) - g(x)| dx} \leq \sqrt{\int_a^b |f(x) - h(x)| + |h(x) - g(x)| dx} = \sqrt{\int_a^b |f(x) - h(x)| dx + \int_a^b |h(x) - g(x)| dx} \leq \sqrt{\int_a^b |f(x) - h(x)| dx} + \sqrt{\int_a^b |h(x) - g(x)| dx}$ (since $\sqrt{a+b} \leq \sqrt{a} + \sqrt{b}$) $= d(f, h) + d(h, g)$.

10. True since we may take a sequence of partitions where P_n has $2n$ equally spaced points, and choose rationals in each subinterval.

11. It is not true. $f(x) = \dfrac{1}{x}$ on $[0, 1]$ is an example. $\int_a^b f(x)dx$ doesn't exist since it is unbounded. But $\int_c^1 f(x)dx$ does exist since for any c strictly between 0 and 1 since $f(x)$ is continuous on $[c, 1]$.

Section 7.5.

1. Any upper sum is $c(b-a)$, so the infimum of the upper sums is $c(b-a)$. Similarly the supremum of the lower sums is $c(b-a)$.

2. Any upper sum is $\leq M(b-a)$, so the inf of the upper sums, which is the integral, satisfies this inequality also. Now do a similar argument for the lower sums.

3. $F(b) - F(a)$, being a Riemann sum, is between the infimum of the upper sums and the supremum of the lower sums, and both of these are equal to $\int_a^b f(x)dx$ since $f(x)$ is Riemann integrable. So $F(b) - F(a) = \int_a^b f(x)dx$.

4. Any upper sum for $f(x)$ on $[0, a]$ is equivalent to the negative of lower sum on $[-a, 0]$. So if we let U be the collection of upper sums for $f(x)$ on $[a, b]$ and L

be the collection of lower sums for f on $[-a, 0]$. $U = -L$. So, $\int_0^a f(x)dx =$ inf $U = -\sup(L) = -\int_{-a}^0 f(x)dx$, and the result follows from $\int_{-a}^a f(x)dx = \int_{-a}^0 f(x)dx + \int_0^a f(x)dx$.

5. We know the function is Riemann integrable since the set of discontinuities has measure 0. Any lower sum is 0 since every interval contains rational numbers. So the supremum of the lower sums, which is the value of the integral, is 0

6. The lower sums are all zero, so the supremum of the lower sums, which is the value of the integral, is 0.

7. a. A simple function is continuous except at a finite number of points, and hence is integrable since a set with a finite number of points has measure 0. The integral over a subinterval where $f(x)$ is constant, is just the constant times the length of the interval. We do this for each interval where $f(x)$ is constant and sum the results.

   b. We know that $\int_a^b f(x)dx = \inf U(P, f)$. Every upper sum $\sum_{i=1}^n M_i \Delta x_i$ is the integral of the step function which takes on the constant value M_i on $[x_{i-1}, x_i]$. If we let U be the set of upper sums, and let S be the integrals of step functions which are greater than or equal to $f(x)$ on $[a, b]$, $\int_a^b f(x)dx \leq s$ for each $s \in S$. So $\int_a^b f(x)dx \leq \inf S$ Since, as we have just pointed out, $U \subset S$, inf $S \leq \inf U$. So we have $\int_a^b f(x)dx \leq \inf S \leq \inf U = \int_a^b f(x)dx$. So inf S, being sandwiched between $\int_a^b f(x)dx$ and itself, must be equal to inf S. If we let $\widetilde{S}$ be the set of integrals of step functions $\leq f(x)$ on $[a, b]$ then a similar proof shows that $\int_a^b f(x)dx = \sup \widetilde{S}$.

Section 7.7.

1-3 . Mimic the proofs for Riemann integrals, only instead of the norm being less than δ, use refinements.

4. Mimic the proof of the uniqueness for the Riemann integral (Exercise 1 in section 7.2) only instead of the norm being less than δ, use refinements.

Section 7.8.

1. Every Riemann-Stieltjes sum, $S(P, f, \alpha)$ is 0 since $\alpha(x_i) - \alpha(x_{i-1}) = 0$. So $\int_a^b f(x)d\alpha = 0$.

2. $\int_2^3 x^3 d(x^4) = \int_2^3 x^3 4x^3 dx = \int_2^3 4x^6 dx = \frac{8236}{7}$.

3. $\int_o^{\frac{\pi}{2}} \cos x d(\sin x) = \int_o^{\frac{\pi}{2}} \cos x \cos x dx = \int_0^{\frac{\pi}{2}} \cos^2 x dx = \int_0^{\frac{\pi}{2}} \frac{1+\cos 2x}{2} dx = \frac{1}{4}\pi$.

4. $f(\frac{1}{2})\cdot$jump at $\frac{1}{2} = 1 \cdot 1$ or 1.

5. $f(0)\cdot$jump at $0 + f(1)\cdot$ jump at $1 = 3 \cdot 3 + 1 \cdot 2 = 11$.

6. $\int_a^b f d(\alpha + \beta) = \int_a^b f(\alpha' + \beta') dx = \int_a^b (f\alpha' dx + f\beta') dx = \int_a^b f\alpha' dx + \int_a^b f\beta' dx = \int_a^b f d\alpha + \int_a^b f d\beta$

Section 8.1

1. We say that a bounded measurable function defined on $[a, b]$ is said to be Lebesgue integrable on $[a, b]$ if there is a number A with the property that for all $\varepsilon > 0$ there exists a $\delta > 0$ such that if $||P|| < \delta$ where P is a partition of an interval containing the range in its interior, and for any choice of points $t_i \in E_i$, $\left| \sum_{i=1}^n t_i m(E_i) - A \right| < \varepsilon$. A is called the Lebesgue integral of $f(x)$ on $[a, b]$.

2. We know that $y_{i-1} \leq t_i < y_i$ for $i = 1, 2, 3, ...n$. For each i, multiply the inequality by $m(E_i)$ and sum the results, and you have the proof.

3. a. From the statement $\sum_{i=1}^n y_i m(E_i) - \sum_{i=1}^n y_{i-1} m(E_i) \leq ||P||(b-a)$, we see that if we take $||P|| < \frac{\varepsilon}{b-a}$, then $\sum_{i=1}^n y_i m(E_i) - \sum_{i=1}^n y_{i-1} m(E_i) < \varepsilon$.

   b. Since $\sum_{i=1}^n y_{i-1} m(E_i) \leq a \leq b \leq \sum_{i=1}^n y_i m(E_i)$, we have that $|a - b| < \sum_{i=1}^n y_i m(E_i) - \sum_{i=1}^n y_{i-1} m(E_i) < \varepsilon$ (when $||P|| < \frac{\varepsilon}{b-a}$). Since we can make $|a - b| < \varepsilon$ for any $\varepsilon > 0$, we have that $a = b$.

   c. Take $||P|| < \frac{\varepsilon}{2(b-a)}$. Then (1): $\sum_{i=1}^n t_i m(E_i) - \sum_{i=1}^n y_{i-1} m(E_i) = \sum_{i=1}^n (t_i - y_{i-1}) m(E_i) \leq ||P||(b - a) < \varepsilon/2$. Similarly we get (2) : $\sum_{i=1}^n y_i m(E_i) - \sum_{i=1}^n y_{i-1} m(E_i) < \sum_{i=1}^n (y_i - y_{i-1}) m(E_i) \leq ||P||(b - a) < \varepsilon/2$.

From $\sum_{i=1}^{n} y_{i-1} m(E_i) \leq a \leq b \leq \sum_{i=1}^{n} y_i m(E_i)$ we get that

$$0 \leq b - \sum_{i=1}^{n} y_{i-1} m(E_i) \leq \sum_{i=1}^{n} y_i m(E_i) - \sum_{i=1}^{n} y_{i-1} m(E_i) < \varepsilon/2 \text{ (by 2)}$$

and it follows that (3): $b - \sum_{i=1}^{n} y_{i-1} m(E_i) < \varepsilon/2$. Now $\sum_{i=1}^{n} t_i m(E_i) - a = \sum_{i=1}^{n} t_i m(E_i) - \sum_{i=1}^{n} y_{i-1} m(E_i) + \sum_{i=1}^{n} y_{i-1} m(E_i) - a = \sum_{i=1}^{n} t_i m(E_i) - \sum_{i=1}^{n} y_{i-1} m(E_i) + \sum_{i=1}^{n} y_{i-1} m(E_i) - b$, (since $a = b$). So by (1) and (3) and the triangle inequality,

$$\left|\sum_{i=1}^{n} t_i m(E_i) - a\right| \leq \left|\sum_{i=1}^{n} t_i m(E_i) - \sum_{i=1}^{n} y_{i-1} m(E_i)\right| + \left|\sum_{i=1}^{n} y_{i-1} m(E_i) - b\right| < \varepsilon/2 + \varepsilon/2.$$

So for any partition with norm $< \dfrac{\varepsilon}{2(b-a)}$, we have $\left|\sum_{i=1}^{n} t_i m(E_i) - a\right| < \varepsilon$.

So $a = (L) \int_a^b f(x) dx$.

4. Both values are 0.

5. $f^{-1}[y_{i-1}, y_i) = \varnothing$ if $c \notin [y_{i-1}, y_i)$, and E if $c \in [y_{i-1}, y_i)$ and both of these preimages are measurable.

6. We have pointed out that subsets of measure 0 are measurable and have measure 0. Since the preimage of every half open interval is a subset of E, the preimage has measure 0, and hence is measurable, which of course, makes the function measurable.

7. Suppose that E is measurable. Then $f^{-1}[y_{i-1}, y_i) = \varnothing$ if and only if $0, 1 \notin [y_{i-1}, y_i)$, and E if and only if $1 \in [y_{i-1}, y_i)$ but $0 \notin [y_{i-1}, y_i)$, and E^c if and only if $0 \in [y_{i-1}, y_i)$ but $1 \notin [y_{i-1}, y_i)$, and X if and only if $0, 1 \in [y_{i-1}, y_i)$, and all of these preimages are measurable if E is measurable. Conversely, if χ_E is measurable, then $\chi_E^{-1}([1/2, 3/2)) = E$ is measurable.

8. $f^{-1}[y_{i-1}, y_i) = \bigcup_i (f^{-1}[y_{i-1}, y_i) \cap E_i)$ and all the sets on the right are measurable and so their union is also.

9. For any a, we have that $f^{-1}([a,\infty)) = f^{-1}([a,a+1) \cup [a+1,a+2) \cup [a+2,a+3).....)$

$= f^{-1}([a,a+1)) \cup f^{-1}([a+1,a+2)) \cup f^{-1}([a+2,a+3))......$ Each of these sets is measurable by the definition of measurable function, and so their union, which is a countable union of measurable sets, is measurable. Taking complements, we get that $f^{-1}(-\infty,a)$ is measurable. Now, $f^{-1}(-\infty,a] = f^{-1}(\cap(-\infty,a+\frac{1}{n})$ $n = 1,2,3...$, so $f^{-1}((-\infty,a]) = \cap f^{-1}((-\infty,a+\frac{1}{n}))$, which is measurable being a countable intersection of measurable sets. Now $f^{-1}(\{a\}) = f^{-1}((-\infty,a] \cap [a,\infty))$ which is measurable, being the intersection of two measurable sets.

10. a. It takes on only a finite number of values.

b. Every step function greater than or equal to $f(x)$ on $[a,b]$ is a simple function greater than or equal to $f(x)$ on $[a,b]$. That is, $S \subset E$. So, inf $E \leq$ inf S. Also, since every lower sum is the integral of a simple function less than or equal to $f(x)$ (that is, $L \subset \widetilde{E}$), sup $L \leq$ sup $\widetilde{E}$.

c. Observe that a step function is continuous except on a set of measure 0, and so is measurable and the same is true for f by Lebesgue's Theorem. We have, using Exercise 7 in section 7.6 and part b of this question, that $(L)\int_a^b f(x)dx =$ inf $E \leq$ inf $S = (R)\int_a^b f(x)dx =$ sup $L \leq$ sup $\widetilde{E} = (L)\int_a^b f(x)dx$. Thus, the Riemann integral is sandwiched between on the right and left by the Lebesgue integral, and so the two must be equal.

Index

$A \times B$, 8
$A + B$, 27
arbitrary union, 81

Bolzano Weierstrass Theorem, 48
bounded, 21, 73
    alternate definition on R, 22
    alternate defintion, 77
    converges, 51
    equivalent definitons, 73
    function, 158
bounded above, 21

$C[0,1]$, 65
    is complete, 84
cA, 28
Cantor set, 154
Cauchy
    sequence is bounded, 59
Cauchy Schwartz Inequality, 65
Cauchy sequence, 58
    of real numbers converges, 59
    converges, 60
Cauchy-Schwartz-Bunyakovski inequality, 65
CBS inequality, 65
closed, 80
    arbitrary intersection, 81
    if and only if it contains its limit points, 83
    lmits of convergent sequences, 84
closed
    subset of complete space is complete, 84
closed interval, 2
closure, 131
compact, 111
    closed subset of, 113
    continuous image of , 118
    finite set is, 111
    if and only if closed and bounded , 118
    implies closed, 117
    in R closed an bounded is, 114
    not equivalent to closed and bounded in metric spaces, 115
compact
    is complete, 145
complement, 3
complete, 84
    R^k is complete, 85
Completeness Theorem, 23
composition, 13
conditional statement, 19
continous
    interpretation in terms of balls, 100
continous at every irrational number, 105
continous function
    is measurable, 206
    theorem and relation to sequences, 103
continuity
    in terms of closed sets, 108

in terms of open sets, 106
continuity at isolated points
    at isolated point, 100
continuous, 98, 99
    polynomials are, 104
continuous
    on a set, 103
continuous function
    relation to sequences, 103
converges
    inplies Cauchy, 73
    sequence, 31, 70
converges uniformly
    if and only if Cauchy, 144
    integrals converge, 179
countable, 149
    finite set is, 149
    R is not, 150
    set of rationals is, 150

decreasing
    sequence, 42
definite integral, 167
Δx_i, 157
DeMorgan's Law, 6
difference, 2
Dini's Theorem, 143
discontinuous
    everywhere, 105
discrete metric
    every Cauchy sequence is eventually constant, 84
    no limit points, 82
domain, 8

Extreme Value Theorem, 119

finite intersecdtion property, 121
finite intersection property
    compact sets, 121
function, 8

geometric series, 153
greatest lower bound, 22

half closed interval, 2
half open interval, 2
Heine Borel, 113

image
    of a set, 9
    of an element, 8
increasing
    sequence, 42
increasing
    function, 192
index, 3
infimum, 22
inner measure, 223
intersection, 2
isolated point, 100

least upper bound, 22
Lebesgue integral, 203
Lebesgue sum, 203
Lebesgue sums, 207
Lebesgue's theorem, 168
    corollaries of, 170
Lebsgue's Dominated Convergence Theorem, 211
limit
    definiton includes right and left , 92
    does not exist, 96
    of a function, 90
    of a sequence of real numbers, 31
    of function
        relation to limits of sequences, 95
limit
    of a function between metric spaces, 93
    theorem for functions, 91, 96
limit inferior, 51
limit point, 81

INDEX

not a limit point, 82
    sequence of points converging to it, 82
limit superior, 51
Lipschitz condition, 129
Logic Preliminaries, 19
lower bound, 21
lower integral, 164
lower Lebesgue sums, 206
lower sum, 159
    less than or equal to upper sum, 163

measurable, 223
    function, 206
measurable function, 209
measure, 153
measure 0, 153
    any finite set has, 153
    countable sets have, 154
    subset of, 154
    the null set has, 153
    uncountable set with, 154
metric, 63
    discrete, 64
    trivial, 64
metric space, 63
monotonic, 42
montone
    subsequence, 48

nested sequence
    of sets, 122
    compact sets, 122

open, 77, 80
    arbitrary union, 80
    finite intersection of open sets, 80
open ball, 75
    is open, 78
open cover, 110
open interval, 2

outer measure, 223

pairwise disjoint, 5, 204
partition, 157
    norm of, 158
peak, 48
pointwise convergence, 135
    of coniinuous functions
        does not imply continuity, 136
pre-image, 10
product metric, 87

Q, 2

R, 2
R^+, 2
real number
    sequence of rationals converging to it, 105
refinement, 158
    common, 158
Reverse Triangle Inequality, 18
Riemann integrable, 167
Riemann integral, 165
    alternate approaches, 181
    continuous almost everywhere, 168
    evaluation of, 167
    evaluation of , 172
    exists, 167
Riemann sum, 161
    between upper and lower sum, 162
Riemann-Stieltjes integral, 193
    evaluation, 195, 199
    properties, 194
Riemann-Stieltjes sum, 193
R^n, 67

sequence
    converges to L
        geometric interpretation, 34
    covergent, 48

of real numbers, 30
    strictly decreasing, 42
    strictly increasing, 42
    unique limit, 71
sequence
    converges, 74
sequentially compact, 123
set, 1
    elements of, 1
sets
    equal, 5
simple measurable function, 215
Solutions to Most Exercises, 231
square root algorithm, 43
Squeeze Theorem, 41
step function, 183, 198
subsequence, 47
subsequence
    converging, 74
subset, 2
subspace, 68
    open balls in, 76
supremum, 22

The Mean Value Theorem, 129
    applied to antiderivatives, 130
Triangle Inequality, 17

unbounded, 23
unformly continous
    extension to closure, 132
uniform convergence, 138
    implies integrals converge, 142
uniform convergence of continous functions
    implies limit is continuous, 139
uniformly bounded, 211
uniformly Cauchy, 143
uniformly continuous, 124
union, 2

universal set, 3
upper and lower sums
    as we refine, 163
upper bound, 21
upper integral, 164
upper Lebesgue sums, 207
upper sum, 159
usual metric, 64

Venn Diagrams, 3

Z, 2
Z^+, 2

CPSIA information can be obtained
at www.ICGtesting.com
Printed in the USA
BVOW07s0858210817
492585BV00013B/103/P